BIRDS *and*
CLIMATE CHANGE

John F. Burton

DEDICATION

Dedicated to the memory of the late KENNETH WILLIAMSON
(1914–1977) who contributed much to our detection and understanding of the effects of climate change on British and other
European birds, and also to my late wife HELGA (née LACH-
MUND) (1938–1984) who saw the start of my work on this
book, but, sadly, not its final completion, and to WEGA who
inspired me to finish it. (Photograph: David Hosking)

BIRDS and CLIMATE CHANGE

John F. Burton

CHRISTOPHER HELM

A & C Black · London

© 1995 John F. Burton

Line drawings by John Davis

Christopher Helm (Publishers) Ltd, a subsidiary of
A & C Black (Publishers) Ltd, 35 Bedford Row, London WC1R 4JH

ISBN 0-7136-4045-6

Printed and bound by Biddles Limited,
Guildford, Surrey in Great Britain

CONTENTS

FOREWORD

We all talk about the weather. It is a fallacy that only the British are preoc-cupied with the day-to-day, month-by-month vagaries of climate. Throughout Europe, it is a topic of conversation and a constant source of speculation and reflection. June was the wettest since meteorological records began. Snow has never come so late to the French ski resorts. The mistral fans forest fires in Corsica, threatening campers. Floods ruin the root-crop harvests of northern Germany. In southern England, water authorities ban the use of garden hosepipes. It seems that the climate of Europe is always in a dramatic state of flux.

For we mortals, such short-term manifestations of the climate shape the pattern of our lives. They dictate our domestic timetable—from planting crops in our gardens to taking weekend breaks and annual holidays. Seasons and climate control our moods and sense of well-being. It is difficult to imagine Scandinavians without their long, dark winters and Spaniards with-out the sun. European culture has been shaped as much by climate as by the nature of the land.

A decade ago, when I was researching and writing my book and BBC tele-vision series The Living Isles, I decided that the changing climate of Britain and Ireland would be its central theme. To my mind, a 'natural history' has to look back in order to fully appreciate what we have today—and explore the reasons why. The story I set out to tell described the changes to the land-scapes and wildlife of the British Isles from ice-age to motorway-age. In that period of less than ten millennia, not only did these islands become cut off from mainland Europe, but their entire flora and fauna was shaped and reshaped by the changing climate—and increasingly by the activities of people. It is a dramatic natural history which has been echoed throughout Europe.

One reference became a constant source of information and inspiration; it became my 'bible' of events in the changing European climatic scene. For several years, John Burton, who was then a close colleague in the BBC Natural History Unit, had been writing and revising his book about climate and the distribution of European birds. He and I had many discussions about the spread of vegetation in the wake of the retreating ice, and about the rate of colonisation by the plant and animal communities that became the native wildlife of the British Isles. The wealth of factual information and informed speculation in John's manuscript was central to the theme of my own book and to the scripts for the television series which has now been screened three times in Britain and in many other countries in Europe and around the world. I am indebted to John for his inspiration.

I hope that, in some way, I have been instrumental in encouraging John to complete his much more thorough work and see it through to publication. No time, he told me, would ever be quite right. Each season, each year brought new observations and revised conclusions. Even that most famous of British naturalists, Charles Darwin, faced a similar dilemma with the publication of his own 'great work'. John Burton's book is masterful in its own way. The changes to the natural world that he so eloquently describes in the following chapters have mostly happened within the time-span of European history—many of them within living memory. The key markers in this intriguing natural history of our climate are events immediately familiar to all of us. I still find it fascinating that at the time when legions of Roman soldiers came to Britain, our climate was so Mediterranean that red grapes grew in southern England, but by the time they eventually retreated and Britain entered the Dark Ages, the grape vines died of cold and birds such as the Bee-eater withdrew to warmer southern climes. Then, a few centuries later, all but the Romans started to return. In the early years of the Middle Ages, our summers became warmer and our fauna and flora shifted its composition yet again. John Burton's book chronicles these climatic changes and reveals how they have fashioned the lives of the native wildlife —and native people—of Britain and the rest of Europe.

His key indicators are of course the birds. John is one of the most knowledgeable and enthusiastic ornithologists I know. Birds—and insects—have been his lifelong passion. No other groups of animals better illustrate the effects of changing climate on the nature of our land. This timely book will fascinate the amateur and inspire the professional. It will be long admired for its breadth of vision and originality of thought.

Peter Crawford
Executive Producer
BBC Natural History Unit, Bristol
March 1994

PREFACE

Britain's climate, in common with that of the rest of Europe, is not static; on the contrary, since the end of the Ice Age it has undergone a wide range of major and minor fluctuations, from the very warm period of the early Middle Ages (approximately AD 750–1250), when vineyards flourished in England, to the 'Little Ice Age' of 1250–1850, when frost fairs were held on the frozen Thames. Soon after 1850 another warm period developed, reaching its maximum in the 1930s. It gradually came to an end after 1950. Within these major climatic trends a number of minor fluctuations have occurred, such as the series of generally fine summers and cold winters in the 1940s.

All these climatic changes, great or small, have had a profound effect on the distributions of plants and animals in Europe and elsewhere in the Northern Hemisphere. For example, as a result of the 1850–1950 climatic amelioration, Britain gained from southern and eastern Europe many new species of birds and insects. Moreover, many other species already present have extended their ranges northwards in Britain and Europe. The current picture, however, is complicated by the fact that the Arctic ice and glaciers began advancing south again after 1950 as the climate became colder there, with the result that many Arctic and subarctic animals have been retreating southwards, and some are still doing so in those sectors of the Arctic zone where the cooling has continued. But since 1975–80 or so, this natural cooling has been increasingly confronted and counteracted by the escalating global warming due to the anthropogenic (man-aggravated) greenhouse effect, a cause of great concern to all environmentally aware people. And so we find that northern Britain is apparently being colonised from the north by retreating northern birds such as the Redwing, Lapland Bunting, Whooper Swan, Wood Sandpiper and Snowy Owl, while in the south new species continue to invade us from the Mediterranean, aided now by the growing greenhouse warming.

As Michael J. Ford pointed out in his book *The Changing Climate: Responses of the Natural Flora and Fauna*, published in 1982, 'the location of the British Isles at the eastern limit of a large ocean means that any changes in the strength of the westerly [atmospheric or wind] circulation are quickly registered as changes in the oceanicity of the British climate. Thus the British Isles constitute a particularly sensitive area in which climatic changes and their effects are indicative of the trends prevailing throughout middle latitudes'. This is reflected extremely well in the reactions of birds and insects in the British Isles to these changes which, in many cases, are more marked than in continental Europe, and therefore explains to a large extent why events in the British Isles feature so prominently in this book. The other, less important reason is, of course, that most of my life has been spent in Britain, so I am naturally more familiar with the details of changes in the birdlife there. Nevertheless, I have endeavoured to provide a balanced account of the effects of climatic change on the distributions of birds in Europe as a whole.

In this book, then, I have sought to trace and explain in a readable and uncomplicated way these perplexing changes in bird distributions in Europe and the North Atlantic sector (particularly since the 19th century), relating them not only to climatic changes, but to other environmental factors such as alterations to habitats and effects on food supplies.

Whatever ultimately happens as a result of the anthropogenic greenhouse effect, we are still living today in an Ice Age, in fact in one of the warm interludes between the extremely cold periods (glaciations) when the ice sheets expand outwards from the polar regions. Within the warm interglacials intermittent climatic oscillations occur which alternately produce long-term or short-term periods of greater warmth or coolness. A knowledge of the Ice Ages, combined with an understanding of their effects on flora and fauna, is of obvious benefit in appreciating the reactions of wildlife to climatic changes both in the past and in the present. I have set out the background to this in Section One, especially Chapters 2 and 3. Chapter 5 and the whole of Section Two are concerned with the relatively long climatic amelioration (warming) which commenced around 1850 and lasted for about a century. Compared with the times before 1850 this period is fairly well documented ornithologically, and I have thus been able to provide in these chapters a reasonably detailed analysis of the effects of this important and quite recent climatic phase on the breeding distributions and populations of European birds, and to some extent of North American birds too. In Section Three, I have made a similar detailed review of the response of European birds to the natural climatic deterioration (cooling) which followed the 1850–1950 amelioration, and to the growing anthropogenic greenhouse effect global warming that is now counteracting it. As well as discussing the causes of this worrying greenhouse effect, Chapter 13 takes a look into the future and considers whether birds and other wildlife, sensitive indicators as they are of change, play a role in predicting climatic changes. Then follows a series of appendices (1 and 2.1 to 2.8) summarising the distribution changes in European breeding species since 1850, indicating whether they appear to be responding to climatic cooling or warming and in which directions they are advancing or retreating. Finally, a contribution by the late Kenneth Williamson on some effects of the climate on the evolution of migratory birds is also included as an Appendix to the book, making a fitting memorial to his highly valuable work in this field. Maps and diagrams are included wherever possible to enhance the text, plus 25 drawings of relevant species by John Davis.

Although aiming primarily at committed naturalists and interested laymen, I have tried to provide an accurate and useful analysis and review of the subject for serious zoologists and climatologists as well. I hope very much that it will form a basis, however imperfect, for much future research into this absorbing field of study.

John F. Burton
January 1994
Heidelberg, Germany

ACKNOWLEDGEMENTS

As described in the Introduction (Chapter 1), my original stimulus to study this subject and eventually write this book arose from early discussions with the late Dr Bruce Campbell OBE and the late Kenneth Williamson, and I am greatly indebted to both of them for that. I only wish that they could have lived long enough to see the outcome and that it would have satisfied them. I am indeed grateful to Kenneth Williamson's widow, Esther, and son Robin for generously allowing me to draw upon the draft of an unfinished book Ken was writing on the same subject, and to include one of his draft chapters as an Appendix to my book. I am also indebted to Dr David W. Snow and the late Robert Spencer for helping to make this possible.

Although she was not directly concerned with its writing, I must express my appreciation of my late wife Helga who gave me much encouragement and forbore the lengthy periods when I closeted myself in my study. The shock of her sudden and unexpected death in October, 1984, and its effects on the lives of my immediate family and on my health, was the main reason for the long delay since then in completing this book at a time when it was almost finished. That I began revising and updating it in 1989 is largely thanks to the tremendous encouragement and support given me by my long-time friend Wega Schmidt-Thomée, who typed (word-processed) the vast bulk of the final version. Elke Jantz-Kurzer and Susanne Gross did the remainder and I am grateful to them, too, as also to my former BBC Natural History Unit colleague Wendy Dickson, who typed the whole of the first draft. Another former BBC Natural History Unit colleague, and television producer, Peter Crawford, read the original draft and I am very grateful to him for his comments and encouragement and for writing the Foreword. Dorothy Vincent read through the final manuscript extremely thoroughly and made many improvements to the text for which I am especially happy to express my deep gratitude.

Others to whom I owe thanks in one way or another for their help and encouragement are Dr Hamish Batten and his wife Audrey, Prof. Dr Hans-Heiner Bergmann, Dr Linda Birch (librarian of the Alexander Library, Oxford), K.W. Brewster, Michael Bright, Neil Curtis, Dr Roger L.H. Dennis,

Prof. Dr G Mauersberger, Chris Mead, Dr Denis F. Owen, Robin Prytherch, Eric Simms DFC, Dr John Sparks (past Head of the BBC Natural History Unit), David J. Tombs and his wife Margaret, David Tomlinson, Nigel Tucker, Mrs Joan Whittock (former BBC librarian in Bristol) and Michael G. Wilson.

I am particularly glad to acknowledge the painstaking work of Robert Hudson in tracing the great majority of the references in Appendix 3 which were missing from Kenneth Williamson's original manuscript.

An invaluable source of information on changes in the distribution and movements of European birds which I have drawn upon freely, but not specifically mentioned in the Bibliography, is the biannual report 'European News' in the magazine *British Birds*, compiled by Dr J.T.R. Sharrock from information supplied by numerous national recorders. I am greatly indebted to him and them for their magnificent work.

John Davis of the Society of Wildlife Artists is responsible for the beautiful drawings of appropriate species of birds which enliven the text, and I am most grateful to him for them. I would also like to thank the following authors and publishers for kindly permitting me to reproduce text figures, photographs and other illustrations included in this book (page numbers are given in parentheses): Dr J.C. Coulson, Editor of *The Ibis* (Figs 68a & 68b p351, Figs 69a & 69b p352); Frau Nicole Debon (Fig 10 p49); Dr R.L.H. Dennis and E.W. Classey Ltd (Fig 3 p26, Fig 6 p29); Dr Andy Gosler, Editor of *Bird Study* (Fig 50 p195, Fig 57 p231, Fig 70 p353); Dr John Gribbin and Bantam Press (Transworld Publishers Ltd) (Fig 2 p23, Fig 58 p238); David and the late Eric Hosking (Frontispiece, Fig 67 p345); Dr Brian S. John (Fig 4 p27); Methuen (Reed Elsevier Publishing Group) and the late Prof. B.P. Beirne (Fig 3 p26, Fig 6 p29); Dr C.M. Perrins, Chief Editor of *The Birds of the Western Palearctic* (Oxford University Press) for permission to use maps in that splendid work as the basis for Figs 17, 21, 22, 25, 28, 29, 32, 38, 44, 47, 50, 53 and 64–66; Dr Oliver Rackham (Fig 11 p59); Dr John Sparks (Fig 8 p39, Fig 30 p141, Fig 59 p240); and David J. Tombs (Fig 9 p40, Fig 16 p86).

Robert Kirk, Editor at Christopher Helm, has taken great care in preparing my manuscript for publication and it is a pleasure for me to acknowledge my sincere thanks to him and to his colleagues. I would also like to express my thanks to Josef Gund of Satzzeichen, Eppelheim, Germany, for preparing the maps and other figures in their final computerised form.

Finally, any errors or shortcomings in this book are, of course, entirely my own responsibility and not due to anyone mentioned above.

Section One

HISTORICAL BACKGROUND

Chapter 1

INTRODUCTION

During the last thirty years or so there has been a remarkable growth of interest in climatic changes, especially where the Northern Hemisphere is concerned; people are increasingly asking about what is happening to the climate. This has probably arisen from a growing awareness among the general public, fuelled by the media, of global warming and the possible threats it poses, together with a realisation that the climate is not as static as they may formerly have been led to believe.

Recent great technological advances have given a new impetus to the study of past climates, and have done much to lead us towards a better understanding of the climatic changes which we have been living through lately. Moreover, they offer opportunities for a better appreciation of those we face in the future.

Although there is a lot of interest in the effects of recent climatic changes on birdlife, especially in the Scandinavian countries and Finland, and numerous papers on the subject have been published in scientific journals, nobody else, so far as I am aware, has made an attempt to draw together and analyse in detail in a single book as much as possible of the vast, widely scattered literature on the subject, with the exception of Michael J. Ford (1982), and the late Kenneth Williamson, whom I greatly admired and had been privileged to know since 1954. Nevertheless, it was only after Ken's untimely death in June 1977, when a first draft of this book was in an advanced stage, that I learned that he also had embarked on a book about the effects of recent climatic changes on European, and especially British, birds. It was brought to my attention by Dr David Snow and the late Dr Bruce Campbell, and with Ken's friend and former colleague Robert Spencer acting as intermediary, I have been most kindly and generously allowed by his widow, Esther, and her son Robin, to incorporate in my book as much as I wished from the seven chapters which Ken had already drafted.

This I have done as far as has been possible, bearing in mind the marked differences between us of style and treatment. Much of what Ken had written had already been published by him in various journals, and so had been drawn upon by me with the customary acknowledgements. One of his draft

16

chapters dealt with an aspect of the subject which I had not proposed to deal with in this book, so I have decided to include it as an Appendix under Ken's name. Incidentally, I am also indebted to Neil Curtis, at the time Natural History Editor with the Oxford University Press, to whom the chapters had been submitted with a view to publication, for his helpful attitude and interest in the matter.

When I first heard of Ken's death, I immediately decided to dedicate this book to his memory, and the subsequent events I have just recounted make that decision even more appropriate. The importance of Ken Williamson's work in this field of study will be obvious from the frequency with which his published and unpublished work is quoted in the following chapters.

As it happens, it was through a discussion in 1962 in the then popular BBC radio series *The Naturalist*, chaired by Ken Williamson and produced by Bruce Campbell, who was then head of the BBC Natural History Unit, that my own interest in the influence of climatic change on wildlife was first aroused. Entitled 'The Arctic in Retreat', the discussion dealt chiefly with the response of wildlife in Europe to the 1850 climatic amelioration. I was invited to speak about the effects of the climatic warming on insects, while Ken dealt with the birds and mammals, and Professor Gordon Manley provided the authoritative climatic data to which we related the changes in animal distributions and so forth. In the course of researching my contribution to this programme, I unearthed so much intriguing data that I decided it would be worth collecting it together and publishing it at a later date with my thoughts on the subject. My original intention was merely to publish a paper concerned only with the insects in one of the entomological journals, but I have so far deferred doing so as I feel that I must research the subject still more thoroughly. However, I came upon so much more information about changes in the breeding distributions of birds related to climate changes, scattered through the literature, that in the end I decided to concentrate on them and collect it all together for publication in book form. In this way I hope I may provide a better service to naturalists and climatologists, as well as attracting the interest of the general reader.

This book then is essentially a study based upon changes in the breeding distributions of European birds which may be due to climatic change. It is the outcome of some thirty years of intermittent research, the greater part of it concentrated in the past seventeen years. I realise that it has gaps and imperfections. I have tried to cover a very large field, even vaster than I thought when I first set biro to paper, and inevitably I have probably overlooked a good deal of published and unpublished data, especially in languages other than English, French and German. In drawing together as much as possible of the widely scattered published data, I have endeavoured to include the interpretations by others of the events recorded and described, while giving my own, sometimes differing, interpretations and conclusions.

Although I have tried to look at climatic events and their implications in the Northern Hemisphere as a whole, most of my detailed information concerns Europe; but I have also tried to take as much account of North America as I can. I can gain some consolation from the fact that the most pronounced influences of recent climatic oscillations in the Northern Hemisphere have occurred in the North Atlantic sector, especially in north-west Europe. So there is some justification for this book being Europe-orientated. As an amateur, writing in my rather limited spare time, it was clear to me that if this book was ever to see the light of day—and already it has taken me three times as long to complete as I imagined—I had to draw limits somewhere.

In the first place, I have felt obliged to limit myself primarily to consulting published data. To have entered fully, as I ought perhaps to have done, into a potentially voluminous correspondence with naturalists in other countries, who might have been able to fill in some of the gaps, would have substantially delayed my book. I hope such people will feel stimulated (or stung!) into action and will publish their data, at the same time bringing to light those publications I have overlooked. Nothing in this respect will please me more. Indeed, I see this book of mine as a first, and I hope helpful, step, leading to series of other books and papers which will advance the study of the reactions of wildlife to climatic change much further than I have been able to achieve.

In the second place, I have had to set a time limit for the incorporation of information published since I commenced writing. One of the dilemmas arising from writing a book of this kind is that, because of the large amount of research involved, it takes a long time to put together, and all the while it is in preparation new articles, papers and books of relevance to the subject are constantly being published. Naturally I have wanted to be as up to date as possible and have therefore endeavoured to incorporate as much fresh information as I could; but one cannot continue revising and retyping earlier chapters indefinitely, so in the later stages of writing this book I have only made revisions where it seemed to me really necessary to do so. Sometimes this has meant adding a footnote. In general, readers will find that I have managed to revise up to the end of 1992, and in some instances up to the end of 1994. On the whole, though, I have drawn the line at adding information which did not materially alter what I had already written, particularly as regards accuracy.

I am primarily a naturalist, not a climatologist, although my studies of wildlife have led me to take a great interest in climate and its fluctuations. For a lot of the background information on the climate I have had to rely therefore on the researches and writings of experts in this absorbing field of study.

As a glance through the list of chapters will show, I have endeavoured to relate this main theme of mine to the climatic pattern of the Devensian or so-called Last Glaciation of the Pleistocene and, in particular, to the Flandrian Period in which we live. I have tried to show that the retreats and advances of animals and plants in the face of the periodic expansions and contractions of the polar ice-caps are still going on to a lesser degree to this day, in response to less extreme variations in the area of polar ice. Sometimes these changes in distribution may be quite extensive, as seems to have happened during the Little Ice Age (the subject of Chapter 4), and as has also clearly been the case during the recent climatic amelioration between about 1850 and 1950.

The effects of that quite recent long-term climatic warming, the intervening period of cooling, and the present 'greenhouse' warming have been my principal concern. I have tried to document in as much detail as possible their influence on European birds, although now and again there are references to other animals. In the final chapters, I have given as much information as I can on the trends that have developed during the forty or so years since the 1850–1950 amelioration ended; to this I have added a little healthy speculation on what may happen in the future.

The reader will find quite a lot of speculation in this book—reasoned, I hope, from the available facts. I anticipate, nevertheless, that some of my views will prove contentious, especially those which do not comply with more or less accepted theories. I hope, however, that my critics will at least find the assemblage of as much data as possible in one work useful, and my arguments and expressions of view thought-provoking. Some are bound to accuse me of seeing climatic change behind every shift in an animal's distribution, although I have paid due regard to other possible factors. Clearly, in some instances I will or may be proved wrong; but I believe that climatic factors have often been overlooked or obscured by other influences, such as the alteration of habitats. If I am instrumental in rectifying such omissions, or at least in bringing about serious consideration of the possibilities of climatic involvement, I will feel that my work has been of some service. When reading papers on changes in distribution or status of this or that particular species, it has seemed to me that too often the discussion of the possible reasons for change has suffered from the species being considered in isolation, rather than as a part of an overall pattern affecting whole groups of species. In many of these cases climatic changes have either been the underlying cause, or at least a possible contributory cause.

19

Chapter 2

THE ICE AGE AND ITS EFFECTS ON WILDLIFE

A knowledge of the history of the last great ice age and its effects on man and wildlife is of enormous help in understanding the effects on them of recent climatic changes, and in predicting the likely effects of those changes which are occurring at the present day.

THE CAUSES OF ICE AGES

Although in this book I am not primarily concerned with the causes of the ice ages, about which in any case complete agreement has not yet been reached, it is as well to mention briefly some of the main theories. One of the plausible explanations is that they are caused by slight shifts in the position of our Solar System during its orbit around the Galaxy, or by changing conditions in the path it takes. For instance, McCrea (1975), developing work by Hoyle and Lyttleton (1939) and Simpson (1957), suggested that ice ages may occur when the Sun passes through a lane bordering a spiral arm of the Galaxy containing dense interstellar clouds of dust. The infalling dust affects the Sun's radiation output and produces temporary variations which can initiate ice epochs by increasing precipitation rates. His calculations were supported by some quite independent findings by an Australian, Professor G.E. Williams. Taken together their calculations suggested that ice epochs occur twice in every orbit of the Galaxy by the Solar System (i.e. about once in every 150 million years) and last on average about 50,000 years.

This dust lane theory receives some support from the well-known fact that great volcanic eruptions ejecting huge amounts of dust into the atmosphere can temporarily upset the weather pattern, taking up to three years to settle. During the summer of 1783, for instance, an eruption in Iceland, lasting five months—the largest in historic times—produced a persistent fog over the whole of Europe and much of North America. This reduced the penetration of the Sun's rays and their warming effect on the Earth to a considerable degree, though allowing the escape of the Earth's own heat because of its greater wavelength. Thus, although the summer was very dry, and in July sultry, the ensuing months were much cooler than usual. The winter of 1783–84 set in early and was a particularly severe one, while the following summer was abnormally cold. Parson Woodforde (Beresford 1935), the 18th century English diarist, noted in late June and July 1783 the 'very uncommon hazy and hot weather' with 'the Sun very red at setting', while his contemporary Gilbert White, the famous parson-naturalist, described it thus in Letter LXV of his classic *The Natural History of Selborne* (1789):

'The summer of the year 1783 was an amazing and portentous one, and full of horrible phaenomena; for, besides the alarming meteors and tremendous thunder-storms that affrighted and distressed the different counties of this kingdom, the peculiar haze, or smokey fog, that prevailed for many weeks in this island, and in every part of Europe, and even beyond its limits, was a most extraordinary appearance, unlike anything known within the memory of man. By my journal I find that I had noticed this strange occurrence from June 23 to July 20 inclusive, during which period the wind varied to every quarter without making any alteration in the air. The sun, at noon, looked as blank as a clouded moon, and shed a rust-coloured ferruginous light on the ground, and floors of rooms; but was particularly lurid and blood-coloured at rising and setting. All the time the heat was so intense that butchers' meat could hardly be eaten on the day after it was killed; and the flies swarmed so in the lanes and hedges that they rendered the horses half frantic, and riding irksome. The country people began to look with a superstitious awe, at the red, louring aspect of the sun; and indeed there was reason for the most enlightened person to be apprehensive; for, all the while, Calabria and part of the isle of Sicily, were torn and convulsed with earthquakes; and about that juncture a volcano sprung out of the sea on the coast of Norway.'

The famous eruption of the Indonesian volcano Krakatoa in 1883 likewise produced dust clouds which drifted around the Earth and diffused to cover most of the globe. It gave rise to similar effects to the dust clouds of 1783, including a succession of glorious, fiery sunsets which, as Peter Francis (1975) has pointed out, inspired this passage in Tennyson's poem 'St. Telemachus':

'Had the fierce ashes of some fiery peak
Been hurl'd so high they ranged about the globe?
For day by day, thro' many a blood red eve,
In that four-hundredth summer after Christ,
The wrathful sunset glared against a cross ...'

So impressive have the temporary effects of volcanic eruptions been on weather over large areas of this planet, that there are some scientists who believe that veils of dust produced in periods of excessive volcanic activity may have been responsible (or at least in part) for the ice ages. Detailed meteorological records have only been kept since the middle of the last century, and the incompleteness of the geological record until recently did not help the scientists' ideas very much; there were too many tantalisingly large gaps in our knowledge of the periods preceding the ice ages, when unusual volcanic activity might have occurred. Fortunately, however, new evidence of such activity has quite recently come from the Deep Sea Drilling Project (DSDP), which has produced strong indications that there was indeed an upsurge of volcanic activity coinciding with the start of the Quaternary ice ages in the Northern Hemisphere, over two million years ago.

It has also been postulated that ice ages might be connected with continental drift. The theory of continental drift was discounted by geologists when it was first put forward by the German scientist, Alfred Wegener, but recent research has not only established it as a geological fact, but has also provided much information about the time-scale involved. Between 450 and 200 million years ago the southern part of the huge mass of what was then a supercontinent is believed to have drifted across the South Pole. As it is now known that ice-caps form more readily over land than over the sea, it is thought by the proponents of this theory that the passage of so great a land mass across the pole may have resulted in successive glaciations over what is now North Africa, South America, South Africa and Australia (Gribbin 1975a).

Finally, and most convincing of all in my view, small shifts in the angle of tilt of the Earth, and the manner in which it wobbles, over very long periods during its path around the Sun appear to be sufficient to cause glaciations. This has been strongly supported by recent research in several quarters. This, incidentally, might well explain the cooler climatic phase we who live in the Northern Hemisphere began to experience from about 1950, although cyclic variations in sunspot activity might also be involved as discussed below; because at the same time as our climate was becoming cooler, in the 1950s and 1960s, there was accumulating evidence that the Southern Hemisphere was actually warming up. For instance, in New Zealand, following the coldest thirty-five years in its recorded history, a climatic amelioration began in 1935, becoming particularly marked from 1950 onwards. Moreover, there was also evidence of similar temperature rises from about 1945 in south-east Australia, Scott Base in Antarctica, Campbell Island, and Orcadas Island in the South Orkneys (Salinger 1976). Thus the natural (i.e. not anthropogenic) changes in the Southern Hemisphere seem to mirror those of the Northern Hemisphere, and indeed it may be that it is the Arctic that calls the tune for the whole of the world's weather machine. It is in such high latitudes as the Arctic that the effects of the alterations and wobbles in the Earth's path round the Sun are felt most strongly.

22

Alterations in the magnitude of the eccentricities, or precesses (wobbles), inherent in the Earth's elliptical orbit around the Sun, and in the angle of tilt of its axis, as well as its precession[1], are known to be the causes of the long-term cyclical variations in the amount of solar radiation falling at certain seasons of the year on certain northern latitudes, and are therefore considered to be critical in determining climatic change (Figure 2). They led Milutin Milankovitch, a Yugoslav astronomer and geophysicist, to make complicated calculations of their overall effects, and thus to form more than half a century ago his initially controversial astronomical theory of the cause of ice ages. He suggested that glaciations occur when the total solar radiation for the half-year containing spring and summer at the critical northern latitudes is at a deep minimum value, and that interglacials occur when the radiation is at a high maximum value (Weertman 1976).

For long unfashionable, Milankovitch's theory was suddenly revived in the late 1970s as a result of new, supporting evidence produced by various

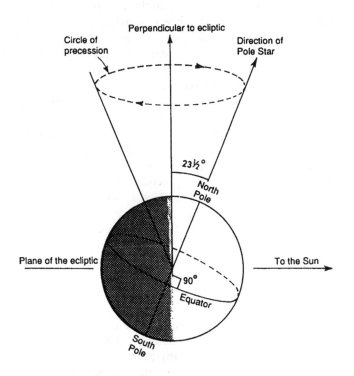

FIGURE 2 *The Earth is tilted by about 23.5 degrees out of the perpendicular to a line joining the Earth to the Sun. Changes in the angle of tilt and in the way the tilted Earth wobbles, or precesses, over thousands of years explain many features of climatic change and ice age cycles (from Gribbin 1990). (Courtesy of Dr John Gribbin)*

drilling projects; this arose from examinations of the carbon and oxygen isotope fluctuations in the ice cores and deep sea cores, the latter of which contained the remains of plankton and other small sea animals from the sea bed sediments. Using modern techniques it proved possible to date the various sediments and to measure the extent of global ice cover at the time. Furthermore, Johannes Weertman (1976) presented calculations which he believes show quite convincingly that Milankovitch's solar radiation variations are large enough to induce ice ages, although only when precipitation in high latitudes is much greater than it is at present. Moreover, further analyses and calculations by others show that Milankovitch's results really do seem to fit the oscillations of the Last Glaciation and the period since then. For an assessment of these I am much indebted to John Gribbin's recent book *Hothouse Earth* (1990). As he wrote there: 'The Milankovitch Model ... is now supported by an impressive weight of evidence and is fully accepted as the key to understanding ice age rhythms.'

But although these studies revealed that all the Milankovitch rhythms really do exist, they did not explain exactly how they control the Earth's climate. The answer to this problem was supplied in the 1980s by Swiss scientists investigating the varying concentrations of carbon dioxide in air bubbles trapped in an ice core, and by American scientists using a different technique to study its concentrations in a deep sea sediment core. Both teams found precisely the same pattern of carbon dioxide variation, the proportions of it in the air bubbles being at their lowest during the glaciation maximums and at their highest in the interglacials. In analysing their sedimental core, the Americans were able to cover a span of 340,000 years—more than three complete glacial/interglacial cycles; they discovered that changes in the Earth's orbit preceded changes in carbon dioxide levels, and that changes in the latter in turn preceded changes in climate, as shown by variations in ice cover (Gribbin 1990). As Gribbin aptly puts it: 'Somehow, the astronomical changes cause the biological pump to increase or decrease its activity, and this produces a change in the strength of the greenhouse effect which is then a key element in switching the world into or out of an ice age—and which also helps to explain ups and downs of temperature on lesser scales, *within* an ice age or an interglacial. Carbon dioxide amplifies the changes that the Milankovitch process is trying to produce.'

Gribbin goes on to say 'The effect almost certainly operates through changes in the workings of the carbon pump at high latitudes. High latitudes feel the effects of the Milankovitch rhythms more strongly than tropical regions do, and it is also at high latitudes that the deep ocean water breaks through the surface and makes direct contact with the atmosphere.' This seems to me to explain why the effects of climatic change are so much more striking at high latitudes in the Northern Hemisphere, with the result that the reactions of the flora and fauna are more pronounced and obvious, as detailed in later chapters of this book.

It has been suggested that cyclical variation in sunspot activity might be responsible for marked climatic changes and even the onset of ice ages. As well as the well-known cycle of approximately eleven years from one minimum number of spots to the next minimum, recent research has revealed that there are two additional solar cycles of 80 and 180 years respectively. Cycles of these same two longer periods have also recently been discovered in the variations of temperature revealed by the Greenland ice cores, and from studies of tree rings in very long-lived trees from Japan as well. Sunspot cycles also vary in strength (number of spots at maximum intensity) from one cycle to another, and can remain comparatively weak over long periods. They were very weak during the coldest decades of the Little Ice Age between 1650 and 1710, but they strengthened subsequently and have remained fairly strong up to the present (Gribbin 1990). It therefore appears that the relatively small variation in the amount of heat put out by the Sun during sunspot activity could be sufficient to cause climatic variations on the Earth of the magnitude of the Medieval Warm Period, the Little Ice Age and, presumably, the climatic amelioration in the Northern Hemisphere between about 1850 and 1950.

THE LAST ICE AGE

Whatever their cause, ice ages have been coming and going for many millions of years, and presumably will continue to do so; it is believed that about 650 million years ago the entire world was covered by ice. The ice age which concerns us most in this book is the last one, because most, if not all, of the fauna and flora of the once glaciated regions of the Northern Hemisphere stem from this period.

The Pleistocene epoch, the name given by geologists to the most recent ice age, began, it is thought, nearly two million years ago. Although popularly known as the Ice Age, it was not an epoch of continual extreme cold. Far from it. There were, of course, long cold periods in which the polar ice sheets expanded outwards and covered extensive areas of the Earth, but they were interspersed with warmer periods when the ice retreated. Indeed, some of these periods had a much warmer climate than we are experiencing in modern times. However, at the height of the expansions of the polar ice sheets in the Pleistocene, a third of the world's land surface was ice-covered compared with about a tenth today.

During the Pleistocene there were at least three main expansions of the polar ice sheets or, as they are generally called, glaciations. In between, there were major warm periods, or interglacials, when the ice retreated. But even during the glaciations it was not all unrelieved cold and gloom: each glaciation consisted of one or more glacial phases of varying intensity, interrupted by relatively warm phases when the climatic optimum approached, or may even have surpassed, that which we experience in the same latitudes at present.

EPOCH	PERIOD	APPROXIMATE DATE B. P. [i]	DOMINANT VEGETATION TYPE	VEGETATION TYPE	ICE SHEETS
FLANDRIAN (= Holocene) INTERGLACIAL (So-called Post-glacial)	Sub-Atlantic (Iron Age of human culture & Historic Period)	2,500 yrs.	Rather cool	Mixed woodland, grassland and arable	Absent
Start of significant human influence on the natural environment	Sub-Boreal (Neolithic & Bronze Ages.) of human culture	5,000 yrs.	Rather dry	Mixed woodland	Absent
	Atlantic (Mesolithic Age)	7,000 yrs.	Warm and oceanic	Mixed woodland	Absent
	Boreal (Mesolithic Age of human culture)	9,000 yrs.	Warm and oceanic	Mainly pine and Hazel woodland	Absent
	Pre-Boreal (Mesolithic Age)	10,000 yrs.	Becoming warmer	Henthland, Birch and Hazel woodland; some pine	Mainly absent
LAST GLACIATION OR WEICHSELIAN (= DEVENSIAN in British Isles) (Early Mesolithic & Upper Palaeolithic Ages of human culture)	Glaciation (Valley Glaciation, Post-Allerød or Younger Dryas Time) (Beirne's 3rd Glacial Phase)	11,000 yrs.	Subarctic	Tundra and heath Birch in south	On high ground
	Late Glacial or Allerød Insterstadial Phase (Beirne's 2nd Interstadial)	13,000 yrs.	Mild subarctic and oceanic	Grassland and heathland with some Birch & pine	Partly absent
	Glaciation (Beirne's 2nd Glacial Phase)	30,000 - 40,000 yrs.	Arctic	Tundra	Extensive
	Upton Warren Interstadial Phase (Beirne's Ist Interstadial)	45,000 yrs.	Warm	Woodland	Mainly absent
	Glaciation	60,000 yrs.	Arctic	Tundra	Extensive
Beirne's Ist Glacial Phase	Chelford/Brørup[ii] Interstadial	65,000 yrs.	Warm	Woodland	Mainly absent
	Glaciation	75,000 yrs.	Arctic	Tundra	Extensive
LAST INTERGLACIAL or EEMIAN (= IPSWICHIAN in the British Isles) (Middle Palaeolithic Age of human culture)		125,000 yrs.- 135,000 yrs.	Warm	Woodland	Absent
SAALIAN (= WOLSTONIAN in the British Isles) GLACIATION (Middle Palaeolithic Age of human culture)		250,000 yrs.	Arctic	Tundra	Very extensive
Earlier Glaciations and Interglacials of the Middle and Lower Pleistocene (Lower Palaeolithic Age of human culture)					

FIGURE 3 *The upper Pleistocene succession in north-west Europe (adapted from Beirne 1952; Dennis 1977).*
Notes: (i) BP = before present; (ii) Chelford Interstadial apparently not recognised by Beirne (1952).

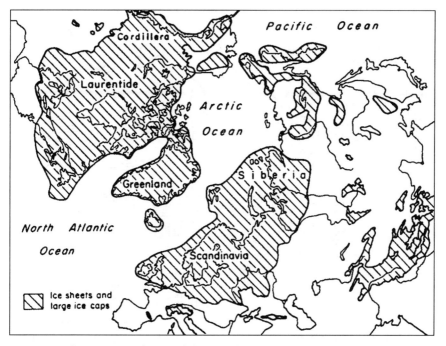

FIGURE 4 *The main ice sheets of the Northern Hemisphere at their maximum extents in the Pleistocene (from John 1977, with permission).*

The climate of the major interglacials was much warmer, and at times subtropical conditions existed in the British Isles and at equivalent latitudes in North America. For example, the last major interglacial (the Ipswichian or Eemian) extended over about 50,000–60,000 years; at its climax, conditions were much warmer than at the present day, and such animals as the short-tusked elephant, hippopotamus, rhinoceros, lion and spotted hyena inhabited Britain. Coope (1975) found that of a collection of twenty-one identifiable species of beetles made near Ipswich, Suffolk, from deposits dating from the thermal maximum of this interglacial, six species (29%) are nowadays unknown in Britain, being confined to southern Europe.

The Last Glaciation (known as the Devensian in Britain and the Weichselian in north-west Europe), which began some 75,000 years ago, consisted of at least three glacial phases—advances of the ice sheets of varying severity, alternating with warm interstadial phases when the ice retreated. The maximum extent to which the ice advanced during the First Glacial Phase is apparently not yet known precisely everywhere, but was more extensive than the subsequent glacial phases or readvances which occurred up to about 10,500 years ago.

At its maximum, the advance of the ice sheets during the First Glacial Phase covered approximately all of the British Isles, except the southern, midland and eastern counties of England and the southern third of Ireland;

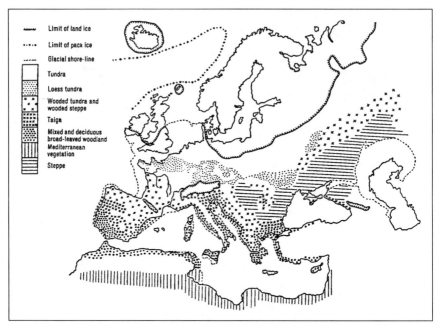

Legend:
- Limit of land ice
- Limit of pack ice
- Glacial shore-line
- Tundra
- Loess tundra
- Wooded tundra and wooded steppe
- Taiga
- Mixed and deciduous broad-leaved woodland
- Mediterranean vegetation
- Steppe

FIGURE 5 *The maximum advance of the ice in Europe during the Last (Devensian or Weichselian) Glaciation (from R.E. Moreau).*

it also covered most of northern Europe as far south as northern Denmark, northern Germany and Poland. The last readvances of the ice between 12,000 years ago and its final retreat some 10,000 years ago were much less severe and quite short-lived, the maximum duration being less than 1,000 years and most probably in the order of 500–700 years. During these final assaults of the ice before the climate really warmed up and the Flandrian Period, in which we live today, began, the ice sheets did little more than cover the highland regions of Scotland, north-west England, Wales and Ireland, while the lowlands were largely covered by tundra.

It will help our understanding of the effects on the fauna and flora, and the subsequent discussion of them, to examine these alternating cold and warm phases of the Last Glaciation in greater detail.

When it commenced about 75,000 years ago, the climatic cooling was apparently gradual. The broad-leaved forests of the latter part of the Ipswichian (Eemian) Interglacial were slowly replaced by boreal forests of birch, hazel and coniferous trees, and subsequently by tundra (John 1977). Then some 65,000–60,000 years ago the climate warmed up again for a few thousand years, this warm phase being known in Britain as the Chelford Interstadial and in Continental Europe as the Brørup. It became mild enough for coniferous forests to develop again, but eventually tundra returned as the climate deteriorated and re-established an icy grip for another 15,000 years or so.

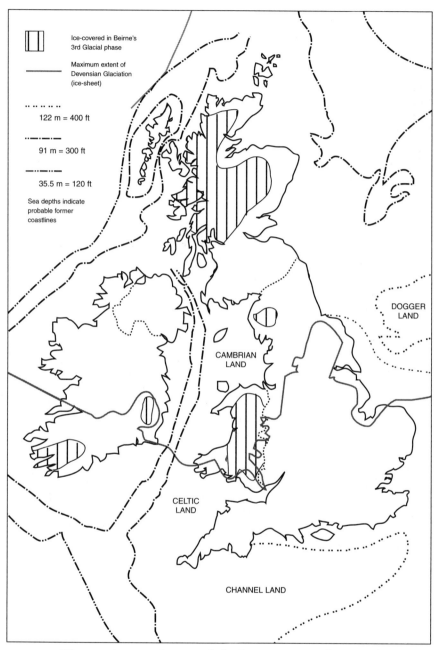

FIGURE 6 *The maximum advance of the Devensian ice sheet in the British Isles and Beirne's glacial refuges (adapted from Beirne 1952; Dennis 1977).*

This was followed about 45,000 years ago by another warm phase, the Upton Warren Interstadial, when it became so warm at its maximum about 42,000 years ago that the ice completely disappeared from Britain, and southern England became warmer and drier than it is today, with average July temperatures of 18°C. However, such warmth in the Upton Warren Interstadial lasted only for a short time, not more than about 5,000 years, before the climate gradually cooled during the ensuing 10,000 years, to be followed by the next cold glacial phase of the Last Glaciation. The period of maximum warmth of the Upton Warren Interstadial was too short-lived to allow the re-establishment of broad-leaved forests, or even the development of coniferous forests of any great size. However, it did allow the return of tribes of the Aurignacian period of human culture into the Russian steppe-like landscape.

A later interstadial between about 13,000 and 11,000 years ago, known as the Allerød Period, was mild and oceanic in character; it became at its maximum as warm as it is today, but lasted only some 2,000 years or so. Although trees, mostly birch, began to reinvade southern England from the European mainland, woodlands did not really develop on a large scale in the British Isles; instead heaths and grasslands formed the dominant type of vegetation.

When the ice sheets finally retreated about 10,000 years ago, the present so-called Post-glacial Period began; but as this is now believed to be an inter-glacial period, or possibly even a relatively minor interstadial phase, it is generally referred to these days as the Flandrian Period or Flandrian Interglacial. The climate gradually warmed up and, as it did so, the sea level rose steadily, until the climatic optimum was attained at the end of the Middle Stone Age and the beginning of the Neolithic Era (New Stone Age), some 7,000–6,000 years ago.

So to sum up, the Pleistocene was an epoch of tremendous and complex climatic changes with big fluctuations in sea level, which rose when the ice sheets melted in warm periods and dropped when the ice re-formed. Moreover, within the major cold phases there were comparatively minor fluctuations of varying duration when warmer interludes brought a respite from the icy grip of the polar ice sheets.

EFFECTS ON ANIMALS AND PLANTS

Not only did the alternately advancing and retreating ice sheets leave their mark on the land, but the climatic fluctuations which gave rise to them inevitably had profound effects on the animals and plants inhabiting the affected areas and, indeed, the Northern Hemisphere generally. These were repeatedly caused to evacuate old haunts and colonise new ones in more favourable areas, then to recolonise their former haunts when conditions returned to something like their original state.

To put it in its simplest form, each time the ice sheets advanced during the Last Glaciation, for example, northern species retreated before them and colonised areas further south in Europe, while southern species retreated even further south. When the ice sheets retreated, the northern species returned northwards and recolonised their old haunts. The territory they vacated was reoccupied in turn by the southern species moving north in their wake. And so they have see-sawed back and forth as the ice sheets advanced or retreated. Of course, the process may have taken many years, as species cannot colonise new areas or recolonise old ones until the climatic and environmental conditions are suitable.

Since the final retreat of the ice sheets marked the conclusion of the last glacial phase, southern species have slowly recolonised the areas they were forced to evacuate. Indeed the process may be considered to be still going on, held up now and then by the occasional cold periods or 'Little Ice Ages', such as that from AD 1250 to 1850, and speeded up by the warm ones, like that from AD 750 to 1250.

Only fairly recently (about 1950) we emerged from a warm phase which lasted about 100 years and which reached its maximum between 1920 and 1940. It hastened the northward spread of many species whose centre of distribution is mainly in the Mediterranean. Since 1920 we have gained as breeding species in Britain at least nineteen species of moths and eleven species of birds which have been expanding their ranges northwards or westwards in Europe. Although a new climatic deterioration from about 1950 to 1975 began applying the brake, it was some time before the advance of these species showed signs of slowing down. For example, from 1972 Cetti's Warbler colonised southern England with remarkable speed, while at the same time the post-1950 deterioration began to stimulate the colonisation of northern Britain by subarctic species of birds, such as the Great Northern Diver, Goldeneye, Snowy Owl, Wood Sandpiper, Temminck's Stint, Redwing and Lapland Bunting.

Thus a study of the effects of the climatic fluctuations of the Last Glaciation on the faunas and floras of those distant times not only enables us, by comparison, to understand the significance of some of the recent and current changes in the geographical distributions of animals and plants, but may also provide some fairly early indications and confirmation of the approach and magnitude of climatic changes. Furthermore, as Roger Dennis has remarked to me (in litt.), a study of present changes in the distributions of animals and plants allows a fuller appreciation of those that would occur on a far larger scale during a glaciation.

One of the most stimulating and interesting accounts of the effects of the Last Glaciation on animals is Professor B.P. Beirne's reconstruction (1952) of the origins and history of the British fauna. Although he based his reconstruction upon a sequence of glacial and interglacial phases deduced by F.E. Zeuner and now generally regarded as out of date, and despite the fact that some of his deductions are much open to question in the light of subsequent studies, I nevertheless believe that we owe a great debt to Beirne for his admirable and thought-provoking analysis. In any case, I believe that some of his conclusions and ideas may yet prove to be more or less correct.

Beirne based his conclusions upon evidence gleaned from the pollen record (for vegetational changes) and from palaeontological, historical, ecological, taxonomic and zoogeographical evidence. He postulated that the present British fauna is made up of the survivors of at least six main periods of invasions corresponding with various climatic periods. He correlated these invasions with three cold glacial phases of the Last Glaciation (arrival of species with Arctic or northern centres of distribution) and two warm interstadial phases plus the subsequent warm Holocene (= Flandrian) Period (arrival of species with temperate or southern centres of distribution); his correlations, however, do not appear to fit the modern picture of events at all closely. I believe the divergence is largely one of inaccurate dating. Thus, although Dennis (1977), for instance, seems to correlate Beirne's First Interstadial with a period of time spanning the early Devensian (Weichselian) Glaciation and the Upton Warren Interstadial, I think it should have been correlated with the Upton Warren Interstadial Complex only.

In the following account, I have opted to follow Beirne's theories, but have modified them in the light of more recent studies such as those of G.R. Coope (1965, 1975) and A.J. Stuart (1974). I have also largely drawn upon examples from the British Isles, as they more or less parallel events on the continent of Europe, though it is difficult nevertheless to correlate them exactly with similar events there; in any case, that is beyond the scope I have set myself in this chapter.

As I have already described, following the final waning of the very warm Ipswichian (Eemian) Interglacial, the cold glacial phase which began about 75,000 years ago was checked after some 10,000 years by a temporary amelioration of climate—the Brørup or Chelford Interstadial—lasting only some 2,000 years or so. At its maximum this interstadial apparently did not attain a high degree of warmth. Coope (1965, 1975) concluded from his investigation of fossil beetle and other insect assemblages that the climate of the Cheshire Plain of England was similar to that of south-central Finland at the present time, i.e. with average July temperatures of about 15°C, a degree or two lower than now. It was apparently rather more continental than during the middle of this century, and the insect fauna was characteristic of cool temperate coniferous forests. At least two insects present in England at that time are unknown today, but still occur in Scandinavia. Thus it would appear that during the Chelford Interstadial the bird and other animal life of Britain would have been like that of the southern half of Fenno-Scandia today, with beavers, elks, giant deer, red foxes and wolves present.

Little seems to be known about the British fauna of the preceding cold phase, although the beetles identified from deposits of this period, such as *Bembidion dauricum* and *Helophorus obscurellus*, indicate, to quote

Professor Coope, 'a harsh climate of arctic severity' and a steppe-tundra land-scape of extreme barrenness with but a scant vegetation. From this, one can imagine a bird life reminiscent of that of Lapland today; while, surprisingly, in spite of the adverse climatic change, large carnivores such as lions, spotted hyenas and wolves apparently survived wherever large herbivores remained abundant, as elsewhere in Europe (Stuart 1974).

With the passing of the Chelford Interstadial, Arctic conditions once more returned to Britain with increased severity, so that tundra must have covered most of the country and north-west Europe. Most, if not all, of the species that inhabited this region during the Chelford Interstadial, plus any species of temperate conditions that had managed to survive from the latter stages of the Ipswichian (Eemian) Interglacial, must have been pushed far to the south (and completely out of the British Isles) and replaced by species characteristic of the tundra which had themselves retreated from the far north, or which had descended from alpine regions of central Europe and migrated westward into Britain. As Professor Gordon Manley (1972) put it: 'the slate was virtually wiped clean'. However, even at its maximum, the ice sheets of this glacial phase did not penetrate anything like as far south as Beirne (1952) believed, assuming, of course, that this glacial phase corre-sponds more or less with the latter part of his First Glacial Phase. If my assumption is correct, he understood the ice sheets to have covered the whole of the British Isles except southern England and the extreme south-west of Ireland; in fact, the exact limits reached by the ice are not completely known. Indeed, there is apparently little if any evidence for the early part of the Devensian (Weichselian) glaciation in the British Isles. Nevertheless, one can assume that even southern England was colonised by such Arctic and subarctic mammals as the musk ox, woolly mammoth, reindeer, Arctic fox, Arctic and Norway lemmings, and mountain hare, and probably such birds as the Snowy Owl, Snow Bunting and Ptarmigan, plus elements from the steppes further east in Europe. Insect fossil evidence from this period suggests an insect fauna typical of the extreme north of Europe.

Animals characteristic of the former temperate conditions would not, however, have been completely eliminated. As the ice sheets advanced and the northern seas froze, so the sea level dropped drastically, perhaps up to 300 feet (90 metres) or more. As a result land formerly under the sea, far beyond the former British and west European coastlines and the reach of the ice, was opened up for colonisation by the retreating populations of warmth-loving species.

These ice-free regions, which had once been beneath the sea, became refuges for survivors from the ice-covered and tundra-covered regions, and these species were later able to expand and repopulate their former territory when the ice finally retreated. Beirne (1952) suggested that 'considerable differences must have existed in the climate of different parts of the ice-free areas' depending upon their distance from the ice fronts, their distance from the sea, and their latitude and altitude. Research in recent years by Professor

C. Lindroth of Sweden and others has shown that a remarkable number of plants and invertebrates are surprisingly tolerant of Arctic conditions, and are able to live in close proximity to present-day ice sheets and glaciers in Iceland, Scandinavia and elsewhere in the Arctic regions; moreover, that a number of species characteristic of temperate zones are able to withstand the subarctic climate in refuges where, for various reasons, the microclimate is warmer than surrounding areas.

So it is possible that the climatic differences between various parts of those ice-free areas that became refuges for retreating temperate-phase species were less marked than Beirne suspected; but nonetheless it was probably less difficult than was previously thought for many non-Arctic species to survive in them despite the comparatively close proximity of the ice sheets.

In general terms, therefore, one could compare the fauna and flora of these glacial refuges with the present-day fauna and flora of parts of Iceland, Scandinavia and northern Britain. It seems not impossible that just as in these regions today we can sit in warm spring or summer sunshine in a green meadow surrounded by a profusion of wild flowers and insects, yet within sight of a glacier, so such scenes may have existed within sight of the glaciers during the glacial phases, and perhaps not too far from the ice fronts either.

Beirne named four main areas in which British temperate-phase plants and animals survived the glaciations: the Celtic Land to the west and south-west of the British ice sheets, the Cambrian Land occupying what is now the Irish Sea and St. George's Channel, the Channel Land which is now occupied by the English Channel, and the Dogger Land to the south of the ice sheets in what is nowadays the North Sea. All these refuges were geographically continuous with each other, but differed to some extent in the composition of the plants and animals which colonised them during glacial phases. Their climate and extent also varied according to the relative advances of the ice sheets and the height of the sea during the different glaciations (see Figures 4–6).

With the start of the remarkably rapid amelioration of the climate from about 45,000 years ago, known in Britain as the Upton Warren Interstadial Complex and which more or less corresponded with similar interstadials in north-west Europe, the Arctic-alpine animals which had retreated southwards and colonised areas far to the south would have commenced, in the wake of the retreating tundra and ice, the recolonisation of their former haunts in the north, and would thus have disappeared from southern Britain and parallel latitudes on the European mainland. Similarly, the improving climate would have encouraged the gradual return from the Celtic Land, and other refuges, of those temperate-phase species which had previously been expelled from their pre-glacial haunts by the Arctic conditions.

Thus, as Professor Coope (1975) has shown, the Arctic insect fauna of the English Midlands was quickly replaced by a considerable number of warmth-loving species; climatically-contrasting groups of species followed one another in such rapid succession that they imply a rapid and intense sequence of climatic change. The presence of certain southern species indicates that for perhaps a thousand years or so the average July temperatures in England rose to as much as 17–18°C—a degree or two higher than they are today. He therefore suggested that the climate of England at that time would have been as continental as that of the north German plain or southern Sweden today. However, despite its warmth, the Upton Warren Interstadial, as indicated by its insect fauna and the pollen counts, was open and almost treeless. Coope reasons that this was due to its short duration, probably not more than 5,000 years; not long enough to allow the return of forest trees from those areas far to the south to which they had been obliged to contract their ranges.

Furthermore, since this interstadial was much shorter, and at its maximum was cooler than the Ipswichian interglacial (when, it will be remembered, hippos, rhinos and short-tusked elephants roamed southern Britain), only the species characteristic of that period's less warm phases would have returned. Apparently those spotted hyenas, lions and ground squirrels which had returned did not finally disappear from the British fauna until the end of the Upton Warren Interstadial. Nevertheless, the fact that its climatic optimum was warmer than now would have encouraged the recolonisation of Britain, at least for a time, not only by warmth-loving insects, but by other animals which would find it too cold here nowadays. For example, southern Britain, at any rate, might well have been inhabited by such typically Mediterranean birds of today as Red-rumped Swallow, Bee-eater and Calandra Lark. This seems all the more likely when one considers the speed with which some Mediterranean species have spread north during a mere thirty years or so of climatic amelioration in the 20th century.

As the climate cooled again and the warm Upton Warren Interstadial came to quite a quick end, some 40,000 years ago, so these southern and temperate-phase species must have retreated southwards once more, and eventually would have been replaced by species characteristic of Arctic, alpine and steppe regions spreading in from the north and east. Coope (1975) believes that it probably took about 2,000 years for conditions to return to Arctic severity from the optimum warmth of the interstadial. He states that 'the insect assemblages dating from this period of climatic deterioration show a progressive increase in the numbers of exclusively eastern species, implying that as the summers grew colder there was a corresponding increase in climatic continentality'. For the next 20,000 years or so the summer temperatures averaged around 10°C. Then, by about 25,000 years ago, Arctic tundra became fully re-established in southern England, and a new glacial phase began. The climate apparently became less continental, and the ice sheets spread further south than hitherto in the Devensian (Weichselian) Glaciation, and reached their maximum (see Figure 6) about 18,000 years ago. Once again the sea level dropped, perhaps by 300 feet (90 metres) or more below its present depth, so that huge expanses of land reap-

FIGURE 7 *Snowy Owl*

peared off the former coast to become refuges for retreating species, where the climate was mild enough. Incidentally, during the climatic optimum of the Upton Warren Interstadial, the sea may have attained a higher level than at the present day.

Much of England south of Yorkshire, as well as the southern quarter of Ireland, would have consisted of tundra with typical low-growing Arctic mosses, lichens and plants such as dwarf willow, with perhaps birch heaths in sheltered parts of the extreme south. The wildlife would presumably have been like that of present-day Lapland, with elk, reindeer, Arctic fox and Arctic lemming, as well as Snowy Owls, Snow Buntings and Ptarmigan. In addition, musk ox, woolly mammoth and woolly rhinoceros were also to be found. The insect fauna was certainly Arctic in character.

Most of the 'refugees' from the Upton Warren Interstadial would probably have clung on in the land adjoining the Atlantic Ocean. Beirne considered that they were concentrated in the region between the present south coast of Ireland and the north coast of France, with others inhabiting what is now the bed of the English Channel. The 'southern' species that survived this new and severely unfavourable climatic period best of all were those that were already adapted to life in maritime habitats, or which were able to adapt to such habitats. This often involved changing their ecologies to a marked extent.

Although adaptation to maritime conditions was perhaps the chief way in which species survived in glacial refuges such as the Celtic Land, others survived by becoming adapted to the heaths and moors which probably covered much of such areas. The ringed carpet moth *Cleora (=Boarmia) cinctaria* is an example of a woodland species which became adapted to heaths and moorland. To this day in the British Isles it is confined to such habitats, where its larvae feed upon heaths, bilberry, bog myrtle, birch and sallow, whereas its continental populations continue to inhabit woods and feed upon various quite different foodplants, such as St. John's worts, mugwort and yarrow.

During the latter part of the main glacial phase of the Last Glaciation (its final climax), some 15,000–13,000 years ago, the climate began to improve rapidly and the ice retreated; but there were set-backs when the retreating ice halted or even readvanced considerable distances. Eventually, however, the ice sheets disappeared from the British Isles and a large part of north-central Europe, marking the onset of the Late-glacial Interstadial phase or Allerød Period.[2] It was a period of chiefly mild and oceanic-type climate, and became at least as warm as our present climate and probably even warmer, though not quite as warm as the climatic optimum of the Upton Warren Interstadial. Most of the British Isles appears to have been vegetated with grasslands and heathlands with a good variety of wild flowers, but tree-birch invaded from the south, and parts of England became quite well wooded with birch and in the extreme south-east by some other trees, such as Scots pine. Here wild elk lived, while the giant deer continued to survive in Ireland and elsewhere.

Although the Allerød was relatively short, perhaps little more than about 2,000 years, many plant and animal species which had been forced to retreat returned from the Celtic Land and other refuges, as they had done during the Upton Warren Interstadial; or they spread back to England from the European mainland via the Channel Land and North Sea land connections, which had not yet been cut off by the sea. Moreover, the invasion seems to have been on a massive scale. Many of the early invaders must have been species tolerant of cold or cool climates, such as we find in northern Britain and northern Europe today—especially those adapted to grasslands, heaths and moors with birchwoods and scrub: birds such as the Meadow Pipit, Twite, Bluethroat, Wheatear, Ring Ousel, Hooded Crow, Merlin and Willow Grouse, with Black Grouse, Hazel Hen, Fieldfare, Redwing, Willow Warbler, Redpoll, Siskin and Brambling in the incipient birchwoods.

However, they were soon followed by temperate-phase and then southern species, for, as Professor Coope's studies of fossil beetles (1975) have shown, the transition from the previous Arctic climate to 'the thermal maximum of the interstadial was remarkably sudden'. Even as far north as the Isle of Man and North Wales, the Arctic beetle fauna was speedily eliminated, and was replaced by an assemblage of species containing many whose distributions today are relatively southern, 'so much so that the northern limits of the geographical ranges of some of them do not extend even as far north as southern England'. Based on this evidence from the beetle deposits in North Wales, an attempt to estimate the actual rate of the climatic amelioration at the start of the interstadial gave a rather unexpected rate of change, according to Coope, of 1°C per decade! Such can be the speed of climatic change.

During the warmest part of the Late-glacial Interstadial (Allerød) Phase, between 13,000 and 12,000 years ago, the British beetle communities from Cornwall to Cumbria and the Isle of Man were characterised by the same warmth-loving, southern species. This is probably when some of our grasshoppers colonised Britain, including the lesser mottled grasshopper *Stenobothrus stigmaticus*, a central European species, now known in Britain only in the extreme south-east corner of the Isle of Man. In Ireland and Scotland, however, the insects appear to have been more characteristic of temperate zones.

If such were the effects of this Late-glacial period of some 2,000 years of climatic amelioration on beetles, and bearing in mind the alacrity with which many 'southern' European birds, moths and other insects have responded to the recent climatic amelioration of little more than a century (the main theme of this book), one is greatly tempted to wonder and speculate about the composition of the rest of the British fauna of that distant time.

The Late-glacial Interstadial (Allerød) started to wane soon after 12,000 years ago, although a more precise date has apparently not yet been determined. Again, based upon evidence from numerous assemblages of fossil beetles from the period between 12,000 and 11,000 years ago, which show a rapid disappearance of southern species from the British fauna, Professor Coope (1975) suggests that the deterioration of the climate was almost as marked and as sudden as the amelioration that began it. Northern species appeared in progressively increasing numbers as the climate worsened, so that 'between 11,000 and 10,000 years ago the insect fauna of Britain was characterised by a preponderance of high northern species, many of which do not extend as far south as these islands today'.

For 600 to 700 years, a final glacial phase (the Younger Dryas or Post-Allerød Time = Beirne's Third Glacial Phase) was then experienced in Britain and elsewhere. Those plants and animals characteristic of Arctic regions which had been compelled by the Late-glacial Interstadial to retreat northwards, now moved south again, and those that had survived in alpine regions further south were now driven out of the mountains by the ice which formed there. They presumably reoccupied much of their former territory as tundra or tundra-like vegetation became re-established and dominant (though less extensively than in the previous glacial phase) over much of the British Isles except the south, as well as in comparable latitudes elsewhere. The glaciers and ice sheets of this glacial phase were, however, largely

FIGURE 8 *Tundra, such as this in Arctic Russia, became dominant over much of north-central Europe, including the British Isles, during the glacial periods. (Photograph: Dr John Sparks)*

restricted in the British Isles to the Scottish Highlands and the highest parts of Cumbria, Wales and Ireland. Birchwoods, although much reduced in lowland Britain, survived in sheltered areas, at least in the south, and even in Denmark.

Once more, the drop in the sea level was sufficient to expose areas of formerly submerged land far enough from the ice sheets to provide refuges for retreating species which had invaded the British Isles during the Late-glacial Interstadial (Allerød Period). Moreover, because of the less extreme climate, it may have been possible for many of them to survive in other areas formerly submerged by the sea which would have been quite untenable in the greater cold of the preceding main glacial phase: i.e. the Irish Sea, St. George's Channel and the southern part of the North Sea. Professor Beirne (1952) believed that many more species survived in the English Channel, and on land off the present west coasts of Ireland and Scotland, than during the main glacial phase. As during previous glacial advances, some species had to alter their habits to survive.

Professor Beirne also mentioned that the growth of dense forests in the Dogger Land, now lying beneath the southern part of the North Sea, may have caused some species characteristic of more open habitats, such as downland and light woodland, to become adapted there to fens and salt marshes. One of the nicest examples, I think, is that of the swallowtail butterfly *Papilio machaon*. On the European mainland it inhabits a wide

39

range of habitats from low-lying marshes and meadows to open woodland, and even high alpine meadows where its larvae feed on several different kinds of plants of the carrot family (Umbelliferae). The slightly different-looking and darker British race, *britannicus*, on the other hand, is entirely confined to the fens of East Anglia where its larvae feed upon a single species of foodplant, the milk parsley *Peucedanum palustre*. The paler continental form of the butterfly, *gorganus*, sometimes migrates to the southern counties of England, and occasionally breeds, laying its eggs upon various umbelliferous plants as on the continent; but nowadays it never really succeeds in establishing itself there, although there is good evidence that it may have done so, at least temporarily, in the past.

The likely history of the species in Britain is as follows: in a warm period during the Last Glaciation, probably the Late-glacial (Allerød) Interstadial Phase, the typical continental race of the swallowtail colonised England, but when the climate deteriorated and the ice advanced once more, it retreated south and south-eastwards. However, a substantial population managed to survive in the Dogger Land, in what is now the southern North Sea, where for a time the type of varied open country to which it was accustomed prevailed (Figure 9). Eventually though, thick forests are thought to have developed, producing an unsuitable habitat for the swallowtail, and this, combined with the increasing cold, caused it to adapt to the coastal marshes and fens, these being the only suitable open habitat available. Here milk

FIGURE 9 *This view of De Grote Peel peat bog in Limburg, south-east Holland, gives an idea of what Dogger Land may have looked like not long before it was inundated by the North Sea after the Last Glaciation. (Photograph: David J. Tombs)*

parsley was probably the dominant umbellifer and became the only food-plant of this, by then, isolated population of the swallowtail.

When the sea rose during the Post-glacial Period, it gradually swamped most of the swallowtail's fenland population in the southern North Sea area, except for the fens of East Anglia, and perhaps those of the Thames estuary; in the first-mentioned restricted area it has tenaciously survived to this day. Its long isolation from the main continental population caused it to evolve into a distinct subspecies.

Such a history might also explain why some species of birds which at present have a southern European or Mediterranean centre of distribution, nevertheless have (or had until recently) breeding outposts in the fens of the Low Countries and England's East Anglia. During the last glacial phase the main populations would have retreated southwards towards the Mediterranean, but isolated populations probably survived in the then rich coastal fenlands of the southern North Sea area. Because these populations did not become isolated, most species being strongly migratory, they presumably did not need to become as specialised as the swallowtail and large copper *Lycaena dispar* butterflies, since the fens retained rich food resources for them as well as an abundance of suitable nesting sites. This probably compensated for the cooler, shorter summers they endured. Examples of such species are Purple Heron, Night Heron, Little Bittern, Bittern, Savi's Warbler and Bearded Tit. As with the swallowtail and large copper, they survived only in coastal fens on the English and Dutch coasts when the sea level of the North Sea rose in the Post-glacial and submerged the Dogger Land between. All these species survived to modern times more successfully in the more extensive fens of the Low Countries than those of eastern England. They probably prospered in both areas, however, during the Post-glacial (Flandrian) Warm Period, some 6,000 years ago, and in the Medieval Warm Period.

Extreme specialisation such as that of the swallowtail butterfly would have been essential for the survival of an isolated population in such an unfavourable climate at the limit of its range, but not for the more highly mobile birds. It is now generally accepted that species at the edge of their range stand a better chance of survival if they evolve specialised adaptations to a coastal habitat. This explains why so many animal species in Britain are confined to the coast, but are widespread in a variety of habitats on mainland Europe.

It is only fair to mention that some writers, such as F. Balfour-Browne (1958) and R.L.H. Dennis (1977), did not accept Beirne's hypothesis and argued quite reasonably that the British race of the swallowtail, like the British and Dutch fenland races of the large copper butterfly, evolved in isolation during the early Post-glacial (Flandrian) Period, and not as isolated survivors of a glacial phase. They believed that these butterflies invaded and colonised the southern North Sea area following the final retreat of the ice sheets after the Post-Allerød Glacial Phase. To put it simply, they consider, if

I understand them correctly, that the rising sea level led to such inundation of the coasts and greatly raised water tables that, by the Boreal Period, marshland had vastly expanded at the expense of drier ground. Thus the populations of these two species of butterflies in the marshland regions were forced to adjust more and more closely to the increasingly dominant fenland habitat; and in time the gradual development of dense forests isolated them so effectively from the populations inhabiting other types of open habitat in surrounding areas that they diverged sufficiently, both ecologically and geographically, to become distinct subspecies, all gene flow between the different populations having been interrupted. They may be right; but I must confess that I am more impressed by Beirne's argument, chiefly on the grounds that I think it would have required more time than has elapsed since the Boreal Period for such marked subspeciation to have occurred. Moreover, I do not believe that conditions would have been so severe in the Dogger Land area during the Post-Allerød glacial phase that the development of mature fenland, or the survival of appropriately adapted butterflies like the fenland forms of the swallowtail and large copper, would have been prevented.

The tendency for populations of species of animals isolated in the British Isles to develop quite independently of their main populations elsewhere, and to evolve specialised ecologies to the extent that they often became distinct races or subspecies, was a feature of the Pleistocene epoch. Some populations even became sufficiently different to warrant being classified today as distinct species (e.g. the Scottish Crossbill *Loxia scoticus* and the robber fly *Epitriptus cowini* (Asilidae)). Thus many of the distinct races of animals and plants of the British Isles, and indeed elsewhere in the Northern Hemisphere, arose from populations which became isolated in this turbulent period of the Earth's recent history.

These effects of the fluctuating climate and the advances and retreats of the ice sheets on the wildlife of the British Isles during the Pleistocene epoch were more or less paralleled elsewhere in Europe, in Asia, and in North America.

In eastern Europe the polar ice sheets are known, at their maxima, to have extended as far south as the Carpathian Mountains and almost to the Black Sea. Further east, in Asia, they stretched more than halfway from the Arctic Ocean to the Caspian and Aral Seas, but much less far south and east in Siberia and eastern Asia generally than might have been expected. Apparently, this was because the cooling during the glaciations was insufficient to bring the snow-line down within the zone of heavy snowfall. In these high mountainous areas of hot summers and severe winters, the level of maximum snowfall is often many thousands of feet below the snow-line. Thus mountain glaciers and ice sheets were often prevented from forming. Those that did form were less extensive than in the lower regions further west, where in any case the moist westerly winds of the Atlantic brought plenty of precipitation to feed the ice domes and sheets.

In North America, the advances and retreats of the ice sheets more or less corresponded with those of Europe. The Last Glaciation, known there as the Wisconsin, covered at its maximum most of Canada (but not Nova Scotia or Newfoundland) and the northernmost states of the USA, such as Wisconsin. The Great Lakes were, in fact, formed from the melt-water of the great ice sheets. The effects of these advances and retreats of the glaciations on the plant and animal life were presumably broadly similar to those in Europe. Although the advancing ice forced the tree-line back, and caused populations of most species to retreat southwards, some populations probably managed (as in western Europe) to find refuge until better times in unglaciated areas on the east and west coasts, and in new coastal lands provided by the drop in sea level consequent upon the freezing of the northern seas. With the return of warmer conditions and the disappearance of the ice, they were able to regain lost territory in the north.

As pointed out by Gribbin (1990), the ice did not retreat everywhere at the same time, and 10,000 years ago parts of northern Europe still had more ice than at the present day, while in North America the ice had barely begun to retreat in earnest. Eight thousand years ago the North American ice sheet was half the size it was during the last glacial maximum some 18,000 years ago, and the Scandinavian ice sheet had only just disappeared. He further states that the ice sheet persisted over Labrador until only 4,500 years ago, and up to that date the proximity of the ice, and cold air blowing from the Arctic, kept eastern North America cool and the summers there chilly.

[1] Change by which the equinoxes occur earlier in successive years.

[2] For simplicity's sake, I have included the Bølling Interstadial in this period. On that part of the continent of Europe adjacent to Britain, the Late-glacial Interstadial does not quite coincide and was divided by a couple of hundred years of cold conditions; there, the early phase of climatic amelioration is called the Bølling, and the later the Allerød, after the locations from which they were first described.

Chapter 3

FROM THE ICE AGE TO AD 1250

The final withdrawal of the ice sheets from the British Isles, Northern Europe, Asia and North America some 10,000 years ago to something approaching their present limits marked the end of the Pleistocene Ice Age, and heralded the so-called Post-glacial or Holocene epoch in which we now live. But since we have no means of knowing whether or not the Last Ice Age has really ended, modern authorities have preferred to rename it the Flandrian Period or Flandrian Interglacial.

The changes in the distributions of the flora and fauna that followed the end of the last glacial phase (Post-Allerød or Younger Dryas Time) of the Last Pleistocene Glaciation broadly paralleled those which occurred in previous warm phases. Indeed, as I have already indicated, it is believed by many climatologists that we may well be living in one of these relatively short-lived interstadials, or at best a major interglacial period. However, the former seems more likely as, in spite of warm oscillations, the climate during the past 5,000 years has become progressively cooler than it was during the very warm Boreal and Atlantic Periods, some 9,000 to 5,000 years ago; so that, judging by the duration and history of previous Pleistocene inter-glacials, we may long ago have achieved and passed the natural climatic optimum of ours, and could be near its conclusion. But, if unchecked, the present warming climate due to the anthropogenic (caused by man) green-house effect could 'buck' the trend; more of that in the final chapter.

A good deal is known nowadays about the climatic and vegetational history of the British Isles and elsewhere in Europe, as well as of North America, since the end of the Last Pleistocene Glaciation. The results of pollen analysis, in particular, have shown us the composition of at least the dominant vegetation over the ages since then, and thus have incidentally shed much light on the changing climate.

The climate of the first stage of the Flandrian Period (Holocene), the Pre-Boreal, was at first subarctic, characterised by tundra-type vegetation, but as it improved rapidly the partly waterlogged tundra, with its many lakes and pools, gradually dried out and was invaded by heaths, birch, juniper and willows, followed later by Scots pine and other trees, such as hazel.

44

Animals typical of the tundra, such as reindeer, Arctic foxes and lemmings, Ptarmigan, Snowy Owl and the polar fritillary butterfly *Clossiana polaris* would have begun to retreat northwards, and from the low to the high ground. The woolly mammoth had become completely extinct by the beginning of the Flandrian, although it seems to have survived in Siberia almost until that date (Stuart 1974). Following the retreating tundra species, and in keeping with the spread of heather moorland and birch, juniper and willow scrub, would have been animals characteristic of these habitats, such as the Redpoll, Redwing, beaver and Frigga's fritillary *Clossiana frigga*. Already the summers may have become as warm as those of the present day.

Although by the end of the relatively short Pre-Boreal phase (around 1,000 years) the ice sheets had completely vanished from Britain and some other parts of the Northern Hemisphere, the sea, although rising, was still low, and much land then remained beyond present-day coastlines. For instance, eastern England as far north as Yorkshire was still connected to the European mainland by a broad, low-lying plain extending north-east to Denmark. However, the sea had already begun to encroach upon these areas of land in which, it will be remembered, many animals and plants characteristic of temperate conditions had sought refuge from the severe climate of the last glacial phase. Presumably at least some of these would by then have commenced the return to their ancient haunts in response to the ameliorating climate, along with other species invading from the European mainland by the remaining land bridges.

Apparently, by the late Pre-Boreal birds such as the Crane, White Stork, Red-breasted Merganser, Great Crested and Little Grebes, Lapwing and Buzzard, and mammals such as the badger, hedgehog, aurochs, roe deer, red deer and beaver had already reached as far north in England as Yorkshire, since their remains have been recovered from a Mesolithic human settlement of this age at Star Carr (Clark 1954; Stuart 1974).

With the onset of the Boreal Period (some 9,000 years ago) the climate became appreciably drier and warmer in summer (about 2°C warmer on average than at the present day), though it remained colder in winter than now. The sea level rose in earnest and, aided and abetted by land subsidence in some regions (e.g. the Baltic and North America north of the Great Lakes), encroached upon much hitherto dry land, greatly enlarging, for instance, the Baltic and North Seas. Eventually, it rose so high because of the increasing warmth that, at the end of the Boreal, the land connection between the British Isles and the European mainland was finally broken about 7,000 years ago, and the sea circulated all round the British coasts. This free circulation of the salty water from the Atlantic Ocean was probably, in Sir Dudley Stamp's view (1946), a cause of the abrupt change in north-west Europe from the dry and warm continental-type Boreal climate to the wetter, but nevertheless still warm, climate of the phase which followed—the Atlantic Period—during which the Post-glacial Climatic Optimum (the terms Flandrian or Holocene Warm Period are now preferred by most climatologists) was soon attained, some 6,000 years ago.

The dry warmth of the Boreal Period led to the virtual disappearance of the tundra from the British Isles and large areas of northern Europe, and from western North America too, except on the highest ground. It continued to be replaced by the northward advance of heathland and birchwoods, but in the south these were in turn being successfully challenged for supremacy by the spread of woods of hazel and Scots pine. For a time, birch and hazel were co-dominant, but pine gradually became more abundant and extended its range northwards. By the end of the Boreal Period birch was on the decline, and pine was steadily replaced from the south by the advance of elm and oak. Indeed, the final stages of the Boreal saw the establishment of the oak forest which has remained dominant in the British Isles right up to the present day. The Middle Stone Age human inhabitants, being hunters rather than farmers, made insignificant, if any, impact on the woodlands of that time.

The Boreal also saw the large-scale return of plants and animals from their southern refuges, continuing the process which had commenced in the Pre-Boreal Period. Moreover, the rapid improvement in the climate encouraged the arrival of additional species which spread as far north and west as their ecological limitations would allow. In fact many spread much further north in Britain and Europe than they occur today. It is tempting to speculate about some of the southern species which may have colonised Britain during this time, but one can only guess that birds such as the Hoopoe *Upupa epops* and Bee-eater *Merops apiaster*, and butterflies such as the Bath white *Pontia daplidice* and scarce swallowtail *Iphiclides podalirius* did so. However, the survival of such southern European plants in the Breckland district of East Anglia as the Breckland or Spanish catchfly *Silene otites* and the spiked speedwell *Veronica spicata* are indicative. It is also possible that some of the so-called Lusitanian plants and animals still surviving in southern Britain and Ireland may actually have arrived during the warmest part of the Boreal—species such as the spotted slug *Geomalacus maculosus*, silver-barred moth *Deltote bankiana*, Dartford Warbler *Sylvia undata*, lesser horsehoe bat *Rhinolophus hipposideros* and strawberry-tree *Arbutus unedo*.

It is less speculative to deduce from the nature of the vegetation that developed in the British Isles that the vast majority of the fauna and flora were broadly the same as exist in such habitats in modern times.

During the two and a half thousand year span of the warm, but moister Atlantic Period, alder, elm, lime and oak, which had become well and truly established in the British Isles and elsewhere in temperate Europe, spread and flourished so much and so rapidly (especially alder), that together with the pines further north they formed dense forests over most of the land surface. And that included the mountains, because in the Climatic Optimum trees clothed their sides up to far higher altitudes than at present. The same was true of North America, at least the eastern half, where the thick forests were dominated by hemlock and oak.

As one might expect with such widespread and dense forests, the dominant fauna and flora consisted of species characteristic of, and familiar to us in, our present-day woodlands, whether they were the pine-dominated woods of the north or the broad-leaved ones of the south.

The moist climate of the Atlantic Period led to a resurgence of the bogs and fens which had been invaded by pines as they dried out during the preceding, drier Boreal Period. But the rising sea level (around 4,000 years ago it was about three metres higher than it is now) led to the loss of other extensive areas of fen and marsh along the low-lying coasts. In Britain, for example, large parts of the East Anglian fens and the Somerset Levels were inundated by the sea, while the filling up of the North Sea saw the end of the fens and salt-marshes of the Dogger Land. All that remains to remind us of this glacial refuge are the quantities of 'moorlog'—masses of loose peat which have been dredged up from time to time by trawlers fishing over the Dogger Bank, now up to 170 feet (52 metres) below sea level. This has been found to contain the pollen and other remains of birch trees, as well as plants characteristic of present-day fens. Some of the finer plants and animals, such as the large copper butterfly *Lycaena dispar* and the fenland race of the swallowtail butterfly *Papilio machaon britannicus* must have been exterminated, except in those outer limits of their range which remained above sea level in eastern England and the Low Countries (Beirne 1952).

As James Fisher remarked (1966*b*), the surface peatlands, especially those of the fens, are the most rewarding source of information about the fauna of Britain in the Atlantic Period, a time when the Neolithic inhabitants began to practise agriculture seriously. He lists Great Crested Grebe, Dalmatian Pelican, Bittern, Mallard, Teal, Wigeon, Red-breasted Merganser, Smew, Greylag Goose, Mute Swan, Crane, Moorhen, Coot and Woodcock among the bird denizens of the fens of those distant times.

About 4,500 years ago, when the Bronze Age culture began supplanting that of the Neolithic, the climate became drier, but the transition appears to have been less marked and clear-cut in the British Isles than elsewhere in Europe and in North America. Nevertheless, it is well established that in at least part of the Bronze Age the climate was much drier than before or since. The summers of this climatic phase, the Sub-Boreal, were generally distinctly hotter than nowadays, while the winters were drier. The peat bogs of Britain and Europe were invaded to a considerable extent by pine and yew, and ceased to expand. Scots pine and birch spread elsewhere at the expense of elm, lime and oak, and even ascended to 3,000 feet (900 metres) up the mountainsides of mid-Scotland, compared with about 2,200 feet (660 metres) today, while on the higher, limey ground beech extended its range far beyond its present natural limits in Britain. In North America hickory joined oak as the dominant trees.

There was undoubtedly a long period of drought in the Sub-Boreal Period which led not only to the drying out of the peat bogs, but also to the drying up of lakes or at least a marked reduction in their size. The drought also caused the human inhabitants to change their way of life. For a long time Neolithic Man had found it not only possible but preferable to settle in the more open, well-drained uplands, rather than in the heavily forested lowlands where bears and wolves prowled. Although the hills were relatively well wooded with beech, ash, oak and yew, the droughts of the Sub-Boreal climate led to an increase of open grassland at the expense of woodland; moreover, they caused the agricultural Bronze Age people to forsake their dwellings on the uplands, and descend to the lower ground where they commenced clearing the forests on a scale greater than ever before. Here they created cultivated areas around their villages more permanent than those of their Neolithic predecessors, who merely felled and burnt a patch of woodland, cultivating the cleared ground for a space of only a few years before moving on and repeating the process elsewhere.

So by this time the story of climatic change, as unfolded by the pollen records, becomes more complicated to unravel owing to man's greater impact on the vegetation. Pollen from his crops and their associated weeds, increasingly noticeable in the pollen record from Neolithic times, gradually became more abundant, with the result that by the Bronze Age it rivalled the quantities of pollen deposited by the forest trees. And man's increasing impact on the vegetation also began to have an effect on the fauna—an effect, to some extent at least, independent of changes instigated and controlled by the climate, although ultimately it was the climate that set the limits. Although information provided by the sub-fossil record is still too scanty to enable us to form a clear picture of the changing wildlife of Bronze Age times, we can be reasonably sure that the flora and fauna of the cultivated areas were beginning to bear some resemblance to that of present-day farmlands.

For instance, it is fairly certain that the birds and other wildlife of the more open forest, and the woodland edge and glades, found the cultivation to their liking, providing as it did regular and improved sources of food. Among birds, for example, ground-feeders such as the Blackbird, Song Thrush, Dunnock, Robin and Wren must have found that food was more easily accessible in the regularly disturbed ground, while others such as the Chaffinch, Woodpigeon and Rook must have been quick to adapt as well. The Bronze Age agriculturalists' herds of domesticated animals—swine, sheep, goats, cattle, geese and the like—would have proved attractive as direct or indirect sources of food for kites, Ravens, crows, Jackdaws, Starlings, Pied Wagtails and Swallows, to name but a few. The vast majority of these early farmland birds, therefore, originated in the forests; but as the cultivated areas and pastures became more extensive they would no doubt have been also colonised by such typical species of open country as the Skylark and Lapwing (Williamson 1977a,b).

The dry warmth of the Sub-Boreal summers would probably have meant the continued survival, and perhaps even flourishing, in southern Britain of

FIGURE 10 *White Storks at the nest near Waren, Mecklenburg, east Germany. (Photograph: Nicole Debon)*

species which arrived during the Boreal Period, and which nowadays have far more southerly distributions in Europe. Indeed, the dry summers would have particularly suited various species of sun-loving grasshoppers and crickets, whose abundance may well have supported here a healthy popula-tion of Hoopoes, shrikes, Great Bustards and White Storks, all of which greedily feed upon them. Certainly the climate was warm enough to continue to support widespread colonies of Dalmatian Pelicans in the English fens and elsewhere in north-west Europe, a bird which one now has to travel to the Balkans and the Middle East to see.

However, the dry warmth did not last. About 600–500 BC, some 2,000 years or so after the commencement of the Sub-Boreal Period, the climate (with some geographical differences) underwent a rapid deterioration to a cooler, wetter and windier phase, and the glaciers readvanced. Although the winters remained quite mild, because of the amount of cloud cover, the summers are believed to have been as much as 3 or 4°C cooler than during the Sub-Boreal, and up to 1°C cooler than the present. Thus began the Sub-Atlantic Period, which corresponds roughly with the Historic Period in which we live today, and which had, generally speaking, the type of climate we experience nowadays; as we shall see, however, this was not rigidly so.

The much increased rainfall led to the reactivation of the dormant peat in the existing bogs and fens, and the rapid formation of new peat bogs else-where, especially on poorly drained uplands which had formerly been well wooded. The expanding bogs in fact swallowed up large sections of forest which had developed during the Sub-Boreal both in Europe and North America.

On better drained land, the mixed, broad-leaved forests, with their dominant oaks, continued to flourish, but the wetter climate encouraged the growth and spread of alders and willows in low-lying areas. Birches also increased, but beeches and Scots pines declined and the tree-line dropped by as much as 1,000 feet (300 metres). In North America, oak also remained dominant, with an increase in the moisture-loving chestnuts, hemlocks and spruces.

As there have not been any really major climatic changes since the start of the Sub-Atlantic Period, the vegetation of the British Isles, Europe and North America has remained substantially the same today as it was then. Of course, over the centuries the increasing activities of man and his growing numbers have modified it to a striking degree; this has tended to mask the evidence of the effects on the flora of such minor as well as relatively large climatic changes as have occurred in the meantime. Thus the task of detecting and analysing their effects has been made all the more difficult. Nonetheless, I will endeavour to indicate, in the remainder of this chapter and in the next, some of the changes which have been deduced. Most of these examples will, however, refer to the British Isles.

Early in the Sub-Atlantic Period, Britain was invaded and eventually overrun by Celtic peoples, who brought with them iron tools and weapons whose use they helped to spread, although they were not the first to introduce them into the country. Thus armed with efficient and powerful implements, the Iron Age tribes set about forest clearance with more gusto and success than ever before. However, living as they did in a cool, wet, deteriorating climate, the prolonged dry spells and droughts of the Sub-Boreal being a thing of the past, they preferred to settle on the driest land, and they recolonised the hills, especially those of chalk and limestone, which long before had been the home of the Neolithic people. There they felled the beech and oakwoods, either ploughing the cleared land or using it as pasture to graze their growing flocks of domestic animals. All over Britain today the remains of their numerous hill forts and fields provide visual evidence of their dependence upon the uplands. The main centres of population were, however, upon the chalk plateaux of southern England, and according to O.G.S. Crawford and A. Keiller (1928) practically the whole of Salisbury Plain, Cranborne Chase and the Dorset uplands were under the plough or used as pasture in Romano-British times.

Sir Arthur Tansley (1968) pictured the natural vegetation of Britain in the late Iron Age, when the Romans invaded Britain, as follows:

'While there was fairly extensive cultivation on the chalk and also on the loam soils of the south-east, most of the English lowlands, for example the Weald and the Midland Plain, were covered with oak forest, mainly uninhabited and harbouring wolf, lynx and bear, besides numerous deer. Oak forest also occupied the sides of the valleys in the hill and mountain regions of the west and north, giving place to pine and birch woods at higher levels and on the poorer and more sandy soils. In the extreme north of Scotland there was little or no oak even at low altitudes, though oak forest filled the bottoms and lined the lower slopes of the larger glens in the Central Highlands. In the lime-stone regions, for example on the Mountain Limestone of the Northern and Southern Pennines, ashwood, the remains of which may still be seen in Craven and the Derbyshire dales, probably covered the sides of the valleys.'

But although the British Isles were predominantly wooded, there were extensive areas of other habitat types; in addition to the upland grasslands and farmland. For instance, the Romans would have found a good deal of estuarine salt-marsh, bog and fen.

As well as the flora, the fauna of Britain would also have been very simi-lar to that of the present day, although presumably not exhibiting quite so much diversity because of the predominance of broad-leaved forest over much of the country. Of course, a number of species occurred then that have since become extinct or greatly reduced in numbers because of the subse-quent destruction of their habitats or from other causes. In particular, the drainage of fens and salt-marshes between then and now has caused the disappearance or virtual disappearance of several species which we know from sub-fossil evidence were widespread in such habitats in Iron Age/Romano-British times. The most important evidence so far comes from finds made on the sites of the famous Late Iron Age lake villages of Glastonbury and Meare in Somerset, where the remains of beavers, wild boar, pine martens, Dalmatian Pelicans (including fledglings), Greylag (?) Geese, Cranes, Bitterns, White-tailed Eagles, Red Kites, Ospreys, Montagu's Harriers and Red-crested Pochard have all been identified.

Remains of the Dalmatian Pelican were found in profusion at the Glastonbury lake village on the shore of Meare Pool and, as these included the bones of fledglings, there was clearly a well-established and flourishing breeding colony. When it died out, and from precisely what cause, is guess-work; but, apart from predation by man, it seems likely from its south-east-erly distribution in Europe today that an unfavourable change in the climate may have been primarily responsible. We have already seen that this species was apparently widespread in the warm and drier Sub-Boreal Period; but it may well have declined and disappeared from many of its old haunts during the cool Sub-Atlantic (Iron Age) Period, with the result that the lakes and bogs of central Somerset were its last stronghold before the arrival of the Romans. The lake villages appear to have been abandoned during the 2nd century BC because of flooding, and this in itself might have finished off the pelicans. Otherwise they may possibly have lingered on through a warmer oscillation between AD 50 and AD 400, and perhaps even survived the major inundation of the sea in AD 250, until climatic conditions deterio-rated yet again after AD 400 and finally put paid to them.

Another present-day southern European and Middle East bird whose bones have been recovered at the lake village of Meare is the Red-crested

51

Pochard. Although nowadays it has a breeding outpost as near England as the Netherlands (of which more will be said in a later chapter), it was unknown there as a breeding species until at least 1910. It seems therefore that, like the Dalmatian Pelican, it was the deteriorating climate which caused its subsequent contraction of range and disappearance from Britain. Incidentally, the drainage and reclamation of the bogs and marshes could not have been responsible for the loss of either of these species, as little of significance was done until the 18th century.

The cool, very wet phase between about 600 BC and AD 50 led to the flooding not only of Somerset's Iron Age lake dwellings, but also those of Lake Constance (the Bodensee) in central Europe. Here the water level rose by as much as ten metres or more. The Alpine glaciers advanced to such an extent that they more or less attained the maximum expansion achieved in AD 1650, and closed the Alpine passes to human traffic until amelioration around AD 700 opened them up again. Furthermore, the general storminess and wretchedness of the climate of this period has been suggested as the cause of the migrations of the early Celts and Teutons from the western part of the German plain about 120 BC (Brooks 1949).

In Russia, the wetter climate encouraged the southward spread of the forests, particularly beech and hornbeam, which were favoured by the lower summer temperatures then prevailing in those latitudes. The Mediterranean region and North Africa also experienced more rainfall than nowadays, especially in the summer months.

Gradually, however, the improvement from AD 50 made itself felt, and the climate became steadily warmer and drier until around AD 400 when it deteriorated again. Interestingly, this warm phase corresponded remarkably well with the entire period of the occupation of Britain by the Romans— they arrived in force in AD 43 and withdrew their legions in AD 410 when the climate was becoming less and less tolerable. Of course this may have been pure coincidence, as it is usually argued that the Romans relinquished their hold on Britain in order to concentrate their forces for the defence of the heart of the Roman Empire—indeed, Rome itself. But perhaps it is also significant that the attacks on the Empire by the barbarian hordes from northern and eastern Europe and Asia coincided with the increasing severity of the climate in those regions; it may well have spurred on their efforts to escape its rigour at the expense of the Romans. Lamb (1966) mentions that increasing drought in Asia has also been suggested as a possible cause for the barbarian invasions.

Climatic deterioration coupled with barbarian invasions from Asia may, in addition, have been a factor behind the rising frequency of raids by the Saxons on the coasts of south-east England, which, aided by the evacuation of the Roman garrisons, led eventually to their settlement there, and the retreat of the Romano-British population to the west and north-west.

By the time the Saxons had conquered and settled much of England in the 7th century, the climate was again showing signs of improvement over much of Europe. Around AD 750 it was distinctly drier and warmer, and this state of affairs persisted until about AD 1215.[1] Indeed, it became so warm that this period has become known as the 'Little Climatic Optimum', or by modern climatologists as the 'Medieval Warm Period'. Judging by the effects on the distributions of animals of the most recent long-term amelioration, lasting only a little over a hundred years (1850–1950) (see Chapter 5), many species of animals, notably birds and insects, must have extended their ranges considerably further north and west during such a comparatively long warm phase; no doubt this included some which would be regarded as exotic if seen so far north today. On the other hand, other species characteristic of colder zones probably retreated northwards.

Unfortunately, our knowledge of the wildlife of the earlier part of this warm oscillation, during the Dark Ages, is even more limited than is our knowledge of most other topics in that mysterious period following the retreat of the Romans. However, excavations at Jarlshof in the southern Shetland Isles, north of mainland Scotland, have revealed that the bird life of this archipelago between the 8th and 11th centuries was much the same as it is now, except that the Black Grouse and Magpie may have been among the breeding species at that time. The presence of these two basically woodland-edge birds suggests, as James Fisher (1966*b*) pointed out, that sheltered valleys in at least the southern part of Shetland must have been naturally wooded, owing to the remarkably mild climate prevailing then—a climate so mild that the Vikings were able to colonise Iceland and Greenland with great success.

It may be a source of surprise to most people living outside Scandinavia to learn that once upon a time Viking settlements flourished not only in Iceland, but in Greenland too. At the time of its discovery by the Viking chief Eirik (Erik) the Red, in AD 982, Greenland truly deserved the name he bestowed upon it, for the south of the country at least was indeed green. However, the authors of the old Norse 'Greenlander Saga' appear to have been suspicious of his motives in thus naming it, and considered it a device to sell to others the idea of following him there and settling. But in fact we now know that the climate of Greenland was much more benign then than nowadays. Voyagers to Greenland from Iceland in those early days followed Eirik the Red's route due west until they sighted the Greenland coast, whereupon they coasted southwards and round the tip of the great island until they reached his settlement at Østerbygd. The landmark they looked for on the sea crossing from Iceland was a mountain then known as Black Mountain, because for much of the year it was relatively free of snow and the bare black slopes were conspicuous. But with the approach of the climatic deterioration known as the Little Ice Age, which began to be felt in Greenland some 250 years before it was felt in western Europe, and about 150 years earlier than in Iceland, the mountain gradually changed from black to white as it became increasingly covered with snow and ice, even in the summer (Bryson 1975).

However, before that occurred (and with it the tragic end of these Norse settlements which is described in the next chapter), Greenland was justly named from the lush pastures which were such a feature of its southern coasts. Herds of cattle and sheep flourished, and more than 300 farms were established. Even grain was grown for a time. The Norsemen were prosperous and energetic enough to build a cathedral (Gimpel 1973).

Paradoxically, although it remained habitable (but only just) throughout the worst of the Little Ice Age, Iceland received its name from its first settler, Floke (Floki) Vilgerdson, a Norwegian farmer. He, unlike Greenland's Eirik the Red, was defeated in his attempt at settlement in about AD 865 by a very bad winter, when he lost all his cattle and saw 'a fiord filled up by sea ice'. Unfortunately he had chosen to try during a short cold spell. Not long afterwards, in AD 874, his successor Ingolf Arnarson was blessed by more favourable conditions and succeeded where Floke had failed, for he arrived at the beginning of a rapid climatic amelioration corresponding with the Medieval Warm Period (Little Climatic Optimum). He was soon followed by others, with the result that by AD 930 the settlement or landnam was completed. It prospered over the next 400 years, until the Icelanders were afflicted by increasing cold, frequent famines and outbreaks of bubonic plague.

In England the great warmth of the Medieval Warm Period (Little Climatic Optimum) is indicated, among other things, by the remarkable prevalence and wide distribution of vineyards. According to Professor Lamb (1966), many are known to have been in continuous operation between AD 1000 and AD 1300 for between thirty and a hundred years, some of them for even longer. Most were about one or two acres in size, but quite a few were between five and ten acres or even more. Although the majority were centred on such climatically favourable districts as north Kent, the Thames valley between London and Oxford, Essex, the Cambridgeshire Fens, south Hampshire, Somerset, Gloucestershire and Herefordshire, a few existed even as far north as Lincolnshire and south Yorkshire (54°N). The wine those in southern England produced was considered to be almost equal with the French in quality and quantity; and indeed, at least on one occasion, the French wine producers attempted to have them closed down as one of the conditions in a peace treaty with England, but to no avail.

For the successful culture of the grape vine, Lamb (1966) has pointed out that the climate needs to be free from late spring frosts and too much rain; it requires sufficient sunshine and warmth in summer and, in regions where the summer is only just warm enough, in the autumn too; and a dormant winter in which serious frosts are rare. The success of the English medieval vineyards between AD 1000 and AD 1300 therefore gives us quite a good indication of the climate prevailing at the time; as Lamb states, it implies summer temperatures perhaps 1 to 2°C above those of the present and a general freedom from May frosts. That favourable conditions prevailed in England long before AD 1000 is supported by the existence of good vineyards in Saxon times.

Elsewhere in Europe, wine was also produced much further north during the Medieval Warm Period than now; for example, there were vineyards in southern Norway, Lithuania (55°N) and the former East Prussia. Furthermore, in Germany vines were cultivated up to 200 metres or more above the present highest altitude (Lamb 1966).

To return to the Saxon period in England: during the early phase of the Medieval Warm Period, the Saxon colonisers greatly extended their clearances of the lowland forests, shunning the hilltops so favoured by their Neolithic and Iron Age predecessors. These uplands, therefore, to a large extent reverted to woodland. With the spread of arable farming amid the fragments of unfelled woodland, wildlife typical of woodland-edge habitats became adapted to the new, more diversified habitats, and presumably flourished. Gimpel (1973) suggested that the Medieval Warm Period slowed down the natural expansion of the then still very extensive forests of western Europe, and 'must have eased the work of the pioneers who cleared forest lands to make way for the plough'. He also mentioned that pollen analysis has shown that in some regions of Europe the forests retreated naturally.

Tansley (1968) summed up the contribution of the Saxons to the English environment with the apt comment that it was they 'who began the clearance of lowland forest which ultimately turned England from a mainly forest-covered into a largely agricultural, but predominantly pastoral country'. However, at the height of the Saxon domination of England there still existed 'thick and inaccessible' forests, inhabited by deer, wild boar and wolves, like that of Anderida in the Weald, which was so described by the Venerable Bede around AD 731. But by the time the Normans arrived in AD 1066, the woodlands of England, Scotland and Wales had been reduced from the forty million acres of Neolithic times to some ten million. The Danes and Norsemen, who had invaded and occupied much of northern and eastern Britain in between the Saxons and the Normans, mainly settled in land previously occupied by the Saxons. They therefore probably contributed little themselves in the way of new forest clearance.

With the Norman conquest, however, the policy of disafforestation was temporarily reversed, perhaps for a century or more. Incidentally, the conquest of England by the Normans, and their territorial expansions elsewhere, may have been at least to some extent stimulated by the greatly improved climate of north-west Europe releasing in them an excess of energy. Soon after their arrival, the Normans set about establishing Royal Forests in such sizeable tracts of wild and wooded country as still survived. Cultivation within these forests was forbidden under the Forest Law, fines being levied upon all transgressors.

As the area covered by the Royal Forests at their maximum around AD 1150 is believed to have amounted to as much as one third of the land surface of England, there must have been some regeneration of woodland, aided by such natural agents as the Jay, as well as by the invasion of former arable fields by scrub. Presumably there was, in consequence, a decline in at least some of those plants and animals which had slowly become adapted to living on cultivated land.

Meanwhile, the colonisation of southern England by animal species typical today of southern Europe, which probably began during Saxon times, soon after the start of the Medieval Warm Period, was presumably consolidated. For instance, it would appear from the earliest extant version of his Latin to Anglo-Saxon vocabulary *Nomina Avium*, compiled in about AD 998, that the Anglo-Saxon scholar, Ælfric the Grammarian, was familiar with such continental birds (no longer or only relatively rarely breeding in Britain) as the Hoopoe, Golden Oriole, Eagle Owl and Quail. Curiously, moreover, he mentions two cicadas in his list of bird names, presumably because they 'sing' like birds and were not therefore distinguished as insects. The only cicada known in Britain today is *Cicadetta montana*, which is apparently confined to the New Forest, where it is very rare. It is improbable that this was the species know to Ælfric as its song is very high-pitched and not easily heard, except by younger persons; it is more likely that it was one of the noisier, southern European species, such as *Cicada plebejus*. However, it is quite possible that the loudest singers among the bush-crickets or crickets rather than the true cicadas were also encompassed by the name 'cicada'. Of incidental but significant interest in this connection are the plagues of migratory locusts *Locusta migratoria* which occurred more frequently in the early Middle Ages in central and northern Europe than is usual today.

The conservation and regeneration of woodlands and other uncultivated land associated with the Normans did not continue for very long, perhaps only for a century or so. By 1200, the policy was collapsing under the increasing pressure of the needs of England's rapidly growing population. The increased food production made possible by the improved climate—the Medieval Warm Period was at its peak around 1150—led to something of a population explosion in Europe at this time, especially in England and France. Between 1150 and 1200, for instance, the European population increased by 22% (Gimpel 1973). Thus the Norman kings of England and their aristocracy found it more and more impossible to exclude the ordinary people and their domestic animals from the Royal Forests. The growing herds of cattle, sheep, goats and horses, grazing deeper and deeper into the Royal Forests, halted the regeneration of the woodlands by destroying the seedling trees and the wild plants of the woodland floor, thus encouraging the invasion of grasses.

The growth of the human population also made tremendous demands upon woodland timber: for building the homes needed to house the extra people, for the fuel required to warm them and cook their food, and a hundred and one other needs. After all, the whole medieval economy depended upon wood. In the face of such mounting pressure on their Royal Forests, the Norman aristocracy had no option but to yield. Furthermore, the kings of the Norman period were often in financial difficulties and, as the authority on medieval Europe, Jean Gimpel, has emphasised (1973), they were induced to accept vast disafforestation. For instance, he quotes from Stenton (1951) the case of the knights of Surrey, who in 1190 offered 200 marks to Richard I that 'they might be quit of all things that belong to the forest from the water of Wey to Kent and from the street of Guildford southwards as far as Surrey stretches'. He also mentioned that in one year alone, 1204, King John was offered 500 marks and five palfreys for the disafforestation of the 'forest of Essex which is beyond the causeway between Colchester and Bishop's Stortford', 2,200 marks by the men of Cornwall for the disafforestation of the whole of that county, and as much as 5,000 marks by the men of Devon for the reduction of the forests of their county.

Once the Norman kings relaxed their policy, there was no holding back on the decimation of the forests and woodlands right through the Middle Ages. Probably few people today realise just how great the destruction of the forests was at that time, or indeed how early in the Middle Ages the shortage of timber was felt. It is not easy to imagine how populations that were relatively tiny, in comparison with those of European countries today, could devastate the once vast forest cover on such a scale and so rapidly with their primitive tools to meet apparently simple needs. Yet, to quote Jean Gimpel again: 'the population explosion of the Middle Ages played havoc with the environment of western Europe. Millions of acres of forests were destroyed to increase the area of arable and grazing land and to satisfy the ever increasing demand for timber, which was the main raw material of the time'.

Thus, the rising cost of wood resulted in the importation of timber from Scandinavia to the depleted parts of western Europe, as well as the increased use of coal as a fuel. The first shipments of conifers from Norway reached Grimsby on England's east coast in 1230; thereafter the trade escalated, so that by the 14th century it was already considerable. By the end of the 17th century the supply of English timber was all but exhausted and almost complete reliance was placed upon imports.

The Scottish forests lasted longer, but only in the Highlands. Those of southern Scotland were cleared even earlier than those in England. Pope Pius II, who visited Scotland in the 15th century before becoming Pope, spoke of the country as destitute of wood, and dependent instead on sea coal as a fuel. However, large tracts of unspoiled, primeval forest survived in the Highland valleys as late as the early 17th century, but from then on they were felled and burned at an alarming rate, so that only fragments remain today.

Ireland was well wooded in 1183 when Giraldus Cambrensis went there and noted 'vast herds of boar and wild pigs'. But between then and 1700 the woods were nearly all cleared, so much so that the Irish became almost entirely dependent upon peat as a domestic fuel (Simms 1971).

My readers may feel by now that this digression into the fate of the forest cover of Europe, especially as regards the British Isles, may have become too detailed and strayed a little too far from the main path in the story of the effects of climatic change on the flora and fauna between the end of the Last Glaciation and 1250. But I wish to show as clearly as I can that the alterations in the distributions of plants and animals caused by climatic change becomes progressively more complicated through the impact of man on his environment, especially from Saxon times onwards. A less detailed treatment of the reduction of the natural forest cover of large areas of Europe might have led to an underestimation of the seriousness of this fact, and of its widespread repercussions on the fauna and flora. It is very probable that until about the end of the Dark Ages, or perhaps even later, the dominant common wild plants and animals with which we are so familiar today in the 'civilised' areas of Europe were much less common, for these are typically species of woodland-edge and open grass-steppe habitats. It is also probable that in late Saxon and early medieval times the devastation of the forests was so extensive and so rapid that it took many species a long time to adapt to and colonise the new expanses of grassland and cultivated land. This may be less true of some plants and insects than it is of birds and mammals; but it is likely that much cultivated land in early medieval times had, for example, a less varied bird life than such countryside today. There is some evidence in the writings of early naturalists (e.g. Turner 1544; Merrett 1666; Willughby and Ray 1676) that even as late as the 17th century such characteristic birds of farmland as the Starling, Song Thrush, Jackdaw and Rook were less numerous in Britain than they are at the present time.

A process which helped the woodland birds and other wildlife to adapt to the new farmland more quickly than they might otherwise have done, was the medieval practice of coppicing the remaining woodland fragments. This redeeming, but unconscious, act of nature conservation was induced by the serious shortage of wood available for the whole range of uses made of it in the medieval economy. As Kenneth Williamson (1977a) neatly summed it up, 'coppice-with-standards proved to be the half-way house between forest and farm'. The wildlife of the woodland glades and borders, which were already well adapted to colonising the natural succession of field and scrub layers of vegetation that follow upon the destruction of mature trees or areas of woodland, found the similar succession in the coppices to be to their liking. The transition was made particularly easy by the variety of stages represented in each coppice, from the dense canopy and shade just prior to cutting, to the rich ground flora which appeared immediately afterwards, and then to the intermediate stages as the scrub layer developed and the coppiced trees grew again. Moreover, the rotational system meant that coppices in the right state for occupation by particular species were always available when others became unsuitable; and the variety of coppices in a given area in different stages of growth offered a wide choice of woodland-edge habitats.

Even when isolated from the remaining natural woodland by the surrounding arable land and pasture, the coppices, little disturbed by man for years at a time, provided relatively safe breeding places and cover, especially for those woodland birds which were primarily ground-feeders, such as Blackbirds, Song Thrushes, Dunnocks, Robins and Rooks. These discovered food easy to come by in the new farmland, and seed and fruit eaters were presumably not slow to follow their example.

The hedgerows planted around many of the coppices to protect them from damage from livestock, plus those already in existence as parish or other boundaries, were intermediate in form to the coppices. Indeed Williamson (1977a) pointed out that they were 'nothing more or less than linear coppice-with-standards'. As he went on to say, the extension of the hedgerow network in later years, especially as a result of the Enclosure Acts, opened up huge new areas of grass and arable cultivation to birds which had hitherto been unable to exploit them to the full, because of the distance from cover and the consequent exposure to their predators. Thus it is from a woodland fauna that the vast majority of the typical wildlife species of English and, to a lesser extent, European farmland are derived.

Another reason why I have felt it necessary to devote so much space to disafforestation, and one more directly linked with the climatic theme, is that the conversion of so much woodland into exposed open country is

FIGURE 11 *Ancient pollard oaks at Risby, Suffolk. Pollarding was an alternative medieval practice to coppicing which helped to remedy the shortage of wood at that time by producing crops of shoots suitable for use as poles, etc. (Photograph: Dr Oliver Rackham)*

bound to have had some effect on the climate of the affected areas, more particularly on the micro-climate (i.e. differences within extremely localised areas, often quite tiny in extent).

So, in interpreting the changes in the distribution and abundance of plants and animals, one has to be careful to distinguish between those which are due to climatic fluctuations, those which are due to alterations wrought in the environment by man, and those which are due to a combination of both; and, indeed, those which are due to some other factors. This task has been made all the more difficult by the extreme paucity of detailed information about the fauna and flora of Britain and the rest of Europe during the historic period covered by this chapter. Instead, it has obviously only been possible, in the main, to speculate on the changes which may have occurred, based upon our knowledge of the effects of recent climatic and environmental alterations.

One may, nevertheless, sum up the picture of this post-glacial period by remarking in a general way that, apart from temporary interruptions due to cold climatic phases, it probably saw a steady recovery from the effects of the Last Glaciation, with many southern species continuing right up to the end of the Medieval Warm Period to push north and north-westwards, at varying rates of progress, recovering territory lost in the glaciations, while northern species continued to lose ground and contract northwards.

[1] Different spans of years are given by different climatologists and others for the Medieval Warm Period and the succeeding Little Ice Age, depending upon whether they are looking at it from the perspective of the British Isles, Europe or the whole Northern Hemisphere. In this book I have taken, where possible, the Northern Hemisphere perspective, except where stated otherwise.

Chapter 4

1250–1850: THE LITTLE ICE AGE

Unfortunately, the extremely warm climate enjoyed by our early medieval ancestors did not last much beyond 1200. By 1250 the inhabitants of England and north-west Europe must have been aware that things were not as good as they once were; westerly depressions arriving from the Atlantic became much more a feature of the weather as the 13th century progressed, and the coasts were often lashed by severe gales, especially around the North Sea. The inadequacy of the sea walls of those times often meant terrible flooding in lowland areas. For example, in the Netherlands many towns and villages were drowned by the expanding waters of the Zuider Zee. They have since been revealed during the vast reclamation schemes of this century, which have reduced the status of the Zuider Zee to that of a large lake, now called the IJsselmeer.

In England, severe flooding followed tempestuous storms on the east and south-east coasts. In 1287, for instance, following a violent, night-long storm, the inhabitants of New Romney on the Romney Marshes, Kent, awoke to discover that they were an important port no longer, for the course of the River Rother had been diverted overnight to Rye, five miles to the west.

If western European man began to realise in the 13th century that the climate was deteriorating, his fellow beings further north were already well aware of it. Recent studies (Dansgaard *et al.* 1975), involving oxygen isotope analysis of ice cores from the crest of the Greenland ice sheet, show that long-term climatic changes occurred in central Greenland some 250 years earlier than in western Europe, and in Iceland 100–150 years earlier. During the 13th century the southward advance of the polar drift ice forced the Norse settlers in southern Greenland to shift their trade routes with Iceland and Norway further and further south.

Formerly, as we have seen, traders used to sail directly westwards from the west coast of Iceland until they reached east Greenland, whereupon they hugged the coast southwards until they reached the port of Julianehåb in Østerbygd in the south-west. But by 1342 they departed from a point further south on Iceland's coast, and took a more southerly route which then turned south-west long before reaching the Greenland coast. Sea-ice also began encroaching during the 13th century on the coasts of Iceland, whose population came to depend increasingly upon supplies from Europe for their survival during the ensuing 'Little Ice Age', as this cold period has come to be called.

In the following century, the climate became so bad in Greenland that few ships got through and the Norse colonies gradually died out, isolated from other human contact by the sea of ice. By 1350 the Vesterbygd colony in the Godthaab Fjord was no more, and within a century the Østerbygd colony had also ceased to exist. No one quite knows how the end came since supply ships no longer called. There must have been frequent years of famine caused by the increasingly frozen ground. Eskimo legends blame the Norse demise on the raids of European pirates (Dansgaard *et al.* 1975); while Williamson (1975) suggested that they lost out in competition with the Eskimos, who themselves were being driven further south by the increasing cold and advancing drift ice, but who were better adapted to living off polar cod *Boreogaelus saida*, seals and whales than the Norsemen. It is also possible that they were finally wiped out by bubonic plague, which reached Iceland in 1402 (Dansgaard *et al.* 1975). By 1400 their average height had apparently declined to less than 5 feet (150 cm), and their teeth were badly worn from subsisting on tough vegetable food.

From the foregoing we can safely assume that, along with the retreat of the Eskimos into more southerly latitudes, there was likewise a southerly retreat of Arctic and subarctic animals away from the advancing ice and increasing cold. Not improbably, people living in Britain and similar latitudes in Europe were mystified by the appearance of such obviously Arctic birds as the Snowy Owl. No doubt they were equally surprised by the decrease and perhaps disappearance of other species more characteristic of warmer climes, which had become established during the long period of warmth of early medieval times. We can only guess at these, but they may have included such species as the Bee-eater, Hoopoe, Roller and Quail.

During the 14th century other happenings made the inhabitants of Britain and western Europe even more aware that colder conditions were spreading south. It became, for instance, more and more difficult to cultivate the vine. As mentioned in the previous chapter, from at least AD 1000 to AD 1300 vineyards had flourished in southern England, and even as far north as Yorkshire. At that time the English were in fact a nation of wine drinkers like the French, but after that, as Kenneth Williamson (1975) drily remarked, they had to get used to drinking beer. The increasing frequency of cool springs with late frosts and wet, sunless summers led to the decline of the English vineyards. At the height of their cultivation in the 12th century some of them had occupied ten acres or more.

There were more frequent failures, too, in the harvests, especially of corn, upon which, with sheep, the prosperity of early medieval England had been built. These brought famines and rising prices which, together with the scarcity and high price of wood, reduced a greater proportion of the population to poverty. It is not, surprising to learn therefore, that the population explosion in western Europe slowed down considerably.

From about 1460 to 1540 there was some respite when the climate ameliorated somewhat. Manley (1972) mentioned, for example, that the cultivation of the cherry spread northwards and apparently reached County Durham in the 16th century, even to a height of 244 metres (800 feet). He pointed out, too, that there was a long break—from 1434 to 1540—in reports of the River Thames freezing over in London. The frequency with which the Thames and other rivers in western Europe, such as the Seine, froze provides another useful indication of climatic trends in historical times. However, even this warmer period was not without its bad spells, for further harvest failures, due to poor summers and wet autumns, occurred between 1430 and 1440.

It is probable that this short amelioration temporarily halted the assumed retreat of those species of plants and animals at the northern limits of their range; but, as there is little evidence of any significant improvement of the climate of subarctic regions during this period, it may not have been sufficient to halt the southward retreat of northern species. However, as will be shown in later chapters, the evidence of the most recent comparable amelioration (*c.*1850–1950) suggests that even relatively short warm phases can have marked effects on the distributions of some plants and animals.

By 1550 the respite was well and truly over; then the longest and coldest period in modern historical times was on again in earnest. It is not without good reason that it became known as the 'Little Ice Age', especially the period of maximum cold from 1550 to 1700. There were, of course, milder spells of varying duration before it finally came to an end around 1850; but generally it was a time of frequent cold and sometimes very severe winters and cold springs, and often cool summers. It reached its peak in the late 17th century when hard winters were almost the rule. The winter of 1683–84, for instance, so well described for Exmoor in Blackmore's *Lorna Doone*, was especially hard, and it was in that winter that the most famous 'Frost Fair' of all was held in London on the frozen River Thames. For two months the river was completely frozen, with ice up to 28 cm (11 inches) thick in places, and so many shops and booths were erected on the ice, arranged in orderly streets, that the fair looked like a separate city (Gribbin 1975a). The Thames froze again five years later when another frost fair took place.

Incidentally, I mentioned earlier in this chapter that it became significantly colder in Greenland and Iceland from about 1200. Indeed, it now appears from the studies made by Dansgaard *et al.* (1975) that the Little Ice Age started there at about that time—actually from about 1150 to 1400— well in advance of its occurrence in western Europe. However, it apparently affected North America at about the same time as it did Europe, judging by tree ring records and other data. The preceding Medieval Warm Period (Little Climatic Optimum) also appears to have been in step with a similar warm phase in North America.

Although the period of the Little Ice Age in Europe seems to have coincided with an improved or, at least, not particularly cold phase in Iceland, Greenland and presumably elsewhere in the North Atlantic, it apparently did not prevent Eskimos voyaging so far south that they were frequently encountered between 1682 and 1701 around the coasts of Orkney, according to the Rev. James Wallace in *The Statistical Account of Scotland* (1791–97). Williamson (1975) cited the instances of a kayak together 'with ye shirt of ye barbarous man yt was in ye boat' being given to Edinburgh University about this time by the local College of Physicians; and of another kayak at Marischal College, Aberdeen, which was 'taken at sea with an Indian man in it', about 1700. He also stated that the number of visits of Arctic birds, seals and whales to British coasts was greater than in recent years, and went on to say that the hooded seal *Cystophora cristata* 'was not infrequent in the Hebrides and Ireland and even penetrated to north-west France'.

Apart from such records, information on the possible effects of the 150 years or so of maximum cold of the Little Ice Age is difficult to come by from the writings of 16th and 17th century British naturalists, such as the eminent John Ray (1628–1705). They were still in the process of recording the fauna and flora to be found in Britain, and were certainly not yet able to make comparisons between past and present distributions, unless these were particularly obvious. Unfortunately, their observations throw little light on the fascinating question of whether or not Britain had recently lost, or was in the process of losing, species characteristic of southern Europe which may have colonised her during the Medieval Warm Period. Tantalisingly, we know that in about 1600 a naturalist called Nicholas Carter annotated his copy of Conrad Gesner's *Icones Animalium* (1560) with notes establishing the Hoopoe as a British bird (Fisher 1966b); but we do not know whether it was a regularly breeding member of the British avifauna, or whether it was merely a regular summer visitor that occasionally nested, as it is today.

Fortunately, however, several butterflies characteristic today of central and southern Europe, such as the scarce copper *Lycaena (= Heodes) virgaureae*, were regularly mentioned as occurring in Britain in the books of 17th and 18th century naturalists, although they are not now accepted as belonging to the British fauna. Indeed, some modern entomologists deny that they ever inhabited the British Isles, but it seems quite likely that they really did at one time, but died out during the Little Ice Age. It is possible that they represented what may have been a more southern element in our fauna, which either colonised southern England by cross-Channel migration during the warm Middle Ages, or survived from a much earlier colonisation in perhaps the Boreal Period, just before the sea link with the continent of Europe was broken.

As well as being a cold period, the last two decades of the 17th century were drier than average, and this dryness continued into the 18th century as late as 1750. For instance, 1714 was one of the driest years ever known

(Manley 1952). Moreover, the first 40 years of the 18th century were considerably warmer than the whole of the previous century, especially around 1720, when the amelioration was especially marked. Nevertheless, some severe winters occurred, such as the first two of the new century and 1708–09, which Gilbert White, the famous parson-naturalist, referred to as 'that dreadful winter', though as he was not born until 1720 he could only have known of its reputation. Another very cold winter was 1715–16, and, as in 1708–09, the Thames froze sufficiently to hold another frost fair.

At its peak, the increase in warmth of this relatively short-lived amelioration was perhaps comparable with the 1920s. Its end was marked by a very bad winter, that of 1739–40, when yet another frost fair was held on the frozen Thames. Thereafter, the rest of the century was generally cold with five or six severe winters in every decade (Lamb 1963), and there was no improvement until about 1825. The River Thames froze over no less than ten times between 1700 and 1814, the year of the most severe frost of the century and the last occasion on which a frost fair was held there. Indeed, 1814 appears to have plumbed the deepest depths of the Little Ice Age, which is generally accepted to have ended about 1850, following some 300 years in which cold winters were more common in north-west Europe than they have been since; and when the regularity of wet and sunless summers resulted in some of the remaining parts of the ancient Scottish Highland forest, growing high on the glen-sides, being overwhelmed by encroaching peat-bogs or blown down in the frequent gales. During this long cold phase the glaciers of Iceland, Scandinavia and the Alps advanced further than in any period since the late glacial phase (Post-Allerød) of the Last Glaciation, apparently reaching their maxima between 1745 and 1750 (Manley 1972).

While the Little Ice Age ran its chilly course, man continued his clearance of the European forests, but with growing awareness that things had gone too far, and that some steps should be taken to repair the damage. In Scotland, a start was made on replanting as early as the mid-15th century, and the trend gathered momentum in the 16th century; but it was not until it became fashionable among Scottish landowners in the following century to do so that reafforestation began to balance disafforestation. By this time the idea had spread to landowners in England, Wales and Ireland, too, but by the mid-19th century its popularity had waned with the arrival of the Industrial Revolution and the consequent increasing dependence upon iron. This wave of tree planting was valuable, however, and very welcome because, according to Tansley (1968), the proportion of woods and forests remaining in England and Wales at the end of the 17th century had dropped to below 16% of the land area, the remaining surface being divided principally between grassland (31%), heath, moor and mountains (25%), and arable land (23%).

Another growing feature of the British landscape during the period covered by the Little Ice Age was the enclosure of more and more common-land and arable fields by hedgerows. For a long time the process had been a slow one, so that in 1700 over half the farmland was still unenclosed (as it mainly still is on the continent of Europe), but from then on the rate accelerated, especially with the passing of the Enclosure Acts at the end of the 18th century and in the early 19th century, which saw the end of the medieval 'open field system' in Britain.

Professor Manley (1972) has pointed out that during the latter half of the 18th century the unfavourable climate in Britain caused shortages and, consequently, rising prices, due to poor crops, and therefore it was profitable to undertake further woodland clearance and land reclamation, even high up the hillsides.

Bearing in mind the escalating loss of natural or semi-natural habitats during this period, plus other factors due to the impact of man, such as industrialisation and increased persecution, what were the overall effects of the Little Ice Age on wildlife in Britain and Europe? Unfortunately, the picture which may be formed from the available evidence is an incomplete one, especially where it relates to the years before 1800.

In general terms (and judging by what we know happened, often on a larger scale, throughout the Ice Age), as the climate gradually cooled from the 13th century onwards, species of animals and plants with centres of distribution in the southern parts of the Northern Hemisphere presumably contracted the northern limits of their ranges, leaving small, isolated populations here and there where local climatic conditions favoured their survival. Meanwhile, species with northern centres of distribution expanded their ranges southwards, while at the same time contracting at their northern limits in the Arctic or subarctic regions.

In Britain, for example, by the time the Little Ice Age was at its maximum in the late 17th century, or at least soon after, we may have lost as regular members of the fauna such characteristic species of warmer climes as the Hoopoe, Golden Oriole and Bee-eater among birds, and the purple-edged copper *Lycaena hippothoe*, scarce copper and scarce swallowtail among butterflies.

Of those birds that underwent changes in range during the Little Ice Age which might be explained by the climatic factor, the Cirl Bunting presents a somewhat perplexing case. It was first discovered in Britain, and distinguished from the Yellowhammer, by Colonel George Montagu (1751–1815), the celebrated early ornithologist, in 1800 in south Devon. In 1805 he also received specimens from central Somerset, so it appears to have been well established in south-west England at that time. However, some authorities believe that when he made his discovery it had not long been an inhabitant of England, but had in fact colonised south-west England from north-west France sometime during the previous century. It is typically a bird of the Mediterranean region, where it occupies woodland-edge habitats, so it is rather surprising that it should have spread north-westwards to colonise south-west England during a cold climatic period when the reverse trend was the rule. Assuming that it was indeed a new colonist, and had not merely been overlooked by the earlier naturalists, then the climatic phase most favourable to such an expansion of range would have been the period of amelioration during the first forty years of the 18th century, which reached its peak around 1720, or the shorter and less warm spell of generally good summers and mild winters between 1770 and 1783.

Once it had established a foothold in south-west England, whenever that was precisely, the Cirl Bunting apparently managed to maintain itself there, where winters are generally milder than elsewhere in Britain, during the generally unfavourable climatic conditions of the succeeding years up to the long period of amelioration which commenced about 1850. Then after that year, in common with many other species which will be dealt with in succeeding chapters, it recommenced its northward spread, not only in England but also on the continent of Europe, reaching almost as far north as the Baltic coast. By the late 19th century, from its stronghold in southern and especially south-west England, it had extended its breeding range over most of the suitable areas of Wales, the Midlands and even, though sporadically, as far north as Cumberland and Yorkshire. But early in the present century there were indications of a decline in some parts of its British range (Figure 13); for instance, after 1910 it changed from being a quite common species in the Isle of Wight to a rare one (Parslow 1973). By the end of the 1930s it had largely disappeared from Wales and Herefordshire, while in the north Midlands and northern England it was much less often recorded

FIGURE 12 *Cirl Bunting*

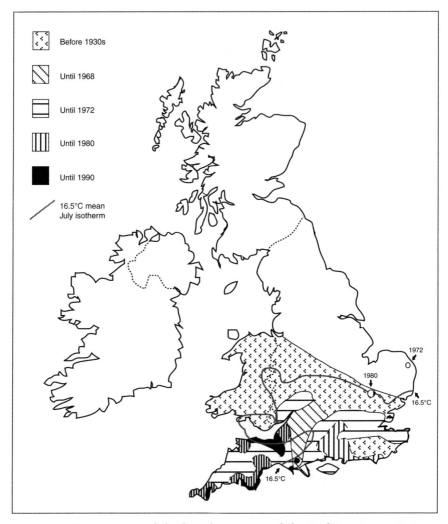

FIGURE 13 *Contraction of the breeding range of the Cirl Bunting in Britain from 1805 to 1990 (adapted from Parslow 1973; Sharrock 1976; Sitters 1982; Gibbons et al. 1993).*

breeding, and after 1950 hardly ever did so. This decline continued quite rapidly, with the result that the species had become mainly restricted to the south of England by 1980, when H.P. Sitters (1982) put the breeding popu-lation at no more than 200 pairs, and considered that it might be as low as 130. He found that by far the highest concentration was in south Devon, where Colonel Montagu first discovered it, and that was still very much the case in 1982 when 136 of the 167 pairs reported were concentrated there. In 1989, all but five of 119 possible breeding pairs were confined to the south Devon stronghold, and four of those were in neighbouring Cornwall

(Spencer *et al.* 1991). A similar retreat southwards has occurred in Belgium, France and Germany (Yeatman 1976; Batten *et al.* 1990).

It is difficult to think of any significant factor other than climatic change to account for the expansion, then contraction, of the Cirl Bunting's range. Losses of habitat have caused local decreases (Parslow 1973), but this alone has not been on a sufficient scale to explain it, nor is there much evidence that alterations in agricultural practice have had any significant effect over the long term, except that the recent practice of sowing cereals in autumn has reduced the number of stubble fields available to Cirl Buntings for winter feeding (see page 271). It is possible that the widespread destruction of hedgerows in some areas may have contributed to the decline, but since 1963, for instance, I have seen it decrease to virtual extinction in the countryside around Bristol where there has been relatively little removal of hedges. Simms (1971) considered that 'a mixture of such climatic factors as sunshine, low rainfall and warm winters together with certain topographical factors' might set the limits on the Cirl Bunting's distribution and population level. I certainly believe that a combination of mild winters and relatively dry summers with plenty of sunshine are important. The period 1909 to 1930 was a distinctly wet phase in Britain, with frequent cool summers and some cold winters and springs, especially between 1917 and 1923 (except for 1921, which was exceptionally dry with a hot summer). This correlates well with the start of the Cirl Bunting's decline, and although 1930–50 was markedly warmer, severe winters were frequent, especially in the 1940s, which presumably inhibited its recovery. The very severe winter of 1962–63 no doubt was initially responsible for the decrease noticed since then. Sitters (1982) also considered that 'it is not unreasonable to suppose that the main reason for its fluctuation over the past two centuries is climatic'.

This consideration of the Cirl Bunting case has caused us to stray into events which come within the scope of the succeeding chapters, to which I must now proceed.

Chapter 5

1850–1950: A HUNDRED YEARS OF CLIMATIC AMELIORATION

General features

At last we have reached a period in history where records of both climatic fluctuations and changes in the distribution of plants and animals are reasonably detailed and accurate, so that it is far more possible to make correlations between the two. However, this is not to say that the picture is a complete one; annoying gaps in our knowledge—at least on the natural history side—will become obvious in the remainder of this book.

As we saw in the last chapter, after the cold and cheerless years of the previous decade or two, there was a slight but steady improvement in the climate of Britain and north-west Europe from about 1825. Indeed the decade 1825–35 appears to have been quite a warm one, but it was followed by several wet summers in the 1840s. A succession of bad harvests led to the notorious potato famine in Ireland and, in a large degree, to the repeal of the Corn Laws in Britain. Nevertheless, from about 1850 the general natural improvement became distinctly more noticeable, with a run of dry summers up to 1865 and milder winters until 1878, when a series of harsher winters ensued. Summers became drier again from 1887 and continued so until 1909, while the winters became milder after 1896. There seems to be general agreement among climatologists that the amelioration began in earnest in the late 1890s, and was especially marked in the 1930s. The 1940s are particularly remembered for their fine summers, for example those of 1940, 1945 and 1947. Soon afterwards, about 1950, the hundred years of amelioration began to wane, to the extent that we were entering a new period of natural climatic deterioration or cooling. According to the American climatologist Wallace Broecker (1975), judging by the Camp Century ice core cycles alone the world should have cooled by the mid-1970s to a degree comparable with the early years of the 19th century; by the beginning of the 21st century the cooling would have become more severe. This subject will be discussed further in Chapters 11 and 13.

The comparatively lengthy amelioration from about 1850 to 1950 affected not only Britain and north-west Europe but the whole of the Northern Hemisphere, and indeed most of the rest of the world as well (Lamb 1966). At its height, the earth was on average 0.5°C warmer than it was in 1880, and about 1°C warmer than it was during the Little Ice Age. This more or less brought about a return to the conditions of the Medieval Warm Period (Little Climatic Optimum), and it was reflected in northern Europe and Asia by an extension of the growing season of up to two weeks; by the raising of the tree-line and a northward extension of the limit to which crops will grow by up to a hundred miles; in the behaviour of the prevailing winds; and in a general increase in the amount of rainfall, accompanied by changes in its seasonal distribution.

The most striking evidence of this amelioration was to be found in the Arctic and subarctic where a quite dramatic reduction in the amount of sea-ice occurred. During the 1830s huge amounts of pack-ice began breaking away and drifting into the north Atlantic. By 1850 the seas around Iceland were virtually ice-free, while the limit of permanent ice on both the east and west coasts had retreated to a considerable extent. For example, the average number of weeks per year when sea-ice was present off the coast of Iceland, which was as high as almost thirteen weeks between 1780 and 1790, declined from about nine weeks in the 1820s to just over three weeks in the 1840s, then rose again during a colder phase between 1860 and 1890. Thereafter, it gradually declined to about five and a half weeks by 1920, following which it dropped dramatically to only one and a half weeks between 1920 and 1950. Around Spitsbergen the average open season for shipping lengthened from about three months in the 1880s to seven months by 1940. In the same period the total amount of permanent ice in the Arctic diminished by some 20%.

Such a remarkable degree of melting of the ice was the result of the penetration into the Arctic of warm air, due to the strengthening westerly depressions in the North Atlantic taking a more northerly path than previously; and to a strengthening of the North Atlantic Drift element of the Gulf Stream, which pushed warmer water into the neighbourhood of Spitsbergen, with a consequent rise in temperature of as much as 8°C and more. These depressions were in turn caused by an intensification and northward shift of the atmospheric circulation over the Northern Hemisphere in general, and the North Atlantic in particular, which significantly increased the prevalence and vigour of the westerly winds, and therefore of the north-easterly movement of the North Atlantic Drift as well.

North-west Europe thus came more and more under the influence of the milder maritime climate of the Atlantic sector, especially in the 1920s and 1930s. From 1950 or thereabouts this influence weakened (see Chapter 11). As well as a stronger tendency to mild winters, the prevalence of Atlantic westerlies brought a somewhat increased likelihood of wetter summers with rather lower temperatures. However, during the warmest period of the amelioration, between 1920 and 1940, when the Atlantic depression track was usually further to the north, the summers were prone to be rather dry and warm. This tendency for fine summers continued up to 1949, but during that decade a run of severe winters also occurred.

The rather dramatic changes in the North Atlantic sector naturally attracted most attention, but a significant amelioration in the climate was observed throughout the Northern Hemisphere. As Professor Lamb (1966) pointed out, increases in the strength of the zonal circulation were quite general from about the middle of the 19th century and culminated around 1930. In the Southern Hemisphere, it apparently culminated somewhat earlier—between 1900 and 1910—and brought colder air from the Antarctic to New Zealand, for example.

Although in this book I am primarily concerned with climatic change in the Northern Hemisphere, especially Europe, and its effect on wild birds, it is worth remembering that this intensification of the atmospheric circulation between 1850 (or even earlier) and 1950 was world-wide, and probably occurred in most months of the year.

Until very recently, the accepted view has been that general climatic changes in the world work through a simultaneous contraction or squeezing away from the poles and towards the equator of the high and low pressure belts. However, climatic studies by scientists working in the Southern Hemisphere strongly suggest that what actually happens is a contraction or expansion around the North Pole of the climatic zones covering the whole world. In other words, the changes in the atmospheric circulation around the North Pole dominate the climatic patterns of the whole world, the zones moving northwards or southwards in response to them. It seems that this has something to do with the concentration of the bulk of the earth's land masses around the North Pole as opposed to the South Pole.

In the Northern Hemisphere, evidence of the amelioration between 1850 and 1950 was forthcoming not only from the retreating ice in the Arctic, but also from the behaviour of the glaciers. During this period there was, on the whole, a steady and significant retreat of the glaciers in Iceland, Norway, the Alps and elsewhere, which became very pronounced after 1930. According to Professor Manley (1972), the Icelandic and Scandinavian glaciers had reached their maximum post-glacial extent about 1745–50; then, following a subsequent retreat, they advanced again before the end of the century, and reached a minor maximum between 1790 and 1815. In the succeeding twenty years they retreated yet again, thereafter advancing rapidly once more to reach a second major maximum about 1850.

The nearest approach to glaciers in Britain, the semi-permanent snow-fields of the Scottish Highlands, such as those on the north-facing gullies of Ben Nevis, have shown similar trends. Thus, in the summers of some years in the 1930s they completely vanished for the first time on record. In the 18th century and early 19th century they had been considered permanent.

As mentioned in the previous chapter, there is evidence of time-lags between the onset of long-term climatic changes affecting Greenland, Iceland and western Europe. It seems that the changes occur in Europe about 250–300 years later than in Greenland, and 100–150 years later than in Iceland (Dansgaard et al. 1975). Data from North America, however, suggest that that continent is in phase with Europe, so the time-lag seems to be restricted to the mid-Atlantic longitudes. So it would appear that the 1850–1950 period of amelioration in Europe and North America was experienced already in Greenland between about 1600 and 1750 and in Iceland between 1750 and 1850. But this does not seem to be borne out by what we know of the fluctuations in the extent of the sea-ice around the coasts of Iceland and southern Greenland during the past two centuries. Moreover, a study of annual temperatures in Iceland between AD 900 and the present by Pall Bergthórsson (1969) clearly showed that 1750–1850 was a decidedly cold period there. Although the amelioration became evident in Iceland earlier than in north-west Europe, this was actually by not more than about forty to fifty years. Certainly a marked amelioration began in Iceland soon after 1850, and did not become particularly noticeable in Britain and western Europe until after 1890. Furthermore, it was not until the 1860s at least that the expansion of range in Iceland of a number of bird species began.

It is obvious from the foregoing accounts of the effects of previous climatic changes upon the distributions of wild animals and plants, that new fluctuations were to be expected in response to the warming-up period from 1850 to 1950; and they have certainly not been lacking. It is fortunate that this phase of climatic history has fallen within a period of increasingly detailed observation and understanding by biologists and naturalists. One could, of course, wish for even more detailed records, especially in the first half or so of the period when the reality of climatic change and its significance in relation to animal and plant distribution was not widely appreciated; but at any rate there is enough from which to draw some reasonably concrete conclusions. At least one is not forced to guess quite so much.

In the following five chapters I shall be listing and discussing in some detail many cases of bird species which have undergone expansions or contractions of range, and fluctuations in numbers, which can be correlated either completely or partially with the climatic amelioration of 1850–1950. Some of these instances appear to be quite conclusive and others less so; but I hope that, in the case of the latter, increased interest in this absorbing subject will eventually produce enough evidence to eliminate or reduce doubt.

In general, as one would anticipate, Arctic-alpine species contracted their ranges, while southern species expanded theirs. Thus many species with Arctic or northern-orientated distributions retreated northwards, while others restricted to high alpine regions in central or southern parts of the Northern Hemisphere retired to even higher altitudes. There was also a tendency for some species in the Palaearctic Region to retreat eastwards or south-eastwards, and more deeply into the continental land mass.

On the other hand, southern-orientated species reacted by advancing northwards or north-westwards, while others—a smaller element— expanded westwards. In Europe, a number of southern species previously confined chiefly to the vicinity of the Mediterranean, or with distributions centred there, spread north and west to a remarkable extent and, especially after 1930, with astonishing rapidity.

Some species appear to be more sensitive in their reactions to climatic changes than others, for reasons which are not often easy to detect. Factors which can be regulated by the climate, such as the length of the breeding seasons, the availability of food, the effects of the mildness or severity of winters on survival, the effects of summer temperatures on breeding success, and the influence of altering wind systems on migration patterns, are all obviously important; but other more subtle factors may also be at work which have not yet been identified.

Some species are clearly more adaptable to climatic changes, or more tolerant of them, than other, even closely related, species. For instance, why have the remarkable northward extensions of range in comparatively recent years by Cetti's Warbler and the Fan-tailed Warbler not been paralleled to a similar extent by other warblers inhabiting the Mediterranean region? Presumably ecological competition with similar species already there has something to do with it. It would seem that there were vacant ecological niches available further north for warblers such as Cetti's and the Fan-tailed, unoccupied by close relatives, which had hitherto been denied them by a climatic factor, but which became open to colonisation by them because of the amelioration—in this case, the greater mildness of the winter—whereas other warblers, such as the Spectacled and Subalpine, were prevented from expanding northwards because their niches were already occupied by related species which were better adapted, such as the Whitethroat and Lesser Whitethroat.

Whatever the precise factors are which determine why some animal species respond to climatic change and others do not, it is clear that some respond more quickly than others. Again, the reasons are presumably connected with subtle differences in their ecological requirements, such as the micro-climate and the level of temperature they are able to tolerate. For instance, most species of grasshoppers depend very much upon the level of the temperature to maintain normal activity, and therefore successful breeding. But some species are more tolerant of cooler conditions than others, and are better able to respond to small rises in average temperature. Thus, to take a simplified case, if the average summer temperature in a certain area outside the normal range occupied by two distinct species of grasshoppers A and B should rise by, say, 0.5°C, then grasshopper A (which tolerates a lower temperature than grasshopper B) may well be able to colonise the new area quite quickly, whereas grasshopper B may not be able to do so until the average summer temperature rises by another 0.5°C (i.e. 1°C above the original level). Therefore species A would respond to a climatic amelioration earlier than B.

Apart from such differences in sensitivity of reaction to climatic change between different species, it is not difficult to imagine that there may be a tendency for a time-lag to occur between observed responses in animals and plants, and the beginning and end of a climatic change. Changes in distribution were often not detectable until well after the 1850–1950 amelioration began; in many cases, once the change in distribution had started the same trend continued, even though the natural improvement in the climate came to an end around 1950. Of course, the long run of mild winters in western Europe between 1962–63 and 1977–78, which occurred in spite of the then expanding deterioration in the Arctic, had much to do with the continuing northward spread of various species in this sector.

There seems to be also a tendency, as the above suggests, for species to respond more quickly to the beginning of a climatic change than to its end. This is probably partly due to the fact that plants and animals are often tenacious in holding on to well-established territory, and disappear only slowly as the changing conditions gradually operate against them, whereas they may be quick to exploit new opportunities. Furthermore, the question of competition between related species with similar ecologies is again relevant. Thus, during an amelioration, a northern species may only retreat northwards, in the face of competition from a related species with a very similar ecology advancing from the south, when the climatic change eventually becomes inexorably more and more unfavourable to it, upsetting the balance between them. On the other hand, climatic advantage may sometimes tip the scales very quickly in the favour of an advancing species. So competition between related species (e.g. the 'northern' Brambling versus its 'southern' relative, the Chaffinch) is another important factor in the speed with which they react to climatic change.

Another factor which should not be overlooked in explaining the apparent slowness with which some species stop responding to a climatic change when it has ceased operating in their favour, and indeed may be operating in the reverse direction, is that of human error. We tend to become more quickly aware of a species which is spreading than the opposite, unless the latter is of a spectacular or conspicuous nature. And this, of course, also depends to a large extent upon the number and distribution of interested observers. Thus, in Britain, the decline of the formerly widespread and fairly conspicuous Wryneck during the past century or so did not escape the notice of the large numbers of birdwatchers more or less evenly distributed over the country; whereas the no less dramatic decline of the equally widespread, but less conspicuous, mole-cricket *Gryllotalpa gryllotalpa* was virtually unrecognized until quite recently when entomologists interested in crickets increased in number and became less thinly distributed.

Partly because of these differences in the speed with which species respond to the commencement or cessation of changes of climate, either operating in their favour or against them, we are still witnessing an apparently paradoxical situation in which the 1850–1950 natural climatic amelioration ended long ago, yet many southern species continue to advance and are indeed still making striking gains, while northern species have been advancing southwards in response to a natural deterioration in the Arctic zone, which began about 1950. In Britain, for example, species of birds such as the Mediterranean Gull, Cetti's Warbler, Firecrest and Serin are, at the time of writing, still colonising or attempting to colonise England from the south, while northern birds, such as the Whooper Swan, Goldeneye, Purple Sandpiper, Wood Sandpiper, Snowy Owl and Redwing, have been attempting to colonise Scotland from the north, or have been showing signs of doing so. However, the global warming due to the anthropogenic greenhouse effect apparently lies behind this complex and, on the face of it, contradictory situation. I will discuss the subject as fully as I can in the final chapters of this book.

Section Two

EFFECTS OF THE 1850–1950 CLIMATIC AMELIORATION ON EUROPEAN BIRDS

INTRODUCTION

Perhaps the most obvious consequence of the general climatic amelioration between about 1850 and 1950 was the northward and westward spread of many species of animals whose breeding ranges are centred in the south. Of these the best known and, usually, the best documented examples are, as one might expect, to be found among birds. Indeed, the northerly advances of such species as the Collared Dove, Black Redstart, Cetti's Warbler, Serin and Little Ringed Plover in Europe (and the Northern Cardinal, Dickcissel and American Wigeon in North America) have been quite spectacular. But many other species have made advances almost as marked; while in a number of others, less conspicuous yet nonetheless significant changes in breeding distribution have been detected. In Appendix 1 I have listed all the examples I am aware of which appear to be due to climatic change, or at least partly so. My reasons for including some of these may seem speculative, since direct evidence that they are due to climatic influences may be lacking. Indeed, other factors may seem to be, or are known to be, a more likely cause of their alteration in distribution. Nevertheless, I have included them if I suspect that the changing climate may be involved either as a primary or a secondary factor. Certainly, I have not been afraid to indulge in specula-tion, in the hope that it may induce others to investigate these particular cases in detail, and thereby to arrive, if at all possible, at the full and true facts where these have not been established. To hesitate to include all such species, I have argued with myself, might result in the way in which they fit into the general pattern being overlooked, since causes other than alter-ations in the climate may have seemed to provide a plausible or even a complete answer.

Of course, the c.1850–1950 climatic amelioration did not benefit all species. Many that are adapted to life in northern climes, such as the Whimbrel, Siberian Tit, Snow Bunting and Brambling, were forced to retreat northwards; as we have seen, this was sometimes in the face of competition from advancing, related species with similar ecologies and habitat prefer-ences. Thus, the Brambling retreated before the advance of the Chaffinch, the Whimbrel before that of the Curlew, and the Siberian Tit before that of the Crested Tit. However, although many of these northern species lost territory in the south of their range, the warmer conditions within the Arctic Circle meant that they were able to penetrate further north and expand their ranges in that direction to a considerable extent. They perhaps regained much of the ground they lost during the Little Ice Age.

Another assemblage of species, which might have been expected to benefit from the amelioration, actually retreated southwards or south-eastwards: such birds as the White Stork, Stone Curlew, Bee-eater, Hoopoe, Wryneck, Roller, Lesser Grey, Red-backed and Woodchat Shrikes, to name some of those most affected. Of those that showed the most dramatic contractions in range, the majority are summer residents which depend to a large extent upon capturing those kinds of adult insects most active on warm, sunny days, such as bees, wasps, ants and grasshoppers. Clearly, the mainly wetter summers in north-west Europe, resulting from the increased dominance of Atlantic depressions driven by the intensifying atmospheric circulation, reduced the activity of such insects, and therefore their availability as food to these birds. Anyone who doubts the effect of wet, sunless days on such insects needs only go into a meadow on a hot day in high summer, when it is alive with chirping grasshoppers and other insects, and then visit that same meadow on the first subsequent wet, overcast day, perhaps a day or two later, to realise what a difference it makes to insect activity. Where, on the first visit, hundreds or even thousands of insects swarmed in the sunshine, you will be hard put to it to find more than a few clinging inactively to the grass stems, or lurking among the leaves of other plants. Birds may find it almost as difficult to locate them, and they and their broods may go hungry amongst a plentiful but hidden supply of food. Even birds such as warblers, which are less dependent upon capturing active insects than those just mentioned, and more skilful at discovering them inactive in their hiding places, may suffer in this way. The temporary contractions in range in western Europe between 1850 and 1950 of, for example, the Woodlark, Crested Lark, Savi's Warbler, Icterine Warbler and Cirl Bunting are likely to have been chiefly due to this cause.

Finally, there remains a small group of species whose distributional changes are confusing, and the causes apparently obscure. Climatic change may or may not be involved. Cases from all these groups will be discussed in the course of the following accounts of individual species which I have selected for analysis, either because they are apparently good examples of changes due to climatic influence, or because this may be one of the factors involved. Some cases have been included because I feel that the climatic factor has been overlooked in explaining the reasons for a change in range or numbers. In practice I have included all European species in which I am aware of a noteworthy change in distribution and suspect that climatic change may be involved to some extent at least. (The reader is reminded that for many of the species discussed in Chapters 6–10, information on more recent trends and status will be found in Chapters 11–12.)

Chapter 6

SEABIRDS

The increased vigour of the North Atlantic Drift, and the penetration of its warmer waters into higher latitudes of the North Atlantic than previously, brought significant benefit to at least some of the seabirds. Firstly, this favoured the northward spread of cod, haddock, mackerel, saithe, pilchard and other fish, with apparently, in some species at least, increases in their abundance; and secondly, it increased the length of the period of their availability.

In the case of the Gannet, Williamson (1975) pointed out that there is a clear link between the recent history of this bird on both sides of the Atlantic and the effect on its food supply of the warmer ocean currents associated with the ameliorating climate. It feeds chiefly upon such surface-shoaling fish as cod, codling, saithe, mackerel, pilchard and especially herring, and their greater availability this century in the North Atlantic led to a great expansion in population and the establishment of several new colonies in the British Isles (including the Channel Islands), Brittany, Iceland, Norway, Nova Scotia and Newfoundland. Williamson also suggested that the abundance, even in winter, of shoaling young saithe in Shetland waters must have helped the substantial growth of the gannetries which became established on Hermaness and Noss during the First World War.

Between 1909 and 1974 the world population of the gannet grew from 67,200 to 197,000 pairs—a rate of increase of about 3% per annum (Sharrock 1976). It continued to increase after 1974 at about the same rate, and by 1985 had reached some 263,000 pairs (Wanless 1987), in spite of the cooling of the climate in northern latitudes from 1950 onwards. It remains to be seen what influence the increasing warmth of the greenhouse effect will eventually have.

Climatic improvement was not the only factor involved in the boom in Gannet numbers; the marked reduction in persecution by man since the turn of the century was important too. Before then, both birds and eggs were taken for food on a large scale, especially on the North American seaboard. During the 19th century the world population was more or less halved, at least partly from this cause. Gannets were, however, probably also declining during the Little Ice Age, as a result of the effects of the cold climate on the numbers and distribution of their fish prey, just as they have increased during the climatic improvement of this century.

The remarkable population explosion and range expansion of the Fulmar in the North Atlantic over the past two centuries has been well documented by James Fisher (1952, 1966*a*), who considered that its only breeding stations in the north-east Atlantic in medieval times were the islands of St. Kilda (west of the Outer Hebrides), Grimsey (off the north coast of Iceland), and the isolated island of Jan Mayen, some 330 miles (528 km) east of Greenland and well within the Arctic Circle. Sometime between 1713 and 1753, Fulmars, hailing either from Grimsey or Jan Mayen, advanced southwards and began nesting in the Westmann Islands, off the south coast of Iceland. From the Westmanns they colonised the cliffs all round Iceland after 1753, then between 1816 and 1839 crossed the 400-odd miles (640 km) to the Faeroes and colonised them. Shetland was next in line for a take-over, starting in 1878, followed by the Outer Hebrides from 1887 and Orkney from 1900. By 1903 the Scottish mainland was being colonised, while in Ireland the first breeding Fulmars arrived in 1911. They penetrated south down the east coast of Scotland and reached England by 1922. Nowadays they are to be found nesting on suitable cliffs all round the coasts of the

FIGURE 14 *Fulmar*

British Isles, with an estimated population of more than 305,600 pairs in 1970, and 525,000–530,000 pairs in 1989 (Sharrock 1976; Marchant *et al.* 1990). Norway was colonised in 1924, and more recently the coast of Brittany, in 1956. This fantastic expansion of the Fulmar continues, though now at a somewhat slower rate.

The extraordinary success of the Fulmar was attributed by Fisher to its adaptation to the marine activities of man—initially, to the growth of whaling, which provided plenty of easy food when the captured whales were flensed at sea. With the decline of whaling in the North Atlantic during the mid-19th century, Fisher believed that the Fulmars then learned to exploit the fish offal thrown overboard by the modern trawlers, which conveniently appeared in increasing numbers at about this time. He pointed out that the slowing down of their rate of increase coincided with the period when their numbers had probably almost outgrown the supply of offal discharged by the trawlers. Professor Finn Salomonsen linked the Fulmar's population explosion with the climatic amelioration in the north-east Atlantic in the late 19th century. The climatic cooling in northern latitudes since 1950 might explain the subsequent decline in the Fulmar's rate of expansion, but not the fact that the spread commenced in the 1750s during a decidedly cold phase around Iceland, although an ameliorating one around Greenland. However, it seems likely that the warming up of the North Atlantic had something to do with the speed of the expansion, since the main impetus was apparently between 1870 and 1940 when the climatic amelioration was most marked, leading to the northward spread of the fish mentioned above; this in turn facilitated the growth of the trawling industry in these waters. (See also Chapter 11.)

Among the cormorants of the Northern Hemisphere, the Shag in Europe and the Double-crested and Red-faced Cormorants in North America have increased and expanded their breeding populations since 1920–25. The Red-faced Cormorant, in particular, quadrupled its breeding stations between then and 1960 (Palmer 1962). Although these increases have undoubtedly been triggered off by a relaxation or cessation of persecution by man, it seems likely that they have also been significantly assisted by the improvement in their supply of food occasioned by the strength of the climatic amelioration in the 1920s and 1930s. The Cormorant (or Great Cormorant) of the Old World began a marked population increase and expansion of its range in central Europe in the 1980s, particularly along the south Baltic coast. The reasons for this are not yet clear. The Pygmy Cormorant also began spreading northwards in the 1980s in south-east Europe.

Several of the auks tended to retreat northwards during the amelioration. One of the most notable was the Little Auk, which became much scarcer between 1900 and 1950 along the southern edge of its breeding range, as for example in Iceland, where the well-known colony on Grimsey declined from at least 150 pairs to only ten, and the colonies at Cape Langanes in the north-east corner died out (Gudmundsson 1951). The species has since

FIGURE 15 *Puffin*

become almost, if not quite, extinct in Iceland. However, since the early 1980s there is evidence of a move south again (see Chapter 11).

The decline in the numbers of Puffins in many parts of the North Atlantic since 1850, and particularly since about 1920, has been truly dramatic. Such decreases have been particularly marked in the southern sectors of the Puffin's breeding range, where many colonies have declined from thousands of breeding pairs to a mere handful or even extinction. For instance, the population on Lundy in the Bristol Channel dropped from 'incredible numbers' in 1890 to ninety-three pairs in 1962 and about sixty pairs in 1966, and that of St. Tudwal's Islands off the Caernarvonshire coast of Wales from 'hundreds of thousands' at the turn of the century to none by 1951. Similar pronounced decreases have been noted along the Atlantic coast of Canada, Newfoundland and the United States. According to some observers, the decline has lately taken an even more serious turn, especially in some of the Puffin's traditional strongholds in the north and west of the British Isles, with numbers apparently dropping almost catastrophically in some colonies (e.g. St. Kilda and the Shiants in the Outer Hebrides).

The picture is not all gloomy, however; in some colonies in the north-east of the British Isles, such as the Farne Islands and the Isle of May, increases have been taking place for some years. Moreover, M.P. Harris and S. Murray (1977), who studied the Puffin colonies on St. Kilda from 1974 to

1976, concluded that though there was a decrease during the period 1947 to 1958, or perhaps even a year or two later, 'it is doubtful if there has been a marked change since then'. Indeed, they believed that the reports of a catastrophic decrease around 1970 had been exaggerated. In an earlier paper, in which he surveyed data for most British colonies, Harris (1976) showed that the recent general decline had stopped, at least temporarily, and concluded that the results of the recent study on St. Kilda supported this.

R.M. Lockley (1953) suggested that the decline in the southern populations of the Puffin and all other Atlantic auks may have been due to the climatic amelioration causing changes in the distribution of their marine prey. This seems to me highly likely, as the more northerly track taken by the North Atlantic Drift may well have moved the Puffin's best feeding grounds (and those of other auks) further to the north. For instance, we have already seen that pilchards, a favourite food, were among the fish which extended their range northwards. Nevertheless, this explanation may not fully account for the Puffin's decreases, especially the recent, serious ones (assuming that they did in fact occur), in the north-east sector of the Atlantic. Pollution at sea has been suggested as a possible cause of these, and among other factors advanced as contributing to the long-term decline in the Puffin are growing disturbance, the accidental introduction of brown rats to island colonies, and the pollution caused by the rapidly increasing populations of the larger gulls, especially the Great Black-backed. However, if M.P. Harris is correct in his belief that the general decline has ceased, then this might well be due to the cooling of the north-east Atlantic during the period 1950–70 driving the fish prey of the Puffin further south again. In this connection, it is interesting to note that the more southerly populations of the Guillemot, Black Guillemot and Razorbill likewise showed definite decreases during this century's climatic amelioration, though there has been some evidence of a recovery since 1940 in the case of the Black Guillemot in the Irish Sea. On the Portuguese and Spanish coasts, the decrease of breeding Guillemots has been dramatic since the 1950s: from thousands of pairs to (in Spain) only 38 pairs in 1982 and 13 in 1989. This continuing marked decrease in the south suggests that, apart from the climatic amelioration, other factors such as oil pollution are at work. Voous (1960) pointed out the remarkable exactitude with which the southern limits of the breeding range of the Black Guillemot are governed by the distribution of waters with surface temperatures of 59–61°F (c.16°C) in August. He also noted that the southern limits of the Razorbill, Guillemot and Puffin are bounded by surface water temperatures in August of 59°F (c.15°C), 66°F (19°C) and 62°F (c.17°C) respectively. Thus a northward shift of these isotherms could account for a reduction in the populations of these seabirds in the southern parts of their ranges, although, as we have seen, the influence of this shift probably makes itself felt through its effects on the distribution of their food.

The gulls are a group in which many, if not most, species have increased in numbers and expanded their ranges to a remarkable extent this century.

Some may have been stimulated to do so by the 1850–1950 climatic amelioration, either wholly or partially. Perhaps the clearest examples of species which have been aided by that and subsequent climatic warming are the Little Gull, the Mediterranean Gull and probably the Black-headed Gull.

The Little Gull, which according to Voous (1960) is almost entirely limited in its distribution in the north by the 57°F (*c.*14°C) July isotherm, began advancing westwards from its strongholds in the eastern Baltic region in about 1900. By 1920 it had established colonies in Denmark, by 1942 in the Netherlands, and by 1950 was attempting to nest in northern Germany. In 1968, after a tremendous build-up in the numbers visiting the British Isles as passage migrants and summer visitors, breeding was suspected, but not proved, in the Wash area of eastern England. Finally, single pairs were proved to nest unsuccessfully in the Ouse Washes in that part of England in 1975 and 1978; single pairs were unsuccessful also in North Yorkshire in 1978 and in Nottinghamshire in 1987; it remains to be seen whether breeding will become regular in future.

The Little Gull is one of an assemblage of birds characteristic of shallow freshwater lakes, bogs and marshes that apparently spread westwards and north-westwards in response to increased aridity in their eastern strongholds, especially the steppes, which resulted in many of the lakes there drying up. Other species involved include the Great Crested and Black-necked Grebes, Gadwall, Red-crested Pochard and other ducks; terns such as the White-winged Black and Whiskered, and the Black-headed Gull. The ability of these birds, and in particular the Black-headed Gull, to establish themselves in north-west Europe can no doubt be attributed to the improvement of the Atlantic-influenced climate there since 1850, and not merely to the necessity to leave areas of increased aridity in the east.

Before 1880 the Black-headed Gull was unknown as a breeding bird in Iceland and Norway, and following a marked decrease it was near extinction in the British Isles. However, sometime after 1850 it began spreading north-west and north on a remarkable scale, so that by 1900 it was breeding in Finland, Norway and Sweden. In about 1900 it began increasing in Britain; for example, between 1938 and 1973 the total number of breeding pairs in England and Wales trebled. In 1911 the first breeding records were reported from Iceland, where it is now widespread and common. In Sweden and Finland the breeding population in 1950 was said to be at least a hundred times larger than it was in 1900. Black-headed Gulls from Iceland, and perhaps elsewhere in Europe, are regularly seen in Greenland, and they are now the most frequent bird visitor from Europe to the east coast of North America. Williamson (1975) attributed this to the recent intensification of the polar easterlies. Maybe Voous (1960) was right in suggesting that a future colonisation of North America is not impossible. Apart from the effects of climatic amelioration, which has probably chiefly benefited this largely insectivorous gull by increasing the availability of its food supply in northern Europe, other reasons advanced for its increasing populations are

FIGURE 16 *Black-headed Gull nesting in a Dutch peat bog, May 1980. (Photograph: David J. Tombs)*

a reduction in persecution by man (especially the harvesting of its eggs), improved protection of colonies, and the creation through man's activities of additional habitats, as well as its exploitation, especially in winter, of more varied food sources, again often provided by man.

Voous (1960) considered the Mediterranean Gull, with its very limited breeding range centred on the Black Sea and the north-east corner of the Mediterranean Sea, to be an ancient relic species probably in the course of complete extinction. That may be so, but ever since 1950, perhaps even earlier, it has shown an increased tendency to visit north-west Europe and attempt to breed, usually among colonies of Black-headed Gulls and often pairing up with them. The Mediterranean Gull is not extending its range systematically north and north-westwards as other species have done, but migrants and breeding pairs have been appearing more and more often outside the normal range (Figure 17). For instance it has bred, irregularly at first, in the Netherlands since the 1930s, Hungary and Germany since the 1950s, Estonia since 1966, the former Czechoslovakia since 1967, southern England since 1968, and southern Sweden since 1970. In southern France, a pair or two have been nesting annually since 1965. Voous stated that there is a considerable dispersal outside the breeding season into central and western Europe along the courses of large rivers, and it seems to me that the mainly milder winters since 1930 have encouraged them to remain and breed in these far-flung outposts, especially since the 1960s.

Most of the other gulls of the temperate zone have increased and expanded their ranges to a considerable extent since the end of last century, or even earlier. In every case the climatic amelioration seems to have played at least a part in their success. The most spectacular has been the population explosion of the Herring Gull, perhaps the most adaptable, aggressive and successful of the larger gulls in the Northern Hemisphere. Professor Voous (1960) regarded it as a 'recent' (i.e. since the end of or during the Last Glaciation) colonist of north-west Europe from North America, where it is much more widely distributed than in the Palaearctic, if you assume, as he did, that its Old World counterpart the yellow-legged Herring Gull is really a race of the Lesser Black-backed Gull and not, as most authorities do, a race of the Herring Gull. He believed that the Herring Gull was able to compete successfully with the Lesser Black-back, and partially replace it, by reason of its remarkable adaptability where food and nesting sites are concerned; and also because it is a hardy, sedentary species which occupies its nest sites at least a month before the Lesser Black-backs return from their tropical wintering grounds.

Nevertheless, there are difficulties in accepting this view of the relationships, such as the fact that the population of yellow-legged Herring Gulls in Finland and the Baltic states interbreeds with the pink-legged Herring Gull of the west, and not with the Lesser Black-backed Gull; and the fact that the latter has not been inhibited by the Herring Gull in expanding its range and numbers this century, but has been almost as successful. Incidentally, the

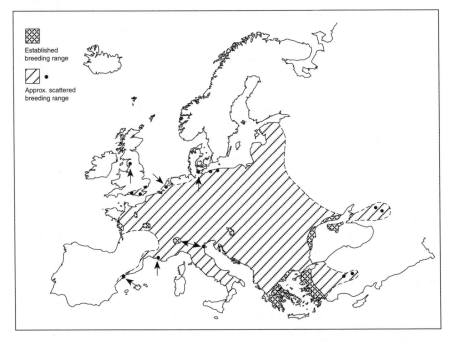

FIGURE 17 *The Mediterranean Gull's expanding breeding range in Europe (adapted and updated from Harrison 1982; Cramp et al. 1983).*

eastern Baltic yellow-legged Herring Gulls apparently colonised that region towards the end of the last century as a result of a north-westward expansion of range caused by the increasing aridity in the areas of the Aral and Caspian Seas. This has already been mentioned as a cause in the similar expansions of other species.

Whatever the true facts are concerning these closely related gulls, the long climatic amelioration which followed 1850 clearly favoured their population growth and extensions of range. The populations of Herring Gulls have grown enormously on both sides of the North Atlantic since about 1900, no doubt aided by the discovery by these powerful birds of new sources of food, as well as by the improved climate (particularly in winter). Their exploitation of the vast amounts of edible refuse provided by man as a consequence of his own immense population and economic growth, and their development of the habit of foraging on his farmland, are important instances of this adaptability. The Lesser Black-backed Gull has likewise benefited from adapting to these increased sources of food, and from the opportunities for range expansion brought about by the ameliorating climate; so, to a lesser extent, have the Great Black-backed and Common Gulls. This increased tendency, since the close of last century, for many formerly coastal gulls to spend much of their time inland has been helped by the increasing provision of large artificial inland reservoirs near big cities; these make ideal roosting places for those gulls which spend most of the daylight hours foraging on city refuse tips or adjacent farmland. It has certainly torpedoed the old countryman's belief that the sight of gulls inland heralds the arrival of stormy weather.

Since they began increasing early this century, Herring Gulls in the British Isles have at least doubled in numbers. Indeed, Parslow (1973) suggested that in England and Wales they might have doubled their population within the previous 20 years. Recent estimates from some of the major British colonies show that they are continuing to grow at between 13 and 15% per annum. Similar increases have been reported since 1920 from elsewhere in Europe, such as the Netherlands and Scandinavia, as well as in eastern North America. Iceland was colonised from the Faeroes in 1932 (first birds arrived 1927); Bear Island, south of Spitsbergen, was colonised in the same year, and Spitsbergen itself from 1950.

Lesser Black-backed Gulls have colonised Denmark from 1920, the Netherlands since 1926 and, undoubtedly aided by the amelioration of the North Atlantic, Iceland between 1920 and 1930. In 1969 this species was reported for the first time on the east Greenland coast, but breeding has apparently yet to be recorded.

Common Gulls have also increased a good deal in northern Europe since about 1870, and have been expanding their breeding distribution southwards as well as northwards. For example, they have colonised the Netherlands since 1908, Dungeness in Kent since 1919 (possibly from the Netherlands) and Ireland since 1934. In the north, they occupied the Faeroe

Islands in 1890, Iceland during the 1950s, St. Kilda in 1963 and Fair Isle in 1966.

The Great Black-backed Gull has certainly benefited on both sides of the North Atlantic from climatic amelioration, and has been increasing in numbers since 1880 and extending its range both northwards and southwards since 1920. In the British Isles the increase since 1880, and especially since 1920, has been very marked, and several new (or formerly occupied) areas have been colonised. For instance, between 1930 and 1956 the breeding population of England and Wales was found to have trebled; but even so, most of the coasts of south-east and eastern England are still without any breeding pairs, though Great Black-backs are numerous there in winter, such as along the Thames as far upriver as London. Elsewhere, the Great Black-back has pushed steadily north to capture Iceland (since 1920), Bear Island (since 1921), Spitsbergen and the Faeroes (since 1930). Along the Atlantic coast of North America it moved the southern limit of its breeding range 720 km southwards from Nova Scotia to Long Island between 1921 and 1942, apparently at the expense of the Herring Gulls in those localities where it has settled. In Iceland, it has been in conflict with the equally large Glaucous Gull (which replaces it in the Arctic), to the latter's detriment, competing successfully with it for nesting sites and food. As well as the climatic factor, other factors contributing to the Great Black-back's expansion are the abundance of edible refuse available to it in coastal towns and at sewage outfalls, and the greatly reduced persecution by human beings compared with last century.

Since the climatic amelioration between 1850 and 1950 had not been to its advantage, the Glaucous Gull was already retreating when competition from the Great Black-backed Gull began. It started diminishing in Newfoundland after 1930, finally disappearing from there as a breeding species by 1959. The big decline in its Icelandic breeding stock began around 1940, and it has since retreated further to the north-west. It has been found, incidentally, that Herring Gulls are now interbreeding with Glaucous Gulls in Iceland, the Kola Peninsula of north-west Russia and also northern Alaska. This probably accounts for the Herring Gull-sized 'Glaucous Gulls' which occasionally appear in southern latitudes and puzzle the local bird-watchers.

The Arctic counterpart of the Herring Gull/Lesser Black-backed Gull complex, the Iceland Gull, may also have withdrawn its southern limit to some extent, but as this has not extended as far south as that of the Glaucous Gull in historic times, it does not appear to have been noticed by ornithologists. In spite of its name, it does not breed in Iceland (although it is the commonest gull in northern Icelandic waters), but is restricted to the southern coasts of Greenland, Baffin Island and elsewhere in Arctic Canada.

The explanation of the Kittiwake's population explosion is rather different from that of the Herring Gull and other gulls. In the southern part, at least, of its circumpolar range, the Kittiwake suffered to a tremendous extent during the 19th century from human persecution. It was shot in its dense breeding colonies on the ledges of sea cliffs for sport and later on, around the 1860s, to supply the insatiable demand of the millinery trade for young Kittiwake wings, which became fashionable decorations for ladies' hats. Reaction in Britain to this dreadful slaughter led to protective measures being taken, and after 1900 the colonies slowly recovered. By the 1920s the population began to expand and recolonise old haunts, and to occupy new ones, particularly towards the south. Since then the rate of increase has accelerated and, as revealed by censuses in 1959 and 1969–70, the population has apparently been multiplying at a rate of nearly 50% every ten years. Breeding colonies are now to be seen around almost all suitable coasts of the British Isles. Big increases have also been reported elsewhere in north-west Europe since 1900, such as in southern Norway and north-west France. Although much of this increase is due to the cessation of persecution and improved protection, and represents a recovery to the Kittiwake's former distribution and status, the continuing recent expansion to the south may have been assisted by the climatic deterioration which set in from 1950 and lasted to at least 1980.

The Whiskered and White-winged Black Terns have already been mentioned as species which have altered their breeding ranges in response to recent climatic changes, but they are not the only terns in the Northern Hemisphere to have done so. The Gull-billed, Sandwich, Common, Little, Roseate and Black Terns have all, to a greater or lesser extent, advanced north or north-westwards since the beginning of the 1850–1950 climatic amelioration, while the Arctic Tern declined and to some extent retreated northwards—at least in southern Britain, Ireland, France and around the southern Baltic Sea—after 1900. It disappeared from the Isles of Scilly, for instance, and many inland sites in Ireland. A truly Arctic bird, breeding far to the north, it replaces its close relative the Common Tern in those regions, and competes with it ecologically along the southern edge of its range, where considerable overlap occurs in north-west Europe. Presumably the amelioration tipped the scales in favour of the Common Tern in the areas of overlap and contact. Curiously, the Netherlands was not colonised by Arctic Terns until 1898; there was a steady increase thereafter, but they are still scarce and far outnumbered by Common Terns. This colonisation is difficult to explain unless it represented a displacement of birds from England. In recent years a serious decrease has occurred in the north British colonies, especially in Orkney and Shetland, particularly the latter where there has been an almost total failure to produce fledged young. Research and other evidence suggests that this is due to food shortage (Batten *et al.* 1990). Whether this results from overfishing by the commercial sand eel fishery on Shetland or is linked to climatic change, or both, has yet to be determined.

In contrast to the Arctic Terns, Common Terns greatly increased in southern Britain and Ireland subsequent to about 1870, undoubtedly putting ecological pressure on the Arctics. Their numbers also increased in northern Britain from about 1890, when they successfully invaded the Arctic Tern's stronghold in Shetland. Such a growth in population is not all due to climatic improvement; in part it is a recovery from the human persecution of the 19th century when they were shot for sport and for the sake of ladies' fashion, and their eggs were collected for food.

Between 1890 and 1940, Little Terns also expanded their populations and range in the British Isles and Europe generally, following a marked decline throughout western Europe in the 19th century, largely due to human depredations. Again, climatic improvement may have aided this recovery, but between about 1950 and 1970 another decrease took place. Apart from increased crow and fox predation, this was largely attributed to escalating human pressure on beaches which are preferred by Little Terns for their nesting sites; and since protective measures have been introduced the situation has stabilised in Britain or even improved slightly, with the British breeding population currently estimated at 1,500–1,700 pairs (Batten *et al.* 1990). In 1960 Little Terns began a colonisation of southern Finland.

The Roseate Tern, which has a distinctly discontinuous distribution over much of its cosmopolitan range, is particularly susceptible to disturbance. It therefore suffered severely from persecution by man during the 19th century, and decreased considerably between 1800 and 1850 in its European stations. However, since north-west Europe is on the northern limit of the breeding range of this warmth-loving tern, it is quite likely that the decline before 1850 was in part caused by the cold conditions of the Little Ice Age. Be that as it may, it is a fact that a strong recovery commenced about 1900 during a period of increasing warmth and greater protection, and lasted until about 1965, since when, in spite of effective protection, the population in the British Isles, France and possibly elsewhere has seriously declined again. Once again this may perhaps be correlated with the climate, this time with the deterioration which set in during the 1950s and lasted until the 1980s, when the warming due to the greenhouse effect began to neutralize it. This and other possible factors of course require detailed study, as with the other sea terns.

The Sandwich Tern is another sea tern whose breeding range is discontinuous and largely restricted to the warmer parts of the Northern Hemisphere, being on its northern limit in north-west Europe. Again, like the Roseate and other sea terns, it declined during the 19th century, ostensibly from the effects of human disturbance and persecution, and recovered this century, although rather later than the other species. For example, in the British Isles and the Netherlands the increase became noticeable during the early 1920s, and in Britain reached a peak of 13,334 pairs in 1982, when the population was considered to have attained its highest level so far this century. It seems to me yet again that the fortunes of the Sandwich Tern, like that of the other terns just discussed, have paralleled the changes in the climate: first the cool conditions of the Little Ice Age, then the 1850–1950 climatic amelioration, and now the present warming caused by the anthropogenic greenhouse effect.

Gull-billed Terns are characteristic of the warm–temperate zones of the world, having their centre of breeding distribution in the neighbourhood of the Black, Caspian and Aral Seas. Elsewhere their distribution is somewhat discontinuous, like the Roseate Tern, and this is particularly true of Europe. Apart from concentrations in the Balkans and southern Spain, isolated breeding pairs and colonies are widely scattered, often well inland, for this is not primarily a coastal species. Strangely, for such a warmth-loving tern, there is a population of around 500 pairs in Denmark, which still appears to be flourishing. Clearly the Gull-billed Tern has benefited from the amelioration in this northernmost outpost. In this connection, it may be significant that, during the especially warm summers of the 1930s and 1940s, pairs nested as far north-west as the Netherlands (first bred 1931, then regularly 1944 to about 1958) and south-east England (pair bred in Essex in 1950, and probably also in 1949). It is also interesting to note that the very large Caspian Tern, which possesses a rather similar world distribution, also increased its most northerly populations—in the eastern Baltic—during the same period.

We now come to the small group of marsh terns, all three European species of which have extended their breeding ranges to the west or north during the climatic amelioration, spurred on, as already noted, by the increased aridity at that time in the Black Sea and Caspian Sea areas. One of them, the Whiskered Tern, has a wide but scattered distribution in the warmer parts of the Old World. In Europe it is largely restricted to the south, but in the course of this century it has been pushing slowly northwards, with the result that isolated breeding records have occurred as far north as Belgium and the Netherlands. The White-winged Black Tern, too, showed signs of expanding westward from the Balkans, and isolated breeding records were reported from Austria, Belgium, southern France, Corsica and Germany in the 1930s. Its continuing spread since 1950 is reported in Chapter 12.

The Black Tern has a more northerly distribution than the other two marsh terns, and is the only one found breeding in the New World, being widespread in central North America. It is also widely distributed and common in suitable marshy localities over much of Europe, and apparently used to be so in the fen districts of eastern England, but ceased to breed regularly by 1850 and altogether by 1885. The causes were said to be the drainage of the fens and the large-scale collection of their eggs; these were undoubtedly very important, but the very wet summers of the early 19th century may have contributed to their downfall through their food supply, which is chiefly insects. Happily, since 1966, Black Terns have returned to nest sporadically in the fen district of East Anglia, helped presumably by the climatic amelioration which allowed them to extend their range north-westwards on the continental mainland. One hopes that they will succeed in establishing themselves. However, there is still little sign that this is happening, despite the new global warming due to the greenhouse effect.

Although not strictly seabirds—indeed, most of them breed inland—it is convenient to treat the waders here, if only on the excuse that we usually associate them with the seashore and coastal marshes. In general, the majority of those species with northern breeding ranges tended to retreat northwards during the 1850–1950 climatic amelioration, at least in north-west Europe, whereas those with southern or central ranges tended to advance northwards or north-westwards. When the climatic deterioration set in, particularly in the Arctic zone, between 1950 and 1980, there was a noticeable reversal of this process which, together with the possible influence of the growing greenhouse effect, will be discussed in Chapters 11 and 12.

Among the species which retreated north during the amelioration may be mentioned the Golden Plover, Dotterel, Turnstone, Greenshank, Whimbrel and Grey Phalarope, whilst the Oystercatcher, Lapwing, Little Ringed Plover, Curlew, Black-tailed Godwit, Marsh Sandpiper, Green Sandpiper, Terek Sandpiper, Avocet and Black-winged Stilt advanced north, north-westwards or westwards. In addition, two species—the Kentish Plover and Stone Curlew—retreated to the south-east.

Taking the last two cases first, it seems that in spite of the greater average warmth, the wetter Atlantic climate of north-west Europe did not suit either the Kentish Plover or the Stone Curlew. Except that the Kentish Plover winters on its breeding grounds in Brittany, both species are summer visitors to north-central Europe. They are characteristic of warm, rather dry, sandy regions and, in the Old World, have their centres of distribution around the Mediterranean, Black and Caspian Seas. The Stone Curlew began declining in England, where it was right on the edge of its range, perhaps as early as 1870, but the decrease has been most conspicuous since 1945. Loss of its natural habitats to the demands of more intensive agriculture and forestry has largely been blamed, as it has been elsewhere in Europe, as well as the increasing unsuitability of those fragments which remain through the invasion of scrub, caused by the disappearance of rabbits as a result of the myxomatosis epidemic of 1954. It cannot be denied that these causes have played an important part in the Stone Curlew's decline, especially as there was a recovery in some areas which went out of cultivation during the agricultural depression between 1918 and 1939; nevertheless, it seems fairly certain that the wetter climate has also played a part, possibly because of its effect on the activity and availability of the lizards, bush-crickets, grasshoppers and other large insects on which the Stone Curlew depends for food. I think it is significant that it has even disappeared from the stony expanse of Dungeness in Kent; and I do not think that can be satisfactorily explained by suggesting increased human disturbance as the chief cause. The story has been the same in the extensive, uncultivated coastal dunelands of the Netherlands, where it began to decrease after 1923 and became extinct around 1958; and also in Austria, the former Czechoslovakia, Poland and Belarus (Byelorussia), where marked decreases have occurred since 1945. By 1991 it was reported to be almost extinct in Belarus, and apparently extinct in Germany, although I would concede that increased human disturbance could have been a factor there. In England the Stone Curlew breeding population dropped from up to 2,000 pairs in the late 1930s to around 150 pairs in 1990.

Kentish Plovers were also right on the north-western limit of their range in south-east England and, since they were always rare birds there, undoubtedly suffered from the attentions of the 19th century collectors. They were also adversely affected by the growth in popularity, in an increasing human population, of seaside resorts; but up to 1920 there was a fairly stable population of about 40 breeding pairs centred on Dungeness. Thereafter, despite protection, a rapid decrease occurred during what was actually (and paradoxically) the warmest period of the amelioration and not the wettest. By the end of the Second World War only a pair or two nested, and they finally ceased to do so altogether in 1957. Like Williamson (1975), however, I have 'little doubt that climatic change has been the root cause of their decline', especially the overall tendency to wetter springs and summers during the amelioration. They may, incidentally, have been comparatively recent colonists of extreme south-east England, as it was not until May 1787 that the species was first recorded in Britain, one being shot on a beach near Sandwich, Kent. The earliest records of probable breeding only date from about 1830 and, more certainly, from 1844. It will be recalled from Chapter 4 that the Cirl Bunting is suspected of having established itself in southern England at about the same time, and some authorities believe that the Hawfinch, too, became an inhabitant of England during this period, since breeding was not known until the early part of the 19th century. I suggested that the amelioration during the first forty years of the 18th century, when it was exceptionally dry, or alternatively the shorter run of generally warm summers from 1770 to 1783, may have favoured the Cirl Bunting's colonisation. I believe that either climatic phase may have led to the Kentish Plover and Hawfinch following suit. Unfortunately, since all three species could easily have been overlooked in those times of poor optical aids, we shall never know the truth.

Elsewhere in Europe there has also been a widespread decline of Kentish Plovers since about 1920, particularly in the north-west, and especially marked since the 1950s. For instance, the Dutch population decreased, with fluctuations, from a possible 3,000 or more pairs in 1900 to between 750 and 1,100 pairs in 1978–79. Kentish Plovers ceased to breed in Norway before the end of the last century, and have decreased and retreated southwards to some extent in southern Sweden. Although human disturbance undoubtedly played its part, the cooler, wetter summers overwhelmingly characteristic of the greater part of the climatic amelioration appear to have been behind the continental decline, as in England. So far, there is no evidence that current climatic trends in north-west Europe have done anything to improve the Kentish Plover's fortunes.

FIGURE 18 *Little Ringed Plover*

Turning our attention now to those waders which advanced northwards during the amelioration of 1850–1950, it seems appropriate to consider first the case of the Kentish Plover's close relative, the Little Ringed Plover, a bird intermediate in size between the Kentish and the Ringed Plover, although resembling the latter much more closely. Considered by Voous (1960) as the southern ecological replacement for the Ringed Plover in inland habitats of Eurasia, the Little Ringed Plover commenced its spread to the north-west soon after 1900, following a recession in the late 19th century. Its typical habitat preferences are for the shingly or sandy shores and beds of partially dried-up inland rivers, and sometimes the dried-up mudflats around fresh-water lakes. However, its range extension in north-west Europe, especially marked since 1968, has been greatly assisted by the provision of such man-made habitats as gravel and mud dredged from rivers and dumped along the banks, refuse tips around cities and industrial sites, and, especially in England, newly excavated gravel-pits which have provided conditions like those of its natural haunts. Moreover, its expansion has coincided with a rapid expansion of these human activities.

In the Netherlands there were only two known breeding records of Little Ringed Plovers before 1900, but by 1925 they were steadily invading that country and its neighbour, Belgium. There were no known instances of breeding in Britain until 1938, when a pair made ornithological history by

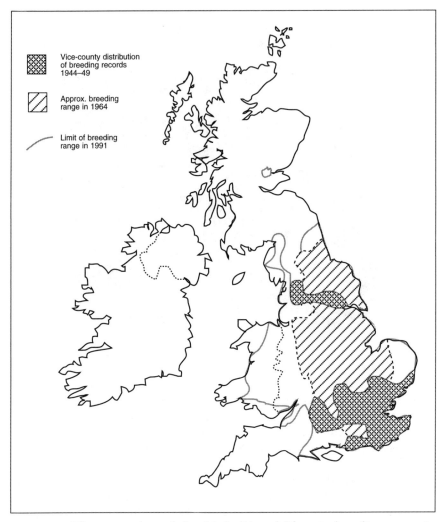

Vice-county distribution
of breeding records
1944–49

Approx. breeding
range in 1964

Limit of breeding
range in 1991

FIGURE 19 *The expansion of the Little Ringed Plover's breeding range in Britain since 1938 (adapted from Parslow 1973; Sharrock 1976; Gibbons et al. 1993).*

nesting on the dry bed of a reservoir at Tring in Hertfordshire. Nothing more happened until 1944, when two pairs bred at Tring Reservoirs and another pair at a Middlesex gravel-pit. From that year on, LRPs, as they are affectionately known by British birdwatchers, have nested every year in ever-increasing numbers, spreading north and north-west, until today there are some 400 pairs breeding as far north as southern Scotland (Figure 19). They are, however, largely absent from almost all of Wales and south-west England, where gravel-workings are far fewer. Although the incidental provision of suitable new habitats by man has undoubtedly aided the LRP's range extension, once again the climatic amelioration is likely to have been the root cause. Unlike the Kentish Plover, it does not seem to have been affected adversely by the wetter climate brought by the increased strength and prevalence of the Atlantic westerlies.

The Lapwing, which has a wide distribution across the Palaearctic Region from the British Isles to China, is usually quoted as a good example of a species which has extended its range northwards since 1900 in response to the climatic amelioration. This has been particularly well documented in Finland. After 1950, with the onset of a new phase of Arctic cooling, this impressive rate of spread, which was most rapid in the 1920s and 1930s, slowed right down and ceased in some areas, such as Finland and Sweden. In fact, in these latter countries, and in Denmark, a big decrease has occurred since 1975 due to severe winters (Hildén 1989). Lapwings shifted northwards elsewhere in northern Europe, notably in the Faeroes, and even started breeding in the south of Iceland. In the British Isles, they increased considerably in northern Scotland and, from 1930, occupied several new islands in Shetland. Ken Williamson (1975) pointed out that the growth of agriculture made possible by the improving climate probably helped the Lapwing and other species associated with man's tilling of the land to expand. Since then, unfortunately, changes in modern agricultural practice in Britain have not suited the Lapwing and it has undergone a marked decrease, especially in the southern lowlands and particularly since 1985 (Marchant *et al.* 1990). This recent decline, then, is largely independent of trends in the climate, although severe winters have some effect. In passing, it is worth mentioning that from time to time large flocks of Lapwings, endeavouring to escape from very cold weather by migrating westward to south-west Ireland or Spain (especially in severe winters such as those of 1962–63 and 1978–79), overshoot and appear along the eastern seaboard of North America, especially Newfoundland and Nova Scotia.

It is also of interest that a close relative of the Lapwing—the Spur-winged Plover—an inhabitant of Africa and the coasts of the Levant and Turkey, has been extending its breeding range north-westwards into Greece and even Bulgaria since the late 1950s. Some other species in this south-eastern corner of Europe have also been spreading north-westwards this century, such as the Olivaceous Warbler and the Syrian Woodpecker.

Although the Oystercatcher was already well established around the northern coasts of Europe, such as Fenno-Scandia and the southern coasts of Iceland, before the 19th century climatic amelioration began, it has increased tremendously in north-west Europe since about 1920. In the British Isles, following a marked decrease in eastern and southern England before 1900, a strong recovery began soon after the turn of the century, and Oystercatchers nowadays breed in almost all suitable localities around the British coasts. This earlier decrease has been ascribed chiefly to persecution by man and this undoubtedly played a part, just as its subsequent relaxation as a result of improved bird protection laws has been an important instru-

FIGURE 20 *Spur-winged Plover*

ment in the Oystercatcher's recovery. Climatic change, however, may also have played a part in both situations, the decline being influenced by the prolonged cold and wet weather of the early 1800s, and the increase by the steady improvement following 1850. Almost certainly, the notable climatic improvement of the 1920s and 1930s in Iceland led to an expansion of the Oystercatcher population around the north and north-east coasts from the warmer south and west coasts where it had long been established (Gudmundsson 1951). Also in this century, Oystercatchers have shown an increased tendency to nest well inland from the coast in western Europe, especially in the British Isles, the Netherlands, north-west Germany, Denmark and southern Sweden. Heppleston (1972) suggested that a behavioural change may have occurred which enables them to exploit habitats that they were previously unable to tolerate. In western Asia, to the north and north-east of the Black and Caspian Seas, they have long nested far inland, but are here believed to be relics from a population which once nested around the shores of the ancient and long since dried-up Sarmatic Sea.

Black-tailed Godwits, which in Britain, Scandinavia and Iceland are on the north-western limits of their wide range spanning the whole of temperate Asia, exhibited a similar big increase after 1920 and a slight northward extension of their breeding range. For instance, they have started nesting in Finland since 1955, in southern Norway since 1969, and in northern

Norway since 1970 (there was also an isolated case in 1955). Between 1920 and 1960 there was a marked increase in the Dutch population despite extensive reclamation work in the marshlands. In Iceland Black-tailed Godwits were originally restricted to the marshy lowlands of the south-west, but in 1913 they began to spread further west. After 1920 they commenced increasing in earnest and spread to various other localities, so that by the 1930s they had also colonised suitable places around the north coast. It is now known that the great increase in the number of these birds wintering in southern Ireland and along the estuaries of the south coast of England since 1940 is due to this remarkable expansion of the Icelandic population. Moreover, it is believed that the pairs which have nested sporadically on wet moorlands in northern Scotland and Shetland since 1946, and in northern Norway since 1955, came from the Icelandic stock, which, incidentally, forms a separate subspecies, *islandica*. It has even been suggested (Harrison and Harrison 1965) that the small but growing population which, since 1952, has colonised the fens of East Anglia in eastern England, is derived from this source, despite their choice of a habitat corresponding with that preferred by the typical race on the European mainland. It does seem more likely, however, that these birds arrived across the North Sea from the Netherlands.

Until their extinction as a breeding species by about 1830, Black-tailed Godwits were fairly common breeding birds in suitable localities over much of eastern England, especially East Anglia, and even as far north as Yorkshire. Their demise is blamed on the drainage of the marshes, followed by extensive shooting and egg-collecting as they became scarcer in their remaining haunts. These pressures were clearly very important and may have been wholly responsible; but I wonder if they offer a full explanation. Although the collectors could certainly have polished off the few remaining pairs once they became a real prize for the cabinet, I suspect that the British population was already dwindling during the unfavourable climatic conditions of the latter part of the 18th and the early 19th centuries. It is interesting to note that its extinction virtually occurred during the notoriously bad run of cool, wet summers between 1809 and 1820, well before large-scale drainage and egg-collecting had any real impact; and that Black-tailed Godwits continued to breed commonly in the Low Countries through the 19th century in the face of similar pressures almost as severe as in England.

I am not convinced, moreover, that the drainage of the fens was the primary factor in reducing the English stock of Black-tailed Godwits—much suitable marshland habitat still remained in the 1830s. Instead, I believe climatic deterioration was the root cause. Thus I am of the opinion that the return of these superb waders to East Anglia since 1952, and more recently elsewhere in England, is another example of range extension due fundamentally to climatic improvement. The rapid success of their return has, in large measure, been assisted by the excellent and unremitting efforts of the Royal Society for the Protection of Birds (RSPB) and of the Wildfowl Trust, but I think they would eventually have recolonised England successfully in spite of these efforts, given the generally enlightened attitude of the British today. Of course, I do not wish in any way to belittle the efforts of these conservation societies; they enabled the godwits to establish and consolidate a strong foothold from which to spread elsewhere.

The Avocet is another species whose return to East Anglia as a breeding bird after almost a hundred years of extinction was probably prompted by climatic amelioration and definitely helped by strict protection, although my remarks concerning the Black-tailed Godwit are probably applicable in its case too. Like that species, Avocets started to decline during the largely unfavourable climate of the latter part of the 18th century and early 19th century; this again was well before the advent of large-scale drainage schemes and the heyday of egg-collecting, although adults and their eggs were harvested for food (Sharrock 1976). The last recorded breeding occurred in Norfolk about 1824, by the Trent estuary about 1837, and in Kent in 1842. The history of the large copper butterfly *Lycaena dispar* in the English fenlands is a parallel case whose extinction was also probably initiated by climatic deterioration and not by extensive drainage.

The belief that climatic deterioration was involved in the Avocet's disappearance from England more than a century ago, and that the subsequent amelioration, most marked this century, was responsible for its return, is supported by the enormous increase in the population inhabiting the north-western countries of the European mainland since 1920, and the consequent extension of its range in the north. For example, since 1920 the Avocet breeding population of Denmark has trebled, while that of the Netherlands has quadrupled. Avocets began to colonise Belgium in the 1930s, and by 1969 there were some 200 pairs breeding there. It is not surprising that prospective colonists of England began crossing the North Sea soon after 1940. The colonisation of Estonia from the south-west has grown apace since breeding was first reported in 1962, while southern Sweden has been recolonised since 1928 and Norway since 1974. This story of expansion in northern Europe is continuing.

The Black-winged Stilt has a characteristically southern breeding distribution (Figure 21) with a normal northern limit contained within the 22°C July isotherm, but irruptions occur northwards now and then far beyond this limit. For instance, during the particularly warm years between 1930 and 1950 several small irruptions took place, and stilts bred for a short time in Austria, the former Czechoslovakia, Germany and Belgium, and even as far north as the Netherlands in 1931, 1935 (over twenty-five pairs), 1936, 1939, 1944, 1945 and 1949 (over ten pairs), southern Scandinavia in 1935, and England in 1945 (three pairs). The only evidence of any permanent extension of range is in France, Spain and Italy; in view of the current warming climate, however, irruptions may become more frequent in due course. One irruption in 1965 saw this species breeding in Belgium, the Netherlands, Germany, the former Czechoslovakia, Austria, Switzerland and the Cape Verde Islands.

100

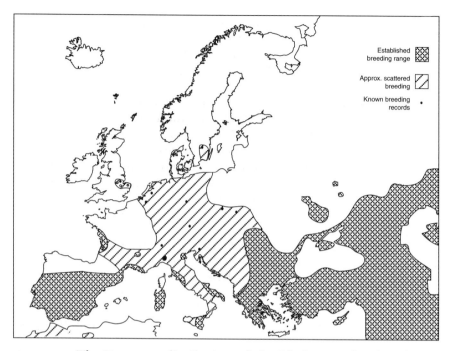

FIGURE 21 *The European distribution of the Black-winged Stilt (adapted from Voous 1960; Heinzel et al. 1972; Harrison 1982; Cramp et al. 1983; Tomkovich 1992).*

Although it has lost ground this century in parts of western Europe, such as Germany and the Netherlands, because of the reclamation of much of its heath and moorland breeding habitats, the Curlew is another wader which benefited from the climatic amelioration; it increased and advanced north-wards along the northern edge of its range, for example in Finland. In this it was assisted by the extension northwards—at the expense of the forests—of man-made grasslands. In Denmark the Curlew was unknown as a breed-ing bird before 1925; thereafter the colonists steadily increased to close on a hundred pairs in 1960, since when the population seems to have become static or has even decreased. Since about 1930 Curlews have also been expanding and spreading in Northern Ireland and northern Scotland. They are also in the process of colonising the Outer Hebrides, though the rate is reported to have slowed down since 1950, as in Denmark and elsewhere in northern Europe, perhaps in response to the short-lived change-over from an ameliorating to a deteriorating climate. In Britain the Curlew is typically a breeding bird of the upland moors of the north and west, but in about 1915 it began to spread into the lowlands of south-east England, breeding on boggy heathlands and in marshes, and on the brecks of East Anglia. Again this expansion appears to have halted since 1950, mainly due, it is thought by most people, to the loss of suitable habitats through drainage and urbanisation. This has undoubtedly happened, as I have seen for myself in Somerset, but I believe that the end of the amelioration has contributed as well. The Dutch population has also suffered severely from habitat loss since the end of World War Two, but has increased in the West Friesian islands (Braaksma 1960).

In Finland there has been a marked expansion northwards of the Green Sandpiper, apparently in response to the climatic amelioration.

In addition to the species so far discussed, which advanced northwards and/or north-westwards, there is a group of northern, boreal species which has been spreading westwards into northern Europe from northern Russia during this century, also apparently in response to the climatic amelioration. It includes two waders, the elegant Terek Sandpiper, which was named after the river in the Caucasus where it was discovered in the 18th century, and the equally elegant Marsh Sandpiper. Since 1930 the Terek Sandpiper has been expanding the limits of its breeding range (Figure 22) from the boreal marshes of northern Russia into the Baltic states and the neighbourhood of Finland, and it started nesting in the latter country in 1957. This westward expansion has continued unabated, presumably in response to the climatic amelioration earlier this century, and has been maintained perhaps by the general global warming caused by the growing greenhouse effect. In conse-

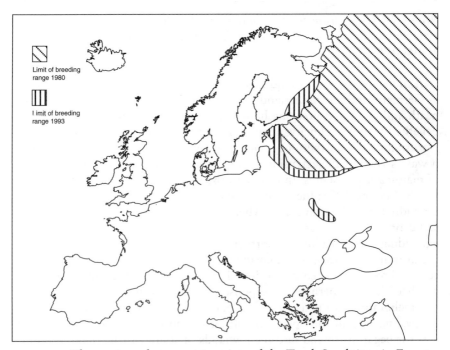

FIGURE 22 *The westward range expansion of the Terek Sandpiper in Europe (adapted from Voous 1960; Makatsch 1966; Heinzel et al. 1972; Harrison 1982; Cramp et al. 1983; Tomkovich 1992).*

quence, the number of occurrences of migrant Terek Sandpipers in western Europe, way off their usual migration routes, has increased in recent years. The majority of these have occurred in spring. For discussion and more recent information, see Chapters 11 and 12.

Earlier this century and in the previous one, the Marsh Sandpiper bred around the Neusiedlersee in Austria and Hungary, an outpost well to the west of the limits of its normal breeding range in east-central Europe, but eventually it died out there. Since the apparent waning of the recent natural climatic amelioration in about 1950, this bird has also been extending its breeding range westwards and north-westwards, rather like the Terek Sandpiper, but perhaps more in response to the greenhouse effect than was the case with that species, which began its spread so much earlier. As the Marsh Sandpiper's spread apparently began well after 1950, I have treated it in more detail in Chapter 12.

We now turn our attention to those waders which retreated northwards because of the ameliorating climate. They include the Golden Plover, Dotterel, Turnstone, Greenshank, Whimbrel and Grey Phalarope (all these species are further commented on in Chapter 11). To take the case of the Whimbrel first of all, we have just seen that its close but larger relative, the Curlew, spread northwards during the amelioration, especially in Fenno-Scandia. The Whimbrel presumably gave ground in the face of ecological competition from the Curlew, the climatic factor, as suggested by Lack (1954), determining the position of the boundary between them. Such competition has been observed in northern Norway; but in Shetland, according to Williamson (1951), it is avoided by the Curlew nesting in the rush-grown valley bottoms and the Whimbrel high on the hillsides. In northern Britain, a marked decrease of Whimbrels was noted during the latter part of the 19th century when the amelioration first began. They ceased to breed in Orkney by the 1880s, and disappeared from several islands in Shetland between 1889 and 1930 (Venables and Venables 1955). As will be described in Chapter 11, this trend has been reversed since the amelioration came to an end about 1950.

A similar situation arose in Greenland between the Grey and Red-necked Phalaropes during the amelioration. Although both species retreated northwards in response to its influence, the more northerly, high-Arctic Grey Phalarope was ousted from many of its most southerly breeding sites by a northward advance of the northern population of the low-Arctic Red-necked Phalarope. At the southern edge of its range, in the Faeroes and northern Britain, the Red-necked Phalarope retreated for the most part between 1930, when the amelioration was at its warmest, and about 1950. Before that a serious decrease occurred during the 19th century, but it is difficult to disentangle how much of this and a subsequent partial recovery was due to climatic change, and how much to the undoubted depredations of collectors. The partial recovery in the early part of this century, which apparently led to a colonisation of Ireland, may well have been due to the beneficial effects of the bird protection acts in Britain, but by the 1920s numbers of breeding pairs were waning in spite of them. The Irish colony, in Co. Mayo, which is on the extreme southern limit of the species' range, was discovered in 1902; in 1905 there were some fifty pairs and in 1929 forty pairs, but thereafter a serious decline set in with the result that by 1966 there were only one or two pairs, and, following a temporary increase after 1967, none in 1972. Similar declines took place in Shetland after 1920 and in Orkney after 1930. From 1950 to the early 1970s, or thereabouts, there was a temporary recovery in some areas in northern Britain, presumably in response to the climatic deterioration in the Arctic region at that period. Everett (1971) and Parslow (1973) mentioned that a few pairs had bred regularly in a part of mainland northern Scotland from 1963, but they disappeared by the late 1980s.

The Turnstone provides a clear case of a northern species which retreated northwards during the height of the climatic amelioration from its southernmost localities around the Baltic Sea. Here its breeding population is considered to be a relict from early post-glacial times, pushed there during the preceding glaciation. By 1917 it had ceased to breed along the north German coast, and it subsequently declined or disappeared from many of its other haunts along the southern shores of the Baltic, for example, Denmark and Latvia.

Greenshanks, too, withdrew northwards to a slight extent in the southwesternmost part of their European breeding range—northern Scotland—between 1900 and 1960, after which there was evidence of a recovery, especially in the Scottish Western Isles, with a remarkable south-westward extension of range to the west of Ireland. Before 1900 the species apparently increased and expanded its distribution in central Scotland, in spite of the fact that it was then being much persecuted by egg-collectors.

The Dotterel is typically an Arctic-alpine species. It certainly has a decidedly discontinuous distribution today, and the tiny alpine populations of central Europe are probably relicts from the Last Glaciation. Breeding was first recorded in the Austrian Alps in 1876 and in the Riesengebirge range (on the border of Poland and the former Czechoslovakia) in 1911; but it was not until 1948 that Dotterels were again discovered nesting in the former location, and 1946 before they were rediscovered in the latter. Then in 1952 the species was proved to be breeding in Italy's Abruzzi mountains. More remarkably, Dotterels were discovered in 1961 nesting below sea level in the newly reclaimed polders of the Zuider Zee in the Netherlands, and they have continued to do so ever since, although proof is difficult to obtain in the huge fields of growing crops. There is even some possibility that they may have nested there for a few years before proof was obtained.

In northern Britain, where Dotterels are restricted as breeding birds to the tops of the highest mountains, they were persecuted by man during the 19th century, and for this reason alone were thought to have been obliged to retreat to the least accessible and most desolate mountain plateaux.

However, the decline continued up to about 1950, well after bird protection laws were introduced and persecution virtually ceased, and before disturbance from tourists became a serious problem. It now seems that, as suggested by Nethersole-Thompson (1973), the climatic amelioration was the prime cause, since there was a distinct withdrawal by the Dotterel after 1900 from some of its southern outposts in England, such as the Lake District, the Cheviot Hills and the Southern Uplands. This process has been reversed with the post-1950 Arctic cooling, and breeding Dotterels have reappeared since 1960 in south-west Scotland, the Lake District and even, since 1969, the Snowdonia district of North Wales. Although the evidence is fragmentary, it is possible that a more or less parallel decline and recovery in the small, alpine populations of Dotterels took place in central Europe during this century.

My final example of the northward retreat of a wader during the 1850–1950 climatic amelioration is the Golden Plover. The southern race of this beautiful bird breeds in the British Isles, southern Norway and Sweden, Finland, Denmark, north-west Germany and, following a long absence, in the Netherlands and Belgium. Throughout the southern part of this range a general decline and withdrawal northwards occurred after about 1830. It is not clear yet whether this contraction of breeding range has halted since the climate in the Scandinavian sector of the Arctic zone cooled, i.e. after 1950 (in Finland there has in fact been a southward advance); but, since this has so far had less effect in Britain and adjacent parts of north-west Europe, it is probably too early to expect any clear pattern to emerge. However, there was some evidence between 1950 and about 1980 of the decline being arrested in some areas of Britain; and even of an advance to the south-west, with the recolonisation, after an absence of more than a century, of Dartmoor (Devon), where the population grew by 1976 to around ten pairs. There were also signs that it was about to return to Exmoor, where it ceased to breed by 1911.

Apart from climatic change, other reasons advanced to account for the Golden Plover's decline include conifer reafforestation in formerly open moorland areas, drainage and reclamation of breeding habitats, and increased predation of eggs. Almost certainly these have been important contributory factors locally, but climatic amelioration has been, I think, the fundamental cause. Even in the Netherlands (where the Golden Plover was once a widespread if not common breeding bird of the heaths and moorlands of the north and east, but became extinct in the 1930s) and in northern Germany and Denmark climatic change was probably the prime, underlying cause of the decline, although land reclamation must undeniably have had a serious effect. Yet, if it were not for some underlying factor such as climatic change, I would have thought from what I have seen that there are still large enough, relatively undisturbed areas of suitable habitat in the eastern Netherlands to have supported at least a small breeding population of Golden Plovers; instead they had virtually disappeared by 1932 with the last breeding case (until the recent return) being reported from Friesland in 1937. In Germany and Denmark the decrease has been serious, and a similar decline has also been reported from Sweden.

In Iceland, where the northern race breeds, Golden Plovers were reported by Finnur Gudmundsson (1951) to be staying much later in the autumn than formerly, presumably due to the climatic amelioration; often as late as December in the south and south-west, and occasionally even wintering.

Chapter 7

WATERFOWL

Under this convenient heading, I propose discussing not only divers, grebes, ducks, geese and swans, but herons, storks, pelicans, gallinules and rails.

Beginning with the divers (or loons, as they are known in North America), I have been unable to discover any significant evidence of a northward withdrawal which could be correlated with the climatic amelioration between 1850 and 1950 for any species, except for the Red-throated Diver (Red-throated Loon); although, as we shall see in Chapter 11, the subsequent climatic deterioration between 1950 and 1980 in the Arctic seems to have led to southward advances of the Great Northern Diver (Common Loon), White-billed Diver (Yellow-billed Loon), Black-throated Diver (Arctic Loon) and Red-throated Diver in some parts of their range.

The Red-throated Diver, in fact, presents a somewhat baffling case, since its changes in distribution seem to have 'gone against the grain' for a typically northern species. Within the boreal and tundra climatic zones it has a circumpolar breeding distribution, roughly contained within the July isotherms of 33°F (0.5°C) in the north and 64°F (17.8°C) in the south. It therefore breeds far to the north, in such places as Ellesmere Island in Canada (80°N and beyond) and the extreme north of Greenland; yet it also ranges as far south as Newfoundland, northern Ireland and south-west Scotland. During the 19th century it was reported to have decreased in its British haunts because of human persecution, but by the latter part of the century it began to increase and spread when, it was claimed, persecution declined and protection laws were introduced. For instance, it steadily increased and spread to new locations in Shetland, Orkney, the Inner Hebrides and south-west Scotland, first nesting on Arran in 1873 and colonising north-west Ireland in 1884. The expansion of range during the past 120 years or so has been considerable, and is still continuing. In Iceland and Finland, too, a marked increase has been reported in recent years. On the other hand, the Red-throated Diver decreased during this period in the Faeroes (Norrevang 1955), and in Norway (Haftorn 1958). Parslow (1973) expressed the opinion that 'had the species been affected by changing climatic factors over this period, one might have expected any change to be reflected first (and most obviously)' in its outpost in north-west Ireland. On balance, it looks as if the 'persecution by man, followed by recovery under protection' theory is the most convincing explanation, but I believe that climatic factors should not be completely ruled out, as they may be having an effect in a way at present unknown.

Of the grebes inhabiting the Northern Hemisphere, three species—the Great Crested, Black-necked and Little Grebes—have, to a greater or lesser extent, expanded their ranges northwards during the climatic amelioration, while one, the Slavonian (or Horned) Grebe, retreated northwards in North America but expanded southwards in north-west Europe.

To take this rather puzzling case first, the Slavonian Grebe, like the Red-throated Diver, with whose recent history it has several features in common, has a circumpolar breeding distribution lying roughly between the July isotherms of 50°F (10°C) in the north and 71°F (21°C) in the south, but breeds much less far north than the diver. However, like the Red-throated Diver it has been slowly extending its range south-westwards this century in part at least of its European range. For example, it was first discovered breeding in northern Scotland in 1908 and has slowly built up a small population (forty-three pairs by 1971, seventy-four to eighty-six pairs by 1990) which has gradually spread to other areas. A parallel increase and southward expansion has taken place in Sweden and Norway (Curry-Lindahl 1959; Haftorn 1958). On the other hand, this grebe disappeared from its breeding haunts in north Jutland (Denmark) by about 1860 (perhaps due to human disturbance), and has similarly vanished as a regular breeding species from a large area of the northern United States and eastern Canada.

Although the Slavonian Grebe's retreat north and westwards in North America may in part have been due to other factors, such as human pressure, it seems likely that the 1850–1950 climatic amelioration, despite being less marked in North America than in Europe, was the chief cause, as it may have been in Denmark. Elsewhere in Scandinavia, and perhaps even in Scotland, most of the increase has apparently occurred since this natural amelioration came to an end. It may have been that the 1908 colonisation of Scotland was a chance occurrence, smiled upon by what was, even during the amelioration, a reasonably favourable climate for it. Certainly, there were signs that these Scottish Slavonian Grebes prospered better in the cooler period between the ending of the amelioration and the recent, noticeable warming influence of the greenhouse effect, as they spread from their headquarters in Inverness-shire to central, and even southern, Scotland. Thus, as with the Red-throated Diver, a climatic cause should not be ruled out; a study of the population trends of both species through the 1950–1980 deterioration, and through the present greenhouse effect warming, may prove illuminating. (See also Chapter 11.)

Of the European grebes, the Black-necked (or Eared) Grebe has undergone the most spectacular expansion of range, having, since about 1870, spread north and north-westwards from eastern Europe right across the rest

FIGURE 23 *Black-necked Grebe*

of the continent to found colonies as far north-west as Ireland and as far north as Sweden. The North American race also expanded north-westwards during this century. As I mentioned in the previous chapter, this grebe and the Great Crested are among a group of birds inhabiting shallow freshwater lakes and marshes which moved north-westwards in similar fashion and during much the same period. One of these, as we have seen, is the Black-headed Gull, and it is interesting to note that in Europe the Black-necked Grebe often nests in colonies in association with this gull. Likewise in America it often associates with the rather similar Franklin's Gull.

Olavi Kalela (1949), a Finnish ornithologist who made an intensive study of the effects of climatic change on birds in northern and central Europe, ascribed the expansion in Europe of this group of species to the gradual desiccation of lakes and marshes in former breeding areas in the steppes adjoining the Black and Caspian Seas, as well as those of Hungary. The intensification of the atmospheric circulation during the 1850–1950 amelioration led to increased precipitation in the region of rising air currents (i.e. in the tropics and particularly in the region of westerly winds, such as north-west Europe); it also led to much less rain in the regions of descending air currents, such as the steppes mentioned above, and the prairies of North America within the Black-necked Grebe's range. Apparently runs of dry

109

years occurred when the desiccation was very intense, and the exodus of the Black-necked Grebes in sporadic and irregular waves fits in very well with this climatic pattern. Thus they colonised lakes in the former Czechoslovakia and Poland towards the end of the 19th century, Wales in 1904, the Baltic states in 1911, Ireland in 1915 (possibly as early as 1906), England and the Low Countries in 1918, Sweden in 1919, Denmark in 1920, Switzerland in 1929 and Scotland in 1930; but the main influxes took place in 1918–20 and 1929–32.

The underlying cause of the expansion of the Great Crested Grebe's range in northern Europe is also considered to have been the desiccation of the steppe lakes and marshes, but the picture was complicated, especially in Britain, by the species' recovery from heavy human persecution in the 19th century; and, during the present century, by the help it has received from the provision of many new breeding sites in the form of gravel pits and reservoirs, resulting from the greatly increased demands by an escalating human population for gravel for building and water for drinking.

Before 1840 Great Crested Grebes were widely distributed in small numbers in Britain, but soon after that date their underpelts became in great demand for use by fashionable ladies as muffs and other useless items of clothing. During the ensuing two decades the fashion led to a virtual massacre of the British population, so that by 1860 fewer than fifty pairs were left. Fortunately the succession of Bird Protection Acts between 1869 and 1880 led to a cessation of the senseless slaughter, and a recovery began. Meanwhile, the practice also declined on the European mainland, and the natural expansion of the species from south-east Europe began, as a result of the increasing aridity in that region.

By 1877, Great Crested Grebes were nesting in Scotland, and during the present century big increases and extensions of range have taken place in Denmark, Norway, Finland and Sweden as well as in England and Ireland. The first full census in Britain in early 1931 showed a recovery to nearly 2,800 birds, while a repeat full census in 1965 put the figure at around 4,500 birds. Ten years later the population had risen to 7,000 adults—the latest count. A growing number of these were by this date spreading from standing water on to canals and rivers (Marchant *et al.* 1990).

In the British Isles and southern Sweden the smallest and commonest European grebe, the Little Grebe, is at the north-western limit of its worldwide range and, as might be expected, is liable to fluctuate in numbers according to the vagaries of both short-term and long-term climatic variations. Severe winters, such as those of 1962–63, 1983–84 and 1984–85, regularly reduce their numbers, since they tend to winter on their breeding grounds, but recovery is usually quite rapid. The long climatic amelioration from 1850 to about 1950, with its mainly mild winters, was therefore distinctly beneficial to the species, at least in the British Isles, where the population increased from 1880 onwards, especially in parts of Scotland and in northern England. The present climatic warming due to the anthropogenic greenhouse effect should also benefit this species as long as sufficient suitable aquatic habitats are available.

Turning to the heron family, one finds that at least five species in Europe—the Grey Heron, Purple Heron, Night Heron, Cattle Egret and Bittern—appear to have benefited from the amelioration, at least somewhere over their ranges. In North America four species—the Little Blue Heron, Green Heron, Yellow-crowned Night Heron and Snowy Egret—also seem to have gained from it.

In spite of the fact that Grey Herons suffer a great deal from the activities of man (whether the effect is intentional or unintentional), they have tended to extend their range north-westwards in Europe, especially in Scandinavia, and northwards elsewhere in the Palaearctic. For example, in the Yakutsk area of Siberia they were unknown as breeding birds before 1916, but are now plentiful there. Herons are also decimated in hard winters: those of 1946–47 and 1962–63, for instance, virtually halved the British breeding population. However, their powers of recovery are such that they are back to normal in two to seven years, depending upon such factors as the degree of severity of the winter and the effect on their reproductive rate of pesticide contamination (Marchant *et al.* 1990).

The Purple Heron's northern breeding limit is restricted almost everywhere in its European and Asiatic range to the July isotherm of 68°F (20°C). Nowadays it has a markedly discontinuous distribution in Europe, breeding mostly in the south, with isolated populations in such areas as the Netherlands, western France, Austria and Hungary. Formerly it was more widely distributed, but it has suffered severely from the steady drainage over the years of the big reed-fens and swamps which are its favourite breeding haunts. Notwithstanding, Purple Herons had increased since the 1930s and spread to new localities. For instance, in 1942 they returned to breed at Lake Neuchâtel in Switzerland, and in 1955 to Bavaria. There were even signs in the 1970s that Purple Herons might attempt to colonise suitable reed-marshes in eastern England from the Netherlands; records in spring and summer increased in frequency, and breeding may possibly have occurred at least once in that decade, but there have been no such records since 1979.

The Night Heron has today a rather similar distribution to the Purple Heron in Europe, and also in southern Asia. Like that species, it was formerly widely distributed in central and southern Europe, breeding as far north as the Netherlands (its most northern limit anywhere in the world), but it lost many of its marshland breeding haunts to land reclamation. In addition, it suffered direct persecution by man. The last Dutch colony at a duck decoy in Zuid Holland died out in 1876 and it remained extinct as a breeding bird in that country, despite direct attempts to re-establish it at the famous Naardermeer in 1908 and 1909, and a general relaxation of persecution. Then after several unsuccessful attempts, the Night Heron apparently re-established itself about 1940; for in 1946 a breeding colony of seventeen or eighteen pairs was discovered in the Biesbosch region of Zuid

Holland (IJzendoorn 1950). Although it is possible that this colony originated from captive birds liberated in Zeeland during the Second World War, it may have been a natural recolonisation, as Hungarian-ringed Night Herons have sometimes been recovered in the Netherlands. Indeed, Hungary would appear to be a more probable source of origin than the colonies in southern France, involving as it does a more likely north-westerly movement.

Although the drainage of marshes, and other forms of human persecution, may have been important factors in the contraction of the Night Heron's European range, the fact that it has successfully returned to breed in the Netherlands in spite of increasing human pressure on its habitats suggests that, as with other herons, climatic change may have been involved. Being, as it was in the Netherlands, on the extreme northern limit of its range, it seems to me probable that the Night Heron population there was a relic from an earlier north-westward expansion during the period of medieval warmth (the Little Climatic Optimum), and that it had been steadily retreating in the north-west part of the European mainland throughout the subsequent Little Ice Age. This may well have been the case for other species, too, such as the Purple Heron and Black-winged Stilt. The Night Heron was able to hold out longest in the Netherlands because of the abundance of suitable habitats, but even there colonies were dying out as early as the cold 17th century. Its final disappearance by 1876 was perhaps the culmination of some 300 years of unfavourable climate, hastened at the end by human persecution, which did not allow the few remaining birds to survive long enough to recover under the beneficial influence of the 1850–1950 climatic amelioration, which was then just beginning. Fortunately, improved protection subsequent to 1900 eventually allowed them to recolonise the Netherlands to a limited extent during the especially warm decades of the 1930s and 1940s.

It is interesting to note in this context the successful establishment since 1950 of a feral breeding colony of Night Herons as far north as the Edinburgh Zoo in Scotland (latitude 55°57′N). These birds, now numbering forty to fifty, fly at dusk to feed along the Firth of Forth. They belong, incidentally, to the North American subspecies *hoactli*, known in that continent, where they sometimes nest or roost in cities, as the Black-crowned Night Heron. This race does not appear to be extending its range in North America, where it is widely distributed and breeds as far north as the southern fringes of Canada, but it has suffered a great deal from land reclamation and other forms of habitat destruction. However, the Yellow-crowned Night Heron has extended its range northwards this century along the eastern seaboard of the United States, presumably in response to the climatic amelioration.

Three other American herons, the Green Heron, the Little Blue Heron and the Snowy Egret, have extended their ranges northwards this century (Palmer 1962), particularly along the Atlantic seaboard, apparently at least helped by the 1850–1950 amelioration. The spread is most marked in the

Little Blue, which breeds mostly in Central and South America; by the 1920s the population had so multiplied that the northward dispersal of the conspicuous white juveniles after the breeding season had become very widespread (Palmer 1962). It nowadays breeds almost as far north as the Canadian frontier on the east coast. To some extent a contributory cause of the expansion was an exodus from Florida because of the destruction of breeding sites by drainage, drought and fire. The Snowy Egret was one of the species which suffered a serious decline because of the demand for its plumage by the millinery trade in the 19th century. Since that threat passed, however, its recovery has surpassed the northern limits of what was its former distribution before the persecution began.

The Great White Egret suffered great losses to satisfy the plumage trade in both the Old and the New Worlds, especially in North America and Europe where the deplorable fashion of adorning ladies' hats and dresses with 'aigrette' feathers flourished most of all. In 1902, for example, it was estimated that 200,000 of these egrets were shot in the breeding season to supply plumes for the London market alone. By the time this senseless slaughter was stopped, they had been practically exterminated over much of their range in southern Europe and in the United States. Even today it has

FIGURE 24 *Cattle Egret*

not fully recovered in the United States, so that it is not easy to say whether the drastic changes in its distribution there this century are entirely due to this revival or to other factors, such as climatic change. According to Palmer (1962), from a low ebb of population in 1902–03 the Great White Egret attained a peak of recovery as a result of legal protection in the mid-1930s, after which there was a gradual decline due to drought, drainage and other developments. In Europe it bred no further north-west than southern Bohemia and north-eastern Austria until 1978, when it began nesting in the extensive marshes of the Netherlands (see Chapter 12).

Without a doubt one of the most dramatic and spectacular of all range expansions has been that of the Cattle Egret, a species well known for its symbiotic association with large grazing mammals, both wild and domesticated. A native of Africa and south-west Europe, it first began its remarkable spread about 1911 when birds were reported in British Guiana (now Guyana), having apparently flown across the Atlantic with the help of the prevailing trade winds. Once it had established a firm footing in Guyana, it suddenly, from about 1930, began colonising other parts of South America, and penetrated northwards to the West Indies and the United States. According to the summary of its subsequent history given by Professor Karel Voous (1960), the Cattle Egret reached Venezuela by 1943, Aruba by 1944, Surinam by 1946, Colombia by 1951, the United States (Florida, New Jersey

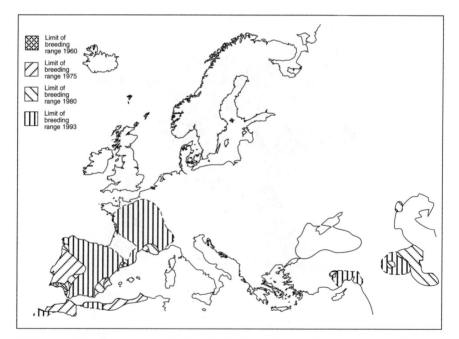

FIGURE 25 *The expansion of the Cattle Egret's breeding range into Europe (adapted and updated from Voous 1960; Cramp et al. 1983).*

and Massachussetts) by 1952, Bermuda and Bolivia by 1953, Costa Rica by 1954, Barbados and Panama by 1956, and Cuba by 1957. Even as early as 1955 several thousand Cattle Egrets were present in Florida, while in the following year more than a thousand nests were found in the Okeechobee area alone.

Cattle Egrets have also become established in many parts of Australia and are still spreading rapidly. Although they had earlier (1933) been artificially introduced into north-west Australia by man, it is thought that the sudden appearance of hundreds of them in Arnhem Land, in the Northern Territory, in 1948 was a natural event. Since 1963 they have been attempting to colonise New Zealand from Australia. In Japan they are advancing northwards in Honshu and now breed with other herons in the Tokyo suburbs. They have also been extending their range northwards in Europe (Figure 25). In the early 1960s they started breeding in the Camargue district of southern France, while in Turkey thirty pairs were discovered breeding in central Anatolia in 1968 (Thiede 1975*a*).

Although the Cattle Egret's world-wide expansion has clearly been greatly helped by its exploitation of the ready source of invertebrate food to be found through associating with man's growing herds of cattle and other domesticated hoofed mammals (as well as with buffaloes, elephants and other wild grazing mammals), it was very probably sparked off by the climatic amelioration. In the first place, the intensifying vigour of the atmospheric circulation after 1850 strengthened the easterly trade winds just as much as the westerlies further north in the Atlantic, particularly between 1900 and 1920, and thus facilitated the crossing of the ocean by Cattle Egrets from West Africa to the northern coast of South America. Secondly, the general warming-up of the climate in higher latitudes, especially in the 1930s, enabled them to push further north in North America, Europe and Asia than they might otherwise have done, the availability of suitable cattle herds notwithstanding. Even the temporary climatic deterioration between 1950 and about 1980 in north-central Europe has failed to deter them.

As with the Avocet, Black-tailed Godwit and Black Tern, which I have already discussed, the virtual disappearance of the Bittern as a breeding species from Britain by about 1950 was largely blamed on the drainage of their fenland haunts, and on the persecution by egg-collectors and shooters which they consequently suffered as they dwindled in numbers and became scarce. However, though these may have been important reasons, once again I believe that climatic influences should not be overlooked. J.T.R. Sharrock (1976) pointed out that the Bittern formerly bred in many parts of England, Wales and southern Scotland, and even in Ireland, up to about 1840. I am not convinced that their disappearance from all these areas, as well as the fens of eastern England, was primarily due to drainage, egg-collecting and shooting. It seems to me more probable that, being on the extreme north-western edge of their range in the British Isles, they had been decreasing because of the worsening climate.

It is well known that Bitterns are badly affected by hard winters, and there were runs of these in the first half of the 19th century (e.g. 1808–20, 1826–30 and 1837–55) which, allied with runs of wet, sunless summers like that of the period 1809–20, might well have been the real reason for the final disastrous decline. Significantly, as observed by Parslow (1973), the main period during which Bitterns re-established themselves in East Anglia occurred when severe winters were fewest—between 1900 and 1939. This was, of course, the height of the climatic amelioration. After 1940 Bitterns slowly spread from East Anglia to other suitable areas in England despite a check imposed by the very bad winter of 1946–47, and by the mid-1950s they were breeding in very small numbers as far west as South Wales and Somerset, and as far north as north Lancashire and south Yorkshire. Of course, rigorous protection greatly assisted their recolonisation; but since 1960, or even earlier, their numbers have seriously declined again, perhaps because of the general climatic deterioration between about 1950 and 1980, plus the very severe winter of 1962–63 and the severe one of 1978–79. In Sweden, where the Bittern is also on the limit of its range in Europe, it became extinct early in the 20th century, but recently returned to re-establish itself over much of the southern half of the country.

The only two storks of the Palaearctic Region, the White Stork and Black Stork, both retreated southwards during the climatic amelioration. It seems probable that the White Stork has been gradually declining in north-west Europe ever since the end of the Medieval Warm Period (Little Climatic Optimum). Although the only known British breeding record is of a pair which nested on Edinburgh's St. Giles' Cathedral in 1416, this suggests, being so far north, that the species nested elsewhere in Britain, especially perhaps in the south-east, during the earlier period of medieval warmth up to the 13th century. We know, for instance, of references to it as an inhabitant of Britain since the end of the 10th century, when the climate was particularly warm, and there is plentiful fossil evidence of its presence from the warm, dry Bronze Age, when it occurred as far north as Shetland; during Romano-British times it certainly occurred in north Hampshire. By 1416, when the climate was deteriorating, the pair in Edinburgh may have been among the last to breed in Britain.

The White Stork has probably been declining over a long period on the mainland of western Europe as well, particularly since 1900. In Switzerland, for example, some 150 pairs nested in that year; fifty years later none were left, while the Danish population dwindled from more than 4,000 pairs in 1890 to only about thirty in 1974. The breeding population of the Netherlands, as revealed by censuses in 1929, 1934 and 1939, rose from 209 to 312 pairs; but thereafter it decreased rapidly to eighty-three pairs in 1950 and to only fifty-eight in 1955. For developments since then see Chapter 11.

In general, the White Stork, and the Black Stork as well, have withdrawn south and south-eastwards into the heart of the continental land mass where

FIGURE 26 *White Stork*

the summers have tended to remain drier and sunnier. The prime reason for this seems to be that these storks depend a great deal upon such large insects as crickets and grasshoppers for food, as well as lizards, which become inactive in cold, wet summers; thus their unavailability results in lowered breeding success, nestling mortality being particularly high in wet summers. Of course, modern agricultural techniques involving 'improvement' of old pastures and marshy meadows will have accelerated this decline by eliminating many prey species. A similar decline appears to have occurred in the isolated eastern population of the White Stork, notably in Japan, which has a more maritime climate. Here it was once a common resident in Honshu, but nowadays only a few pairs linger on in the south-west of the island, near Kyoto.

The Black Stork, a mainly woodland-breeding species, obtains much of its food, such as fish, frogs, newts and water beetles, in more aquatic situations than the White Stork, and would therefore seem to be less vulnerable to wet summers, but nonetheless it likewise vanished from most of its former haunts in western Europe from about 1850. For instance, it died out in Belgium in 1860, in Denmark in 1952, and Sweden in 1954, and disappeared from the whole of western and southern Germany. The isolated population in the Iberian peninsula also dwindled to a serious extent. Curiously though, in Poland the trend has been in the opposite direction, with a slow but steady increase since 1920. This has been attributed to improved conservation, but it may have been due simply to the arrival of retreating birds from Denmark and Germany. However, Black Storks have expanded their range westwards in central Europe since about 1970, a trend which seems in keeping with the recent drier summers associated with the expanding influence of the Scandinavian blocking anticyclone. I will describe this spread in more detail in Chapter 11.

We now come to a consideration of the Anatidae—the swans, geese and ducks. Many species have shown at least some degree of alteration in range which may be correlated with the 1850–1950 climatic amelioration.

Beginning with the swans, the Mute Swan has increased and expanded its range spectacularly since 1940 in north-west Europe, especially in the region around the Baltic Sea. Here it is at the northern limit of its Old World distribution, which extends discontinuously from the British Isles to Manchuria. It has also been spreading northwards along the Atlantic coast of North America, where it was introduced sometime in the late 19th century. In the British Isles a general increase from the end of the 19th century came to a halt in 1960, since when there have been declines, at times rapid, due to an irregular series of hard winters up to 1986. However, in the 1970s and 1980s it was proved unequivocally that a high proportion of swan deaths, particularly along the River Thames and in the English Midlands, was from lead poisoning due to the ingestion of lead weights lost or discarded by anglers. The prohibition of the sale of these since 1987 has significantly

FIGURE 27 *Mute Swan nesting at Wicken Fen, Cambridgeshire. (Photograph: John F. Burton)*

reduced this cause of swan mortality (Marchant *et al.* 1990). Other contributory factors which have been advanced to explain the declines include an increase of casualties through birds, especially immatures, colliding with power lines, and an increase in the destruction of nests by human agency. Since 1987 there have been some local slight increases, but little change overall.

As one would have expected, bearing in mind its much more northerly breeding distribution, the Whooper Swan retreated here and there during the 1850–1950 amelioration along the southern limits of its range, while pushing north along the northern edge. For instance, it increased in Iceland and spread northwards in Fenno-Scandia, while further south in Europe its attempts to breed became less frequent than they were in the Little Ice Age. To take northern Britain as an example, Whoopers regularly nested in Orkney, and probably in Shetland and northern Scotland too, up to the 18th century, when they eventually became extinct. Of course, human persecution may also have helped to exterminate them, but I suspect that climatic 'improvement' was the chief reason. There appears to have been some attempt at recolonisation of north-west Scotland in the 1920s, but this fizzled out soon afterwards. However, since 1970, as will be discussed in Chapter 11, the incidence of summering pairs of Whoopers in north-west Scotland has increased and some of these have succeeded in breeding.

Changes in breeding distribution, at least possibly associated with the climatic amelioration, have been reported for several species of geese in the Northern Hemisphere, such as the Greylag, White-fronted, Snow, Ross's and Canada. All of these, except for the Greylag and some races of the Canada, are northern species breeding more or less in the tundra.

The Greylag Goose's present breeding distribution extends eastwards from Iceland, north-west and eastern Europe right across central Asia to Manchuria. Although it is absent nowadays from most of western and south-west Europe, there seems little doubt that it once bred over most of this region too. Professor Karel Voous (1960) stated that it has nested in historic times in France, Spain and even in north-east Algeria. His countryman, A.L.J. van IJzendoorn (1950), described it as common in the Netherlands in medieval times, as it probably was elsewhere in western Europe then, including Britain. As with various other marsh-nesting birds, the disappearance of the Greylag from these regions has been blamed on persecution by man, the spread of agriculture and forestry into their haunts, and, especially latterly, extensive drainage of the fens and marshes. There can be little doubt that these factors have indeed been very important and may well be the chief causes; but the fact that since the amelioration petered out around 1950 the Greylag has been staging something of a comeback makes one wonder if climatic influences have also been at work. Of course, this is chiefly attributed to improved protection, escapes from wildfowl collections, and, above all, artificial introductions by wildfowling interests, particularly in England and southern Scotland. These are undisputed facts.

The feral population in the British Isles rose from 1,700 in the late 1960s to around 13,000 to 14,000 birds by 1986 (Owen and Salmon 1988). Nevertheless, there have been big increases in some areas, such as Estonia and Iceland, a small increase in the Scottish Western Isles, and apparently natural extensions of range elsewhere in Europe. The Netherlands, where Greylags ceased to nest after 1909, has been recolonised since 1948, with at least 800 pairs breeding in 1990; and it is likely that birds from this source have supplemented the rapidly growing feral populations which have become established in East Anglia, Lincolnshire and east Yorkshire, as well as elsewhere in Britain and Ireland. See Chapter 11 for the latest information.

It is quite possible that the drier climatic conditions of the Little Ice Age suited the Greylag Goose better than the wetter climate which developed in western Europe with the onset of the amelioration during the mid-19th century, and especially during the first part of the 20th century. With the natural return of drier conditions between 1950 and 1980, coupled with improved protection and introductions, it has been able to stage a recovery in its European range, and to continue it under the present, largely drier conditions associated with the anthropogenic greenhouse effect.

Of those geese whose breeding ranges lie generally further north than the Greylag, the White-fronted, which breeds in the tundra zone right round the Arctic, greatly increased and spread northwards in several places during the 1850–1950 amelioration, such as the Canadian Arctic, west Greenland and Novaya Zemlya. In North America, considerable overlapping and inter-breeding between formerly isolated populations also occurred. This also happened among populations of the Greater and Lesser races of the Snow Goose: the latter has apparently been increasing and gradually extending its breeding range north-eastwards into the more northerly Canadian high-Arctic headquarters of the Greater Snow Goose. Moreover, the blue phase of the Lesser Snow Goose, which forms a higher proportion of the breeding population as one travels from west to east in subarctic America, became more numerous up to 1960. Later, when the 1950–1980 climatic deterioration became evident, the blue phase birds appeared to decline. Presumably they are at a greater selective advantage during the periods of milder climate when less snow lies around during the breeding season, and their camouflage is consequently more effective against Arctic foxes and other such predators.

Ross's Goose, whose limited breeding grounds in subarctic Canada were only discovered in 1938, also increased during the same period, especially after 1955, due, it is thought, to the amelioration. They have also been found nesting in small numbers on Southampton Island at the entrance to Hudson Bay, and along the west shore of this immense bay. Since 1970, these Hudson Bay populations have been building up and spreading southwards. From 1972, for instance, Ross's Geese have been found nesting among Lesser Snow Geese as far south as La Perouse Bay (Ryder and Cooke 1973). This recent development may actually have represented a retreat in the face of the 1950–80 climatic deterioration. No doubt the true nature of this trend will become clearer later on as their response to the current green-house effect is revealed.

Over much of its natural range in North America, the Canada Goose spread northwards in response to the amelioration, especially in the eastern part of the Canadian Arctic. In fact, after about 1945 it was even able to establish a foothold in west-central Greenland. Both in Britain, where Canada Geese were introduced in the 17th century, and in Sweden, where they were introduced after 1930, they have been increasing and (particularly in Sweden) spreading northwards over the past forty or so years.

Some of the most marked changes of all in the distribution of the water-fowl of the Northern Hemisphere have occurred among the ducks. Pride of place should, perhaps, be given to the Gadwall, Tufted Duck and Pochard for the most striking alterations apparently associated with climatic change, but they have been significant also in the cases of the Mallard, Black Duck, Blue-winged Teal, Pintail, Wigeon, American Wigeon, Shoveler, Garganey, Red-crested Pochard, Ferruginous Duck, Scaup (or Greater Scaup), Lesser Scaup, Ring-necked Duck, Common (or Black) Scoter, Harlequin Duck, Long-tailed Duck (or Old Squaw), Bufflehead, Goosander, and Hooded and Red-breasted Mergansers.

Starting with one of the three most striking, the Gadwall swept north-west across Europe after 1850, penetrating even to Iceland, while in North America it pushed northwards across Canada in the course of this century. In Eurasia, the Gadwall's centre of distribution lies around the Caspian Sea, where it is a characteristic inhabitant of the extensive freshwater lakes of the steppes. Thus, it was one of the group of species affected by the long peri-ods of drought and increasing aridity in this region since the mid-19th century, and which consequently emigrated north-westwards. The amelio-rating climate enabled Gadwall to establish themselves permanently in their new haunts, and even to increase and expand their range still further. England was colonised from 1850 (helped by artificial introductions), Sweden from about 1870, Scotland from 1909, the Netherlands from 1925, and Ireland from 1933. Remarkably, they reached Iceland as early as 1862, and have since colonised its eastern areas. Moreover, during the course of the first half of this century the Gadwall became one of the two commonest surface-feeding ducks of Lake Mývatn, the other being the Wigeon. Finnur Gudmundsson (1951), the renowned Icelandic ornithologist, who has documented the Gadwall's arrival and spread in his country, was in no doubt that its success is due to the improved climate there, particularly after 1920.

In Britain, the Gadwall's colonisation and spread was assisted by artifi-cial means, but there can be little doubt that its steady northward and west-ward expansion has been primarily natural, as elsewhere in Europe. In North America its breeding range has expanded northwards, especially in eastern Canada, in recent decades (Palmer 1976). It also seems to have done

121

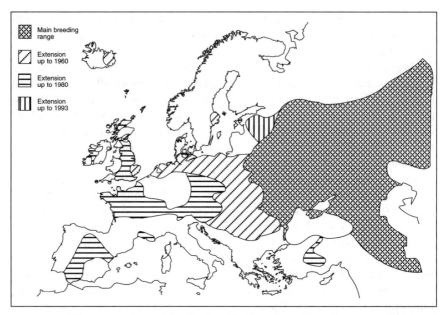

FIGURE 28 *The expansion of the Gadwall's European breeding range (adapted and updated from Voous 1960; Harrison 1982; Cramp et al. 1983).*

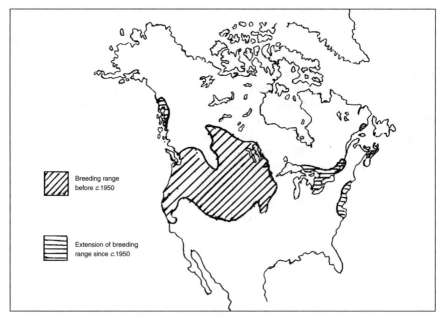

FIGURE 29 *The expansion of the Gadwall's North American breeding range (adapted from Godfrey 1966, 1986; Robbins et al. 1983; Cramp and Simmons 1977–1992).*

122

so in south-west Alaska, while breeding populations were quite recently established on the Atlantic coast of the United States northwards to Long Island, New York. Its advance north was presumably limited by the cooling influence of the Greenland–Labrador current, whereas on the opposite side of the Atlantic Ocean the Gadwall is able to flourish much further north because of the warming influence of the Gulf Stream. As in the steppes of eastern Europe and south-west Asia, it is likely that increasing aridity in the Gadwall's prairie lakeland haunts led to its northward extensions of range in North America.

The Pochard and Tufted Duck have similar histories, and the cause of their north-westward extensions of range appears to have been the same as was the case with the Gadwall—the desiccation of the steppe lakes—as indeed it was also with the Red-crested Pochard and the Ferruginous Duck, though in a less spectacular way.

Today the Tufted Duck enjoys a wide distribution as a breeding species, being found right across the Palaearctic Region from Iceland and the British Isles in the west, to Kamchatka and Sakhalin in the east. Its breeding range also extends far to the north, to the edge of the tundra zone, where it over-laps to a considerable degree with its northern counterpart, the Scaup. However, its colonisation of north-west Europe is a comparatively recent event, although it is highly probable that it originally occurred throughout this region, but retreated south-eastwards and eastwards during the Little Ice Age. It commenced the recovery of its former territory as the Little Ice Age drew to an end after 1830. It first nested in the Netherlands in 1835, but remained rare until 1940 when a big increase occurred which has since been maintained.

The first breeding record in the British Isles occurred in 1849 in Yorkshire, following which the Tufted Duck gradually consolidated its position and bred for the first time in Scotland in 1872 and in Ireland in 1877. There followed a big increase and expansion, especially after 1900, which led to the colonisation of most of the British Isles by 1940. This process was aided in England, according to Olney (1963), by the rapid spread of a staple food, the zebra mussel *Dreissena polymorpha*, from its point of introduction, the London Docks, in 1824. Another helpful factor latterly has been the growth since 1945 in the number of reservoirs and flooded gravel-pits in England.

Iceland was reached by Tufted Ducks in 1895. By 1904 they were breeding in considerable numbers at Lake Mývatn and elsewhere. They continued to increase and extend their range, so that by 1950 they were the second commonest duck on Lake Mývatn (Gudmundsson 1951), and by 1975 the commonest (Williamson 1975). In central Europe expansion has been slower: Silesia was colonised from 1906, Bavaria not until 1930, and Switzerland not until 1940.

The Pochard's spread has not been quite so spectacular, but it is impressive enough. Apparently it did not begin nesting regularly in England until about 1850, but by 1871 it had spread up the east coast to Scotland and to the extreme north by 1921. Ireland was occupied from 1908, and it has since spread over most of the northern two-thirds of that country. It also expanded steadily westwards across central Europe during this time, and has been increasing as a breeding species in the Netherlands and France since 1940. For example, since 1965 it has colonised Brittany, the western-most province of France.

The Red-crested Pochard's westward spread has not been anything like as uniform as the Pochard's. It began breeding around Lake Constance (the Bodensee) in 1919, in north-east Germany (Fehmarn Island) in 1920, in north-west Germany soon afterwards, in Denmark in c.1940 and the Netherlands in 1942. Some Red-crested Pochards may have nested in the Netherlands as early as about 1910, as IJzendoorn (1950) says that birds were regularly shot in the boggy pools of the province of Zuid Holland around that time. Although the situation in England is complicated by the presence of many birds escaped from captivity, some of which are now breeding in the wild, wild birds apparently regularly visit waters near the coast of Essex, where in 1958 a possible wild pair nested at St. Osyth. These birds probably come from the Netherlands. The Red-crested Pochard is still increasing and expanding its breeding populations in central Europe.

In North America the equivalent diving ducks, such as the Canvasback, Redhead and Ring-necked Duck, have undergone similar changes in range, partly, at any rate, due to the drying-up of some of the lakes of the prairies during the amelioration. For instance, the Canvasback, a larger edition of the Old World Pochard, greatly increased and extended its range in Alaska after 1950 following drought on the prairies. However, the general trend this century, especially since 1950, has been a steady decline in population almost everywhere else. Apart from periods of drought, blame for this is attached by American ornithologists to drainage, excessive hunting and even increased predation by the raccoon *Procyon lotor* which has been spreading northwards. A recent retreat of the Canvasback southwards in south-east Canada may be influenced by the post-1950 climatic deterioration in addition to these other factors.

Although the Redhead, another pochard-type diving duck, has long been decreasing overall, due, it is said, to loss of habitat through drainage and drought as well as excessive shooting, it has spread north-westwards in Alaska and north-west Canada since about 1955. Again, prolonged drought is thought to have been the cause of this displacement of population.

The Ring-necked Duck, a pochard which looks very like the Tufted Duck, has undergone the most spectacular increase and spread of all the North American diving ducks. Although it lost some breeding habitat in the south from the desiccation of lakes and drainage, it has made striking gains in territory to the north and east, particularly since 1925 in Ontario, Quebec, Labrador and Newfoundland, as well as around the Great Slave Lake in the North-west Territories and, since 1960, even beyond to eastern Alaska.

Almost certainly the 1850–1950 climatic amelioration was the prime cause. Very probably, the increase and extension of range along the eastern Atlantic coast of Canada, where it has become one of the commonest ducks, explains the greater frequency with which it has strayed across the Atlantic to the British Isles and elsewhere in western Europe since 1955, although a better network of birdwatchers has improved the likelihood of such strays being spotted.

The Mallard population in North America has also increased and spread to a marked degree this century, and to a lesser extent elsewhere in the northern part of its range in the Northern Hemisphere, particularly in Norway from the 1870s. The climatic amelioration seems to have been the prime cause, though the picture is complicated by other factors, such as large-scale releases of artificially reared birds, and other human activities designed to maintain or increase the Mallard population. In addition to the northward and eastward extension of its breeding range in North America, it also spread northwards in Greenland and, at least until recently, wintered there in greater numbers than formerly.

During the same period the American Black Duck has also extended its breeding range westwards and north-westwards to a considerable extent, and now breeds well beyond the Great Lakes, whereas a century ago it did not nest further west than Lake Erie. This advance has taken place in spite of an overall drop in numbers since 1955, following an increase which began in 1948.

The Blue-winged Teal is another North American species which extended its range noticeably during the climatic amelioration, and returned to areas from which it disappeared before 1880. Although the Green-winged Teal, the American race of the Eurasian (Common) Teal, has not apparently altered its range to any significant degree in response to long-term climatic change, the nominate race occurred more frequently on the Atlantic coast of North America during the warm years of the 1930s and 1940s than before or since (Palmer 1976), and seems also to have attempted to colonise south-west Greenland from Iceland, judging from ringing returns.

The spread of the Shoveler over many parts of its very extensive breeding range in the Northern Hemisphere correlates well with the climatic amelioration. From about 1900 it greatly increased and expanded north-westwards and northwards in western Europe, colonising Iceland (from 1931), and many parts of the British Isles where it was formerly unknown or very rare as a breeding species. Since 1950, when the improvement in the climate waned, there seems to have been a halt or even a reversal of the Shoveler's general advance in Europe, although it continued to spread in the eastern Baltic up to about 1970. In North America, where the Shoveler has a mainly westerly breeding distribution, an eastward expansion has occurred, especially since about 1945.

Back in Europe, the Garganey also spread north-westwards during the climatic amelioration, especially in Britain, Denmark, Finland and Sweden between 1900 and 1953, but it slowed down its advance between 1953 and 1975, possibly because of the climatic cooling during that period. In Britain, where it is on the north-west limit of its range, it declined and contracted south-eastwards after 1953 (see Chapter 11).

Although the Lesser Scaup breeds as far north in north-west North America as its slightly larger, but very similar, relative the Scaup (or Greater Scaup), and overlaps with it to a considerable extent (though it is much more of an inland species), it prefers a milder climate. Its general increase this century, and eastward extension and enlargement of range in Canada, were therefore most probably related to the beneficial effect of the climatic amelioration; the Greater Scaup, however, did not benefit, but on the contrary was forced to retreat northwards over much of its almost completely circumpolar range, especially in north-west Europe. This was especially noticeable between 1920 and 1940 when the amelioration was at its height; numbers of breeding birds dwindled, particularly in Iceland, where it was apparently the commonest duck, and in Sweden. Between 1897 (perhaps earlier) and 1913, a few pairs nested fairly regularly in north-west Scotland, but ceased to do so as the climate warmed up. But, as we shall see in Chapter 11, the Greater Scaup renewed its attempt to colonise areas along the southern limits of its breeding range in northern Europe when the amelioration waned after 1950, especially in the southern Baltic.

Other northern ducks which retreated northwards, at least to some extent, during the 1850–1950 amelioration were the King Eider, Harlequin Duck, Long-tailed Duck, Goldeneye and Bufflehead.

Curiously, the Wigeon which, as might have been expected, began to colonise northern Britain during the cooler climate prevailing there during the early part of the 19th century, nevertheless continued its southward expansion throughout the subsequent climatic amelioration, and only came to a halt around 1950 when the amelioration waned. By then it was breeding locally over most of Scotland, and here and there in northern and eastern England as far south as Kent. The part played by climatic change, if any, seems uncertain.

The spread of the Pintail in the British Isles bears some similarity to that of the Wigeon. It first bred for certain in Inverness-shire, Scotland in 1869. After that it spread in sporadic fashion to many parts of Scotland, including Orkney and Shetland. It first nested in England in 1910 (as far south as Kent), and Ireland (Co. Roscommon) in 1917. There are now pockets of breeding Pintails in various parts of both countries. Like that of the Wigeon, the Pintail's expansion seems to have stagnated since the 1950s, and even declined. Elsewhere in Europe it nests sporadically as far south as southern Spain and France. As the Low Countries, northern Germany and Denmark form part of its normal breeding distribution, it would appear that the Pintails breeding in eastern England came from there.

Near the northern limit of its range, the climatic amelioration seems to have enabled the Pintail to increase in such areas as Iceland, along with other surface-feeding ducks, for instance the Gadwall and Shoveler. It has also spread a good deal to the north-east and east in Canada since at least 1940, presumably in response to warmer conditions, colonising areas as far north as Ungava, and as far east as Nova Scotia.

The Goosander and Red-breasted Merganser, the Common Scoter, and above all the Eider (or Common Eider), provide further examples of northern species which advanced, somewhat inexplicably, southwards during the amelioration—all four of them, like the Pintail, commencing their spread soon after 1850.

Following a relaxation of persecution by man in the early 19th century, which is said to have been (and may well have been) a main cause of a decrease in some parts of its range at that time, the Eider started recovering about 1850, and even started the colonisation of new localities. For instance, in Britain it began breeding on the Scottish mainland about 1850 even before protective measures were introduced. Once such protection was effective, Eiders spread in earnest all round the mainland coasts as far south as Walney Island on the west coast of northern England and Tynemouth on the east. The north of Ireland was colonised in 1912. In the Netherlands the first breeding records occurred on the North Sea island of Vlieland in 1906, and here the breeding population built up to some 300 pairs by 1936 and nearly 1,000 pairs by 1948, having survived heavy depredations by man during the Second World War.

Since 1950, however, the rate of increase and expansion appears to have slowed up, except in south-west and eastern Scotland, in north-west Ireland (where Eiders continue to spread south-westwards), and in Estonia. In North America, where the Eider breeds almost all round the coasts of Alaska, Canada and Newfoundland, as well as in Greenland, its increase and spread has paralleled that in Europe. Moreover, between 1910 and 1960, when drift ice greatly decreased in the Davis Strait between Greenland and Baffin Island, Eiders greatly increased in that region. This latter instance is clearly connected with the climatic amelioration, and it is difficult to escape the conclusion that the increases and extensions of range elsewhere are somehow also ultimately bound up with this factor, and are not solely due to the relaxation of persecution by man. Nor is there any real evidence that they are associated with changes in the Eider's food supply, except possibly locally.

The Goosander does not breed as far to the north as the Eider, being largely restricted in the breeding season to the Boreal climatic zone in which it breeds all round the Northern Hemisphere as far north as the tree-line allows, since it nests primarily in holes in trees near water. Here and there, however, such as in western North America, north-west Europe and central Asia, it also breeds southwards into the Temperate zone, especially in mountainous areas.

In the Old World part of its range, the Goosander has definitely spread from the north southwards and from southern Germany eastwards into the

Ukraine since about 1920, probably because of the milder and wetter climate there; the North American population may also have extended its range, although I have not come across any positive evidence of this. But, more remarkably, the Goosander commenced an even earlier spread southwards in the British Isles, in spite of persecution because of its salmon-eating habits. The first British breeding record was in central Scotland in 1871, but there is evidence that it may have nested in Perthshire as early as 1864, and possibly even as early as 1858 in the Outer Hebrides.

Whatever the true facts are concerning the first Goosanders to breed in the British Isles, there is no doubt that after a massive winter influx in Scotland in 1875–76, many stayed to breed. Thereafter they nested regularly, especially after 1890, and by 1900 they were common breeders in western Scotland from Sutherland in the north to central Argyllshire. Their southward spread has continued almost unabated throughout this century, so that at the time of writing they have become widespread in Scotland and in the north of England as far south as north Lancashire and north Yorkshire. Moreover, breeding has occurred in north-west Ireland (Co. Donegal) since 1969, in Wales since 1972 and in Devon, south-west England, since 1980.

From about 1885 the Red-breasted Merganser has also expanded its breeding distribution southwards in the British Isles, in a similar manner to the Goosander. Unlike that species, however, it had long been established in many parts of Scotland and was not therefore a completely new colonist. After 1920 the rate of increase and expansion slowed down, but nevertheless since 1950 it has spread as a breeding species into northern England, and by 1973 had nested as far south as Derbyshire. In Wales, where breeding was first recorded in 1953, the Red-breasted Merganser has steadily consolidated its population in the north-west, and since 1967 it has even shown signs of occupying the Burry Inlet on the south coast. In Ireland, too, this merganser has extended its range considerably since 1900. Elsewhere in Europe, it has greatly increased in Estonia since 1965 (where there are now some 600 pairs or more), while in the summer of 1970 it was recorded in Spitsbergen for the first time. In 1977 breeding began in the Netherlands (in Zeeland), and by 1989 about ten pairs were breeding annually.

So far nobody seems to have offered a very satisfactory explanation for the southward spread in the British Isles of either the Goosander or the Red-breasted Merganser. One theory put forward suggests that the improved summers since 1850 in Scandinavia led to such a high breeding success among the populations of both species that they eventually outgrew the available food supply, and many of them were forced either to emigrate elsewhere or perish through starvation. Another suggestion is that the prolonged hard winters in the late 1870s and 1880s caused Goosanders and Red-breasted Mergansers wintering on the European mainland to move to the slightly less severe conditions in Britain. Here the bad weather, which persisted late into the spring, caused them to stay so long that many of them remained to breed in Scotland. If this theory was correct, it is surprising that the Red-breasted Merganser did not become established as a breeding species in Britain as a result of the many severe winters of the late 17th and 18th centuries. But somehow it does seem likely that climatic change was involved.

In North America, a third merganser, the Hooded, which has, on the whole, a more southerly distribution than the other two (although considerable overlap occurs), has also increased since 1930. As well as advancing southwards to recover, with the help of greater protection by man, much former breeding territory which was lost through unrestricted shooting in the 19th century, it has been spreading north-westwards into Alaska in recent years. This latter trend was probably stimulated by the 1850–1950 amelioration.

Like the Goosander and Red-breasted Merganser, but on a far smaller scale, the Common Scoter colonised Scotland after about 1850, and north-west Ireland in 1905. The first proved Scottish breeding record occurred in Sutherland in 1855. Then in the 1860s the Common Scoter was found to be nesting regularly in Caithness; after 1870 it spread in extremely small numbers to other parts of Scotland, and reached Shetland in 1911. A nest on the shore of Lower Lough Erne in Co. Fermanagh in 1905 was the first one reported in Ireland. By 1917 there were seven breeding pairs in that locality, and they slowly increased to about fifty pairs in 1950, and then more rapidly to some 145 pairs by 1967. Since 1948 other small populations have become established elsewhere in Ireland, so that by the time of the British Trust for Ornithology (BTO) Atlas survey, 1968–72, the total Irish population was 130–140 pairs (in spite of a small decline after 1970); this was in excess of the total Scottish population of 100–115 pairs (Sharrock 1976). (See also Chapter 11.)

Again the reasons for this mid-19th century onwards south-westerly extension of range of the Common Scoter are not certain, but may be basically climatic. Although it apparently represents an extension of the Scandinavian distribution, it is possible that the immigrants came from Iceland, as ringing returns show that many of the birds wintering in British waters come from there and not from Fenno-Scandia or northern Asia.

One of the three truly Arctic ducks, the Long-tailed Duck—the others being the King Eider and Steller's Eider—breeds almost entirely within the tundra zone right across the Northern Hemisphere, and is one of the commonest birds of the Arctic seas. Outside the breeding season it winters regularly as far south as the Pacific and Atlantic coasts of the northern United States, as well as the Great Lakes; in Europe it winters as far south as the southern North Sea, while in the west Pacific it comes down as far as North Korea and northern Japan. In severe winters it may winter even further south, for losses are extemely heavy among those birds which attempt to stay in traditional areas such as the Great Lakes and the Baltic Sea when these become frozen over. Similarly, during cold climatic phases there is a tendency for some Long-tailed Ducks to nest further south than usual. For example, during the Little Ice Age they were commoner in Iceland than during the warm climate of the first half of this century, and they may well have bred regularly in small numbers (at least) in northern Britain during the height of that chilly period. However, the only fairly certain breeding records in Britain date from more or less the end of the Little Ice Age, when pairs apparently bred in Shetland three times in the nineteenth century (including 1848 and 1887), and in Orkney in 1911 and possibly also in 1912. The last report before 1969 of a pair attempting to breed in Britain was in Orkney in 1926.

After the natural climatic amelioration ended around 1950, pairs of Long-tailed Ducks were suspected of nesting again in Shetland (1971) as well as in the Outer Hebrides (1969), but not since then, although a female summered in central Scotland in 1989 and 1990 (Spencer *et al.* 1993). If the post-1950 climatic deterioration continues in the Arctic zone, we may see an attempt by Long-tailed Ducks to recolonise northern Britain. In Iceland, according to Finnur Gudmundsson (1951), Long-tailed Ducks declined to a marked extent after 1900 when the 1850–1950 amelioration made itself felt there, and were largely replaced by Tufted Ducks. On the basis of the recovery in south-west Greenland of Long-tailed Ducks ringed at Lake Mývatn, Gudmundsson suggested that part of the Icelandic population may have retreated to Greenland.

During the climatic amelioration, the numbers of Steller's Eiders appearing in winter as far south as the Baltic and North Seas declined, especially between 1900 and 1950 (see also Chapter 11). That other truly Arctic duck, the King Eider, was also recorded as having nested in northern Britain in the 19th century, during the latter part of the Little Ice Age. A nest with six eggs was apparently discovered by William Bullock on Papa Westray in the Orkneys in the summer of 1812, in the middle of a run of notoriously cold years. The species is also said to have nested in Orkney for two consecutive years in the 1870s (Buckley and Harvie-Brown 1891).

There were few reports of summering King Eiders in northern British waters during the warmer years of this century, but with the climatic deterioration in the Arctic since 1950 they are being reported more frequently, usually consorting with Common Eiders. Indeed there has been speculation that they may begin interbreeding with Common Eiders in northern Britain, as they already do in Iceland. Although Britain, Iceland and Scandinavia all lie outside the usual breeding range of this species, severe weather often causes very heavy mortality, and such weather, when persistent over a number of years, causes some pairs to attempt to breed further south than is normal. In North America, for instance, King Eiders have nested as far south as the Alaska Peninsula and James Bay in Canada. It remains to be seen if this trend will become more noticeable with the Arctic cooling, or whether it will eventually be counteracted by the current global warming caused by the anthropogenic greenhouse effect. See Chapter 11 for more recent information.

Turning from the ducks of the Northern Hemisphere to the crakes and rails, the 1850–1950 climatic amelioration appears to have caused northward extensions of range in the Water Rail, Spotted Crake, Moorhen and Coot. The increase and northward spread of the Spotted Crake, first noted in Britain in the warm springs and summers of the 1920s and 1930s, has been most noticeable since about 1950 in Sweden (although decreases were reported in 1979 and 1991); but there have been new signs since 1960 of a similar increase in Britain, where this generally rare and easily overlooked species, so difficult to prove breeding, was revealed by the 1968–72 BTO Atlas survey to be thinly scattered from the south coast of England to Shetland in the north (Sharrock 1976). Although these later increases seem to have been detected only after the 1850–1950 amelioration had waned, this appears nevertheless to have been the cause, aided by the continuation of generally warm summers in north-west Europe despite the post-1950 current climatic deterioration in the high Arctic.

The Coot seems to belong to that group of marsh- or waterbirds, such as the Black-headed and Little Gulls, and the Black-necked Grebe and Tufted Duck, whose north-westward expansion of range this century was apparently sparked off by the drying up of the steppe lakes of south-east Europe and south-west Asia. Like the Black-headed Gull and the Tufted Duck, the Coot first attempted to breed in Iceland around the end of the 19th century. In fact, a pair bred in 1891, though the next nest was not discovered until 1943; but in between times summering birds were sometimes reported. Since 1943 such birds have been regularly observed there.

On the European mainland Coots expanded their range northwards after about 1900, especially in Finland, but this trend may have slowed down or stopped altogether in recent years. Indeed, in a substantial part of north-west Scotland Coots have disappeared as a breeding species since about 1920. They used to breed in small numbers in Shetland in the early part of the present century, but had ceased to do so regularly by the mid-1950s, and had decreased in Orkney and some of the Hebridean islands. The BTO Atlas survey in 1968-72 also failed to reveal any pairs breeding in Shetland, or even any suspected of doing so. All this may be an indication that the post-1950 tendency to cooler summers in this area is affecting the Coot.

Moorhens, too, extended their breeding range northwards, especially in Fenno-Scandia, during the latter part of the 19th century, presumably in response to the climatic amelioration. They began nesting in Finland for the first time in 1842, in Norway in 1860 and in Denmark in 1865.

Chapter 8

LANDBIRDS: NON-PASSERINES

Having considered the effects of the recent climatic amelioration on seabirds and waterfowl, we can now turn our attention to the remaining orders of birds, which are adapted primarily to life on land.

Starting first of all with the birds of prey (i.e. vultures, eagles, hawks, falcons and owls), we find that of fifty-two species on the European breeding list at least sixteen species appear to have benefited from the amelioration and expanded their breeding ranges, while five species (Hobby, Gyrfalcon, Snowy Owl, Hawk Owl and Great Grey Owl) apparently did not do so, and in fact retreated.

The sixteen species which seem to have benefited are the Black Kite, Red Kite, Short-toed Eagle, Levant Sparrowhawk, Honey Buzzard, Imperial Eagle, Pallid Harrier, Montagu's Harrier, Lesser Kestrel, Red-footed Falcon, Barn Owl, Tawny Owl, Little Owl, Long-eared Owl, Short-eared Owl and Scops Owl.

Black Kites have been increasing and spreading northwards and north-westwards in several parts of their range, for instance in northern France (since 1920) and Finland. Also, since 1955 they have been found breeding as far north as the River Kalix in Norbotten province in northern Sweden. It has been suggested that this increase may be connected with greater mortality among freshwater fish due to eutrophication and pollution during summer in lakes and rivers, for Black Kites, which incidentally are summer residents in the northern part of their range, feed chiefly upon large fish which they find dead or dying. This may well have been a contributory cause in recent years, but it would hardly seem applicable to the steady increase reported as far back as last century. To my mind, a correlation with the climatic amelioration appears a more likely explanation, although they may have gained from reduced competition with Red Kites and other large raptors (which, unlike the Black Kite, have almost all declined as a direct or indirect result of man's persecution), and also from their increased exploitation of the growing numbers of refuse tips associated with the escalating human population.

In passing, it is worth mentioning that the Red Kite, which is normally also a summer resident over almost all of the northern part of its range except Wales, has displayed an increasing tendency to overwinter in recent years, especially in southern Sweden. This, too, may be correlated with the generally milder winters. Red Kites suffered very much in western Europe from human persecution, particularly with the growth of game preservation, and therefore it is very difficult to sort out changes in distribution brought about by this cause from possible climatic influences. However, a temporary recovery in some parts of western Europe, such as northern Germany (where it has become locally the commonest raptor) during the first half of this century may have been inspired by the more congenial climate. Since 1970 the Red Kite has recolonised Denmark, where it died out in the early years of this century, and by 1986 some 20 pairs were breeding.

The Short-toed Eagle is another raptor which over the greater part of its range is a summer resident only, the great majority wintering south of the Sahara. Moreover, it is a highly specialised feeder, preying almost exclusively on warmth-loving lizards and snakes. Since it has been estimated that each family of these eagles requires some 800 to 1,000 snakes and lizards annually for their survival, they are dependent upon regular warm summers to ensure that these reptiles thrive and are active enough to be found easily. One would expect, therefore, the numbers and distributional limits of Short-toed Eagles to fluctuate according to both long-term and short-term climatic changes, and this appears to happen. As usual, of course, the picture has been complicated during the last century or two by the reclamation for agriculture of large areas of heathland and scrubland bordering old forests, where lizards and snakes are particularly abundant.

However, in spite of such extensive reclamation, Short-toed Eagles have recently extended their range a little in France and elsewhere in central and eastern Europe, following a marked recession southwards up to the end of the 19th century. For instance, until about 1850 they bred as far north-west in Europe as Denmark, while in France breeding pairs occurred as far north as Brittany, Normandy and Picardy up to the end of the 19th century. Subsequently, they retreated well to the south of Paris and the Loire by 1920, but since that date they have recovered some lost territory, presumably as a result of the generally better summers of the 1930s and 1940s.

The Honey Buzzard is also a summer resident throughout its breeding range, and a specialised feeder like the Short-toed Eagle. It preys chiefly upon wasps and bumble bees and their larvae, whose nests it digs up. Thus it too is dependent upon dry summers when its food is more easily available, and it is then more widely distributed and numerous. In summers when its prey is scarce, pairs may skip breeding. Fluctuations in the northern part of its range, especially in the maritime north-west of Europe are to be expected in consequence of short-term and long-term climatic changes. In Britain, for example, which is right on the north-west edge of its range, it was considered to be rare in the past, but widespread, breeding as far north as Aberdeenshire and Ross-shire in Scotland, although most frequent in southern England. Its decline during the 19th century has been blamed on human persecution; but since Honey Buzzards are unobtrusive and easily over-looked, while the nest sites are even harder to discover, especially in dense woodland which would have been more widespread in former times, it is unlikely that this was the real cause.

Although there is insufficient data to prove it, I think it probable that the Honey Buzzard was well established and quite common during the Medieval Warm Period, but gradually declined during the 17th and 18th centuries when the Little Ice Age was at its maximum. Therefore it was most likely already a very rare breeding species in England when Gilbert White wrote his famous account of the pair which nested in 1781 in Selborne Hanger, in north Hampshire; but it is significant, I think, that this event took place towards the end of a group of generally warm summers between 1772 and 1783 which may have induced a temporary recovery by the Honey Buzzard. Thereafter the summers deteriorated, especially between 1809 and 1815, and became generally unfavourable to it with the result that it presumably became a very rare breeder.

However, as the climate gradually improved after 1850, the Honey Buzzard may have begun a slow return to Britain which escalated during the much immmproved climate of the 1920s and 1930s; for it was during that period that a few pairs were discovered to be nesting regularly in the New Forest, in addition to scattered pairs breeding occasionally as far north as southern Scotland. And this is still more or less the situation today, except that the resumption of the climatic warming since 1975, and especially since 1980, has led to the Honey Buzzard consolidating its position in Britain. In 1990 up to 19 breeding pairs were reported distributed over 13 counties (Spencer *et al.* 1993), and there may be as many as 30 pairs (Batten *et al.* 1990). Elsewhere in Europe the Honey Buzzard has maintained its status, as in France, or even increased, as in Denmark, since 1930.

Two south-east European raptors, the Levant Sparrowhawk and the impressive Imperial Eagle, have spread slightly north-westwards in recent years, perhaps like the Syrian Woodpecker and the Olivaceous Warbler in response to the climatic amelioration. For instance, the Levant Sparrowhawk has been breeding in Hungary since 1962, while Imperial Eagles have moved further into the former Czechoslovakia and eastern Austria since 1950. They are also appearing more frequently as vagrants outside their normal range (e.g. Scandinavia).

Since about 1940 yet another eastern raptor, the Pallid Harrier, whose main breeding grounds are the open steppes to the north of the Black, Caspian and Aral seas, has been spreading quite rapidly, almost irruptively, westwards into central Europe. It even bred in the early 1950s as far west as Norderney in the German East Friesian Islands and the Baltic islands of Öland and Gotland off southern Sweden; but unfortunately this extreme extension, which seems to have been caused by exceptional weather condi-tions, was short lived, although half a dozen birds were recorded in Öland and Gotland in May and June 1988. However, it appears possible that the main extension of range has been due to the effects of the climatic amelio-ration, although it has also been attributed to the extensive felling of forests and more intensive cultivation of the steppes. It is too early, however, to be sure of the precise reasons for this development.

The similar Montagu's Harrier is exclusively a summer visitor through-out its breeding range in Europe and west-central Asia. Therefore one would expect it to respond more readily to climatic changes at the edge of its range than either the Hen Harrier or the Marsh Harrier, particularly in the wetter, more maritime climate of the British Isles. Such expectations seem to be justified. During the 19th century, and probably even earlier, when the climate tended to be especially cold and wet, it decreased in several places along the north-west limits of its European distribution, but gradually recovered after 1900 when the climate improved. At that time it spread to Denmark from northern Germany.

By 1930 when the climatic improvement was most marked, Montagu's Harriers were breeding in fair numbers over a large part of Denmark and had also invaded southern Sweden; while in Britain a noticeable increase had occurred, with small populations building up not only in the East Anglian stronghold, but in several other southern English counties, and also in South Wales. This increase in Britain became even more striking in the years around 1950, when south-west England became the new stronghold of the species, and pairs nested as far north as central Scotland and as far west as Ireland. However, after 1953, when the post-1950 climatic deterio-ration began to make itself felt, a decline in the British population of Montagu's Harrier occurred, and, following a brief partial recovery in 1967, became so rapid that by 1974 there were only two (possibly breeding) pairs left in the whole country. Since 1976, however, when the greenhouse effect warming began to have an influence in the south, another recovery has taken place, and by 1990 up to 15 pairs were breeding, or attempting to do so, annually in southern and eastern England.

The decrease of Montagu's Harrier in the 19th century has been attrib-uted to persecution by sportsmen, gamekeepers and egg collectors, and also to land reclamation, while its subsequent recovery in the first half of this century has been ascribed to respites from the attentions of gamekeepers during the First and Second World Wars, plus the rapid growth of afforesta-tion. Young forestry plantations are favoured as nesting habitats by Montagu's Harriers until the trees become too big, when they move on to the next suitable new plantation. All these factors may indeed have had an effect, but, as with other species, I believe they are not adequate in them-selves to explain fully the changing fortunes of this fine bird of prey, and that climatic change has, in fact, been the underlying cause. For instance, why should it have decreased so dramatically after 1953 when there has been no apparent increase in persecution by gamekeepers or others, and afforesta-tion continues? Pesticides have, of course, been invoked as a likely reason, but the use of these has been quite effectively controlled since the bad days of the early 1960s. Moreover, the Hen Harrier has been increasing and spreading almost spectacularly during the very period of the Montagu's retreat, and overlapping into the same habitats. Yet interspecific competion cannot be the reason either, for Montagu's Harriers vanished from their southern haunts before the Hen Harriers reached them. In any case their breeding habitats overlap in France as well as elsewhere in Europe. No, climate is, I think, at the root of it all.

Among falcons, there are three small, largely insectivorous species which spend the winter in Africa and migrate north to breed in Europe and central Asia. They are the Hobby, the Red-footed Falcon and the Lesser Kestrel. All rely very much upon catching active insects, so that wet summers are unfavourable to them. Therefore it is interesting to note that the latter two species showed signs of extending their ranges in central Europe in response to the dry warmth of the 1930s and 1940s. On the other hand, probably because of the more maritime climate produced in north-west Europe, with a tendency to wetter cooler summers for the most part, the 1850–1950 climatic amelioration did not, on the whole, suit the more widely distributed Hobby. Like its relatives, the Lesser Kestrel and Red-footed Falcon, it depends to a large extent on catching active flying insects, such as butterflies and dragonflies, which are much more inactive on wet or cloudy, largely sunless, days. Since the climatic amelioration waned around 1950, however, conditions have gradually improved for the Hobby, and with the advent of frequent hot summers from 1975 onwards, as the greenhouse effect began to exert an influence, the species has increased its breeding strength considerably in England and Wales and, locally, elsewhere in north-west Europe (see Chapter 12). It has become clear that the English breeding population has been underestimated in the past owing to many pairs being overlooked on farmland and in woodland, these habitats having been wrongly considered unsuitable for the Hobby (Marchant *et al.* 1990). On farmland in the Rhine plain in Germany, Hobbys frequently nest in old crows' nests quite high up on pylons, as I myself have seen. The increase in the English and Welsh breeding populations since 1975 has led to a marked expansion of range northwards, so that by 1990 pairs were suspected of nesting in Scotland.

The Lesser Kestrel, which was formerly unknown as a breeding species in France, established several colonies from 1947 onwards in the rocky hills and mountains along the Mediterranean coast as far east as the Alpes Maritimes, near the Italian frontier. Presumably this represented a north-eastward extension of its range in Spain, but it subsequently disappeared from its localities in the Corbières near the border with Spain, as none were found there during the French Atlas survey between 1970 and 1975 (Yeatman 1976). The Red-footed Falcon has been spreading slowly westward from its strongholds in the Balkans and Russia, and now breeds west of Munich (Bavaria). Moreover, it has wandered more frequently further to the west in recent years, and odd birds or even small parties are more often seen nowadays in the Low Countries and England.

As mentioned at the beginning of this chapter, the Arctic-alpine Gyrfalcon is another species that did not find the milder conditions of the climatic amelioration favourable, and tended to withdraw its southern breeding limits northwards. Since 1950, when the amelioration was superseded by a cooling of the Arctic, which also affected much of Fenno-Scandia and Arctic Russia, Gyrfalcons have been seen further south again in the breeding season. For instance, in 1988, thirty-seven were reported outside the normal breeding range in north-west Sweden. Moreover, individuals have been reported more frequently in north-central Europe in winter and are staying there longer than they used to do.

Likewise, the Arctic-alpine Snowy Owl and two Boreal zone owls, the Hawk Owl and the Great Grey Owl, also contracted their breeding ranges northwards somewhat during the amelioration, but have shown sign of advancing southwards again since the 1950s. This apparent return south-wards is described in more detail, particularly in the case of the Snowy Owl, in Chapter 11. This species was a regular visitor to Scotland up to about 1890, but was seen much less often as the climatic amelioration took hold strongly in the first half of the present century.

Among the owls, the Barn Owl, which has a primarily tropical and subtropical distribution throughout the world, breeds further north in Europe than anywhere else. Indeed the Barn Owls breeding in northern Scotland are the most northerly of all. Changes in distribution, associated with climatic fluctuations, along this northern limit are to be anticipated, particularly in Europe. In any case, hard winters are known to reduce severely the numbers of Barn Owls in both Europe and North America. It seems likely that they did not breed as far north as they do today during the Little Ice Age, but spread northwards during the improving climate of the late 19th century and the first half of the 20th century. During this period, for instance, they colonised Denmark, apparently from northern Germany, and also increased and extended their range northwards in Scotland.

However, in spite of this general northward spread, a widespread long-term decline has occurred among Barn Owls in some areas, such as south-ern England and Wales, since about 1900. This paradoxical state of affairs has been explained as due to a combination of the effects of severe winters, agricultural changes affecting habitat and food supply, reduction in the availability of suitable nesting sites, direct persecution by man, and poison-ing by pesticides. Since 1950 this decrease has become much more univer-sal, not only in England and Wales, but also in Ireland (especially in the north) and elsewhere in northern Europe; which fits in well with the start of the post-1950 climatic deterioration with its greater frequency of severe winters. Since 1987, however, winters have mainly been mild, probably due to the counteracting influence of the anthropogenic greenhouse effect, and there is growing evidence in some regions that this is helping to reduce Barn Owl winter mortality and increase numbers. Meanwhile, since 1960, it has almost become extinct in Sweden and has declined by over 80% in the Netherlands (Marchant *et al.* 1990).

The Tawny Owl also extended its range northwards in Europe during the recent climatic amelioration. For instance, it colonised southern Finland from 1878, and greatly increased there during the warm decades of the 1920s and 1930s, when it also spread northwards in Scandinavia and in north-west Russia. It also increased considerably and expanded its distribution in Britain (it does not breed in Ireland) and in the Netherlands. Since 1950, however, when the improvement in the climate waned, the Tawny Owl population increase slowed down or even showed a reverse tendency.

The Little Owl, too, seems to have benefited from the 1850–1950 amelioration. It is unlikely that it would have established itself so rapidly in Britain following its successful introductions between 1874 and 1890 and subsequently, were it not for the helping hand given by the steadily improving climate. Indeed, an earlier introduction in Yorkshire, by Charles Waterton in 1842, during a rather cold phase, failed. The Little Owl suffers severely from bad winters; thus it seems to me significant that the introduction attempts which succeeded so well were those after 1870, when the winters became progressively milder. Moreover, the period of really rapid colonisation of Britain, as far north as the Scottish border, coincided with the particularly warm phase between 1900 and 1940 when mild winters were the rule. This hypothesis is supported by the subsequent decline in the Little Owl's fortunes after 1940, when its spread at first slowed down, and then finally stopped almost everywhere as the subsequent climatic deterioration exerted its influence, especially through the severe winters of 1946–47 and 1962–63. As with other species, the picture of events was clouded by other factors, such as losses from toxic chemicals which may have been chiefly responsible for the sudden and heavy losses between 1955 and 1962. After that the British population recovered (allowing for the cyclic-type three- to five-year fluctuations which have now been established for this species), until it reached a peak in 1984, following which it has declined steadily, in spite of mild winters since 1987.

Judging by the Little Owl's widespread distribution on the mainland of Europe, where it occurs as far north as northern Jutland in Denmark, Latvia, and elsewhere along the southern shore of the Baltic, it seems rather surprising that its natural range does not include England. In former times, such as the very warm medieval period (the Little Climatic Optimum), it may well have inhabited England; if so, it probably died out during the severest part of the subsequent Little Ice Age. Richard Fitter (1959) considered it 'almost certainly a British bird *manqué*, that would long ago have been a natural member of our avifauna if only it had spread northwards after the ice in time to cross the ancient land bridge. The British avifauna as it stood in 1875 had a vacancy for a small diurnal, mainly insect-eating bird of prey'.

During the amelioration and probably because of it, the Long-eared and Short-eared Owls both exhibited a tendency to extend their ranges northwards in Europe. Although the Long-eared Owl underwent a marked decrease in Britain this century, possibly because of greater competition from the growing numbers of Tawny Owls (Sharrock 1976; Lack 1986), it spread northwards in Norway and also in Jutland (Denmark). The Short-eared Owl, which breeds over most of northern and central Europe and Asia, as well as North America, began the colonisation of Iceland in the 1920s and today breeds throughout that island.

Scops Owl has a distinctly southern distribution in the Northern Hemisphere, where it is confined to the Old World. In Europe it is common-est in the Mediterranean region, where its soft call is one of the most char-acteristic sounds of warm nights. The further north one travels the rarer it becomes, but it has nevertheless advanced northwards slightly this century, presumably in response to the warmer climate. In France, for example, it has spread north-westwards into Alsace. Tengmalm's Owl also began extending its breeding range westwards in central Europe around 1950; it will be discussed in Chapter 12.

Meanwhile, of the gamebirds, the Quail apparently benefited to a certain extent from the climatic amelioration, and so also, for a time, did the now very uncommon Greater Prairie Chicken of the American grass plains. On the other hand, the Ptarmigan contracted its range upward and northward in the Arctic-alpine and Boreal zones of Europe.

Although the highly migratory Quail apparently reacted unfavourably to the warmer but wetter maritime climate of the British Isles, Denmark and elsewhere along the north-west fringe of its range in north-west Europe between 1860 and about 1942, it generally benefited in central Europe during this period, and maintained its population level in spite of the adverse effects of changes in agricultural methods and the heavy toll exacted by followers of *la chasse*. Since 1942 an upward trend in Quail numbers has occurred in the maritime parts of north-west Europe, including the British Isles; some years proving to be exceptionally good for them, such as 1947, 1953, 1964, 1970, 1979, 1983 and 1989. This trend may be correlated with the increased tendency to drier summers, associated with the growing dominance of the Scandinavian blocking anticyclone in northern Europe since 1940. The Finnish ornithologist, Olavi Kalela (1949), considered the fluctuating fortunes of the Quail in northern Europe to be due to climatic changes.

Although once a fairly common grouse of the North American prairies, the Greater Prairie Chicken is nowadays very rare and local over most of its range. It clearly suffered from a combination of such things as the destruc-tion of its habitat through the expansion of agriculture and severe prairie fires during the breeding season, as well as from intensive shooting in the past. However, in spite of all this, it underwent a northward expansion of range in mid-Canada around 1880 and became fairly common until about 1925, when it decreased and contracted southwards drastically, becoming very rare (Godfrey 1966). The general climatic improvement in the

Northern Hemisphere during this period may well have been involved in this expansion, but it is difficult to explain the sudden decline of the Greater Prairie Chicken after 1925, when the subsequent climatic deterioration had not yet begun (unless agricultural intensification proved too much); its reappearance in southern Ontario about that same year is also difficult to account for. In 1938 it spread to Manitoulin Island on the north side of Lake Huron, and was still present in the early 1960s, although hybridising to a considerable extent with the Sharp-tailed Grouse.

The circumpolar Ptarmigan (or Rock Ptarmigan) is one of the best examples of an Arctic-alpine species which contracted the southern limit of its range northwards during the climatic amelioration. This has been most evident, as one would expect, in the maritime climate of north-west Europe. For instance in Scotland, where it is an inhabitant of the Arctic-alpine heaths, it began retreating north as early as 1830. Before that time it nested as far south as the mountains of south-west Scotland and north-west England, the islands of Arran and Rhum, and, at an earlier date, according to Kenneth Williamson, perhaps even Snowdonia in North Wales. By 1938 it had ceased to inhabit the Outer Hebrides; but since 1950, with the reversal of the climatic trend, it has shown slight signs of expanding again, and there have even been reports of it being seen again in the Outer Hebrides and on Rhum, though this was not confirmed by the BTO Atlas survey (Sharrock 1976). The growing greenhouse effect may be expected to have an adverse effect on the Ptarmigan eventually.

A similar northward retreat, correlated with the improved climate, has occurred in the more southerly Willow Grouse, which is widespread in North America as well as in the Palaearctic Region. Again this has been most noticeable in northern Europe where, for instance, it receded northwards after 1870 in the former East Prussia, Latvia and Finland. The indigenous British race of the Willow Grouse, the Red Grouse, which is probably a relict population isolated during the Last Ice Age, has however shown little or no sign of receding northwards, although, allowing for regular short-term cyclic fluctuations, marked long-term decreases occurred in Ireland from 1920 and in Britain from 1940. But these have been correlated convincingly with factors other than long-term climatic ones, such as a deterioration in heather quality through inadequate management of the grouse moors, overgrazing by red deer, cattle and sheep, and outbreaks of disease. Nevertheless, the run of mild winters since 1986 has seen a recovery in Britain.

One of the puzzles which has confronted European ornithologists has been the cause of the sudden and irregular irruptions of Pallas's Sandgrouse—not actually a grouse, but a relative of the pigeons—from its native haunts in the vast steppes and semi-deserts of central Asia into the countryside of Europe. These incursions were first noted in 1859 and were not infrequent until 1909, when they apparently inexplicably ceased until 1969. That year a small irruption took place, with birds reaching Finland, the Netherlands and the east coast of England, and it was followed by

FIGURE 30 *Willow Grouse. (Photograph: Dr John Sparks)*

another minor one in which birds reached Norway and Poland. The most spectacular of these westward invasions occurred in the late springs of 1863, 1888 and 1908, when parties of sandgrouse were quite common as far west as the British Isles, including Ireland and the Outer Hebrides. Some of these birds nested in their new, temporary haunts, but 30% or more were slaughtered by trigger-happy 'sportsmen'.

In his stimulating book *The Natural Regulation of Animal Numbers*, David Lack (1954) stated that the reason for these irruptions is not known, but he suspected that food shortage was perhaps the cause. In support of this possibility he pointed out that the nomadic tribesmen of the Asiatic steppes have a saying that 'when the sandgrouse fly by, wives will be cheap'. Pallas's Sandgrouse, which superficially resemble sandy brown partridges with long, pointed wings and tails, walk with a curious tripping and waddling gait as they peck at the seeds and shoots of grasses and such succulent plants as those of the goosefoot family. Thus severe droughts, which kill off the vegetation and dry up the watercourses and pools as well, very probably force whole populations to emigrate to other regions, for the loss of their water supply is just as critical as the loss of their food. They need to drink at least once a day to survive, because they are unable to obtain moisture from insects, unlike some other desert birds, since they feed exclusively upon seeds and other dry vegetable matter. Consequently, they must be able to travel long distances at high speed in search of water; and thus they are adapted to a nomadic way of life, having long pointed wings and swift, powerful flight. Moreover, their belly feathers are so constructed that they can absorb water without becoming spoiled in any way. Pairs with young, particularly the males, regularly bring back water for them by soaking these feathers after drinking themselves. The chicks creep under their parents and suck the water-laden feathers rather like piglets suckling a sow.

Another cause of irruptive migrations by Pallas's Sandgrouse is said to be heavy and prolonged falls of snow which freeze their sources of water and make it difficult for them to find their natural food. Indeed westward movements have been observed starting at such times. They have usually reached the east coast of Britain in May or June.

Thus hard winters and/or severe droughts in spring seem to be the usual cause of the irruptive movements of Pallas's Sandgrouse but it remains to be explained why they were so frequent over Europe between 1859 and 1909, and completely absent since then, with the exception of the small irruption in 1969 and an even smaller one in 1990, which reached Norway and Poland respectively. The breeding range of this species extends from the eastern limits of the Gobi desert westwards in a wide band to engulf the Aral Sea and the area around the northern shores of the Caspian Sea, but no further. It will be remembered that this region around the Caspian and Aral Seas became increasingly arid after 1870 and caused a number of species, including the Black-necked Grebe, Little Gull and Tufted Duck, to vacate the area and extend their ranges far to the west in Europe. Between 1859 and 1909, Pallas's Sandgrouse periodically did likewise, but gradually the increasing aridity caused it to contract its breeding distribution much further east. As a result, Europe eventually became beyond the reach of subsequent irruptions. The small 1969 and 1990 irruptions suggest that the recent climatic trends in the Northern Hemisphere may have led to reduced desiccation in the western part of the sandgrouse's range and a gradual increase of the birds there.

Four European species of pigeons expanded their ranges northwards during the amelioration, and one North American species, the Mourning Dove. This common American dove with a long tapering tail and swift, whistling flight, is becoming as familiar to suburban dwellers as it is to those in the country. It breeds almost everywhere in the United States, and also over much of southern Canada and even, it is believed, in south-east Alaska. But in the northern part of its range it is primarily a summer resident and migrates south for the winter. However, because of the milder winters of the first half of this century it developed an increasing tendency to overwinter in those regions, even in the extreme north. It is still not clear whether or not this tendency reversed with the post-1950 deteriorating climate in the Arctic, but apparently northern wintering Mourning Doves have been seen with frost-bitten feet! As well as overwintering more often in the north, they also extended the northern limit of their breeding distribution to the north— in eastern Canada, for instance.

Without a doubt, the most spectacular northward expansion of range has been that of the Collared Dove in the Old World. Indeed, it is one of the most remarkable stories in the history of ornithology. Sometime in the uncertain past, from their ancestral home in India and Burma, Collared

Doves gradually extended their range westwards until, by the 16th century, they had colonised much of Asia Minor. Around 1900 they spread into the Balkans and the invasion of Europe began. For a time, like the Turks, they were held in check by the River Danube, but early in the 1930s they finally made the crossing, and then there was no holding them. They swept north-westwards and north across Austria and the former Czechoslovakia, so that by the end of the Second World War they had colonised most of Poland, Germany and Italy. By 1947 they were in the Netherlands and already within striking distance of England. 1948 saw them invading Denmark and European Russia, from whence they struck into Norway and Sweden. In 1950 they spread across France, within the next few years building up their forces all along the Channel and North Sea coasts for the next phase of their advance—the invasion of Britain and Ireland.

The first pair of Collared Doves nested on the Norfolk coast of England in 1955, and by 1964 the British population was estimated at almost 19,000 birds. Today the total population is certainly in excess of 100,000, although accurate up to date estimates are now extremely difficult to make because of the lack of estimates for the high populations in urban areas (Marchant et al. 1990). However, their rate of increase dropped in the late 1970s, peaked in 1982, and has since more or less stabilised. Collared Doves now

FIGURE 31 *Collared Dove*

breed almost everywhere in the British Isles, except on high barren uplands where the human population is sparse.

The Faeroes and Iceland were more or less successfully colonised early in the 1970s, and soon afterwards birds even reached Greenland. However, although the Collared Dove, not surprisingly, has spread south-westwards since 1974 to colonise completely new territory in Spain and Portugal and across the Mediterranean into Morocco, it has started to decline and retreat from the Low Countries and Scandinavia. These recent trends are discussed in Chapters 11 and 12.

Although the history of the Collared Dove's expansion of range (Figure 32) has been well documented, none of the various explanations so far put forward have yet won general acceptance. The great American zoologist Ernst Mayr has suggested that the cause was a genetic change in those birds which lived on the extreme western limits of the old range, thus leading the species to enter 'a phase of agressive range expansion' (Fisher 1966). Certainly, compared with other doves and pigeons, the Collared Doves of western Europe are noticeably aggressive. Personally, I believe that the climatic amelioration which began in the 1850s and gathered momentum in the 1930s is probably the main cause; its development matches the two main phases of the Collared Dove's spread remarkably closely.

Another factor in its success has been the advantage it has taken of man's agricultural activities. Very much an opportunist and primarily a grain-eater,

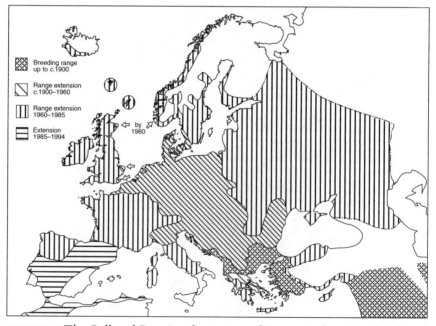

FIGURE 32 *The Collared Dove's colonisation of Europe (adapted and updated from Fisher 1953; Voous 1960; Harrison 1982; Cramp et al. 1985).*

it has little fear of man and is thus able to live in close proximity to him, exploiting any sources of waste grain, raiding his stackyards and ripe corn, and competing successfully with domestic fowls for their feed.

As the human population in Asia has increased over recent centuries and expanded food production, so the Collared Dove has benefited. It is known that it was already spreading outwards from its ancestral headquarters in the Indian subcontinent during the Middle Ages, and that it reached Asia Minor in the 16th century, when it appeared as far west as Istanbul. So, as a parasite of man, its spread may well have paralleled man's own population growth in southern Asia during this same period. Then, although poised at the gateway to Europe for some three hundred years, its progression further west may have been inhibited by the climate. After all, the 16th century saw the beginning of a long cold spell in Europe, characterized by severe winters. But, with the steady improvement of the climate in the 19th century, the stage was set for its invasion of Europe, which had a vacant ecological niche for a medium-sized, grain-eating dove, and whose rapidly expanding human population was, in many countries, wasteful.

Once in a new area, the Collared Dove's prolific rate of breeding enables it to consolidate its foothold and build up its population rapidly. Three, four or more broods of two young are raised in the course of a long breeding season which may start as early as January or February and go on until late November, although late March to late October is more usual. So the milder winters in Europe this century must have greatly facilitated its spread.

A northward advance of the closely related Turtle Dove in Britain, Denmark and elsewhere in north-west Europe from about 1850 may also have been primarily due to the climatic improvement, although it has been most frequently attributed to agricultural change favourable to this bird. During the 1950s the expansion slowed down or halted, and there were even some local contractions in range, particularly in the west. Before that there was a further slight advance in northern Britain, when pairs bred for the first time in Scotland in 1946. Subsequently, Turtle Doves looked like becoming established in the south-east of that country, as far north as Argyllshire and Angus, but eventually failed to do so. Between 1963 and 1979 they steadily, if slowly, increased in Britain, but then fell into a steady decline through the 1980s at least up to 1992, when an increase was detected by the BTO's Common Birds Census. It is to be hoped that this may indicate an upturn in the Turtle Dove's fortunes, perhaps connected with the growing influence of the greenhouse warming. In continental Europe, it has been increasing in Estonia and the other Baltic republics since the 1960s, and is now being seen more frequently in southern Sweden, although breeding has yet to be proved (Marchant *et al.* 1990). Meanwhile, the Danish population has been growing.

R.K. Murton *et al.* (1964) have shown that the Turtle Dove's most important food is the seeds of the fumitory *Fumaria officinalis*, a weed of arable land, and that its distribution in Britain closely matches that of this plant.

They pointed out that fumitory did not arrive in Britain until Roman times, and also that although the Turtle Dove was well known to the ancients, its occurrence in Britain was not mentioned until the 10th century. Moreover, even though fossil and sub-fossil remains of Rock Doves, Stock Doves and Woodpigeons have been found in interglacial deposits, no trace has been discovered of the Turtle Dove. They therefore considered it highly likely that it only spread north in comparatively recent times. Perhaps it invaded southern England as a result of the great warmth of the Medieval Warm Period (Little Climatic Optimum), which would have been well established by the 10th century. It probably retreated south-eastwards during the subsequent Little Ice Age, advancing again when this ended around 1850.

C.A. Norris (1960) suggested that the Turtle Dove may be limited as a breeding bird by the 19°C July isotherm and also by the amount of rainfall, with 102 cm (40 inches) as the upper limit of toleration. It seems to me that the correlation with the 102 cm rainfall limit is a good one, but I am much less convinced by his correlation with the July isotherm. The 15.5°C mean July isotherm looks a more likely limit, but I believe that it is the level of rainfall that is probably most important.

The westward and northward spread of the Stock Dove in Europe since about 1850 has also been quite spectacular. Before that date it was unknown as a breeding species in much of western Europe, including France, the Low Countries and Spain. In Britain it was confined to south-east England; but by the 1860s it was expanding north and west in earnest, so that by 1866 it had begun to breed over the border in Scotland and was rapidly colonising western England. Ireland was reached in 1877, and it is nowadays a widespread nester in that country. On the continent of Europe it began nesting in north-east France towards 1870, in the Netherlands about 1880, and in southern Sweden by 1900. In France its westward expansion reached Normandy in 1925, after which it pushed south to colonise Languedoc in 1936. Meanwhile the colonisation of Spain commenced as early as 1927, possibly from North Africa.

R.K. Murton and others have attributed the Stock Dove's spread in the British Isles to the growth of arable farming in the 19th and early 20th centuries, but, as might now be expected, I favour Kenneth Williamson's belief that the underlying cause was the post-1850 climatic amelioration with which there is obviously a close correlation. However, the growth of agriculture, by improving the Stock Dove's food supply, undoubtedly assisted the rapidity of its increase. But in my opinion it does not alone explain the bird's spread in western Europe as a whole. Between 1951 and 1962 there was a marked decline in Britain that was correlated to the temporarily uncontrolled use of chlorinated hydrocarbon pesticides, which caused similar losses among other birds at this time. Following their ban in 1966, Stock Doves began to recover, reaching their previous population level by 1980. Since then the population has more or less stabilised (Marchant et al. 1990). Meanwhile, on the European mainland, declines have been reported from several areas, including Fenno-Scandia, where climatic cooling may be involved. Elsewhere, misuse and over-use of pesticides may be to blame.

A similar link with the development of agriculture has been advanced for the increase and spread of the Woodpigeon in Britain and elsewhere in Europe since 1850. Murton (1965) associated it in particular with the increased availability in winter of such crops as clover. Once again though the climatic amelioration has played a part, especially in the northward expansion of the Woodpigeon's breeding range. It has pushed northwards significantly since 1870 in Fenno-Scandia and northern Britain. By 1900 it had colonised the Outer Hebrides and Orkneys; it began nesting in Shetland in 1939, and colonised the Faeroes from 1969. Since 1964 breeding has also occurred from time to time in Iceland.

The only cuckoo in the Northern Hemisphere, so far as I am aware, to have significantly extended its range in apparent response to the climatic amelioration is the Great Spotted Cuckoo, a large and impressive species which parasitises magpies and other corvids. Apart from a couple of quite exceptional breeding records in 1885 and 1924, it was unknown as a breeding bird in France until about 1943 when it spread north-eastwards from Spain along the French Mediterranean coast. Today it is well established in this region, especially in and around the Camargue, and is still increasing.

The decline of the Cuckoo in the British Isles from about 1940 to about 1970 has attracted a great deal of attention there, but is very difficult to explain. It may have been connected with a possible decrease due to improved control measures, in the numbers of hairy caterpillars of such moths as the goldtail *Euproctis similis*, browntail *E. chrysorrhaea*, vapourer *Orgyia antiqua*, and lackey *Malacosoma neustria*, which are among its principal prey. Climatic change may also have played a role, though it should be noted that there has been little appreciable change in its status in Scotland and, moreover, it has nested more frequently in Shetland since 1947. However, Williamson (1975) pointed out that by the 1950s the springs in Britain were colder on average than they had been and the summers wetter, as the climate cooled following the 1850–1950 amelioration. The BTO's Common Birds Census indicates a recovery since 1970, especially on farmland, with the population becoming more or less stable during the 1980s. In the early 1990s the trend was upwards again with the population in 1992 reaching the highest level on farmland yet recorded (Marchant and Balmer 1993). This improvement since 1970 coincides with the growing influence of the greenhouse effect in the south, which may be causing more Cuckoos to breed in northern Europe.

The Nightjar is one of a small group of highly insectivorous summer residents in north-west Europe like the Red-backed Shrike and Wryneck, which retreated south-eastwards during the course of the present century. It seems probable that the generally wetter summers of the climatic amelioration have affected it unfavourably, presumably because they reduced the activity of the crepuscular and nocturnal insects upon which it chiefly preys. It was a common bird over much of the British Isles until the end of the 19th century when a decline began to be noticed in many areas, which accelerated after 1930—the peak of the climatic amelioration—and particularly so after 1950, when the climate became more unsettled, especially in the north-west. Similarly, in the Netherlands and northern France the Nightjar also decreased subsequent to 1930. Since then a pronounced and widespread decline has been reported from many parts of Europe, especially in the north-west, where a south-eastern contraction of range has occurred. In the British Isles this has resulted in a withdrawal since the early 1970s from Scotland and northern England in the north, and Ireland and Wales in the west. Nightjars are also arriving later in spring, on average, than in the 1920s, and fewer are rearing second broods than in the 1940s—both indicative of climatic influences (Marchant *et al.* 1990).

According to Parslow (1973) and others, the destruction of the Nightjar's habitat and increased disturbance are the two causes most commonly given, but he considered that another factor, possibly a climatic one, must be operating to account for the widespread nature of its demise. His view that habitat loss and increased disturbance cannot be wholly responsible concurs with mine, although they have undoubtedly played a part lately. If anything, afforestation in recent years should have benefited the species in its early stages, as it has birds such as the Hen Harrier and Short-eared Owl, but this it plainly has not done, except to a very limited extent. Climatic change offers the most likely explanation, and it remains to be seen if the gradual and increasing warmth due to the anthropogenic greenhouse effect will promote a real revival in its fortunes, assuming that the loss and fragmentation of prime Nightjar habitat is arrested. In fact, the preliminary results of a repeat BTO census in Britain in 1992 have shown a recovery of at least 52% since the 1981 census, especially in south-west and north-east England, and Wales. There was even some recolonisation of some previously lost areas (Morris 1993).

The amelioration seems to have benefited most of the swifts in the western Palaearctic. The Swift (or Common Swift), which is found breeding almost everywhere in western Europe, except for the extreme north and north-west (e.g. Iceland), increased and spread westwards in Ireland from 1932, and has increased in northern Britain. An interesting effect of the climatic improvement has been the increased tendency of Swifts to arrive earlier in Britain and depart later after the breeding season than formerly. Nowadays, they arrive nearly two weeks earlier and leave nearly a month later than they did in the early 1930s.

The Alpine Swift, an inhabitant of chiefly rocky mountainous country in the Mediterranean region, south-west Asia, and southern and eastern Africa, gradually extended its European range northwards from at least 1930, presumably because of the warmer summers. For instance, it now breeds as far north as Freiburg in south-west Germany (since 1952), and not far from Dijon in east-central France. Yeatman (1976) stated that in the 19th century it was unknown in Provence, but is now widespread in suitable localities.

Similarly, the Pallid Swift, which has a largely coastal distribution along the Mediterranean seaboard of Europe, has been advancing northwards since about 1930. It was first reported nesting in Corsica in 1932, and was found breeding for the first time in continental France in 1950 at Banyuls, since when it has colonised many parts of the coastal zone, and has penetrated inland as far as the neighbourhood of Toulouse.

In 1966, to the surprise of the ornithological world, four pairs of White-rumped Swifts were discovered in the Sierra de la Plata in southern Spain, nesting, as is their habit, in the disused nests of Red-rumped Swallows; previously this species was thought to be confined to Africa south of the Sahara. Breeding has subsequently been shown to be regular in the extreme south of Spain and, since White-rumped Swifts were seen in this region as early as 1962, they have probably been nesting there at least since that year. In 1968 the species was proved to be breeding in the Atlas Mountains on the African side of the Straits of Gibraltar, but it is not known whether these Moroccan and Spanish colonies represent an extension of range from much further south in Africa or are part of a relict population long isolated north of the Sahara. However, there is little doubt that the similar Little Swift, with which the White-rumped Swifts nesting in Spain were at first confused, has been extending its range northwards in Morocco since the 1920s. Since 1925, when it began to colonise Rabat, it has spread almost 140 miles (224 km) to the north to occupy Tangier, where it is now a common and noisy bird. It seems to be only a matter of time before it joins the White-rumped Swifts across the Straits in Spain and becomes an addition to the list of European breeding birds, as individuals are already (in the early 1990s) being reported there.

That gorgeously colourful bird so evocative of the Mediterranean region, the Bee-eater, has also been pushing northwards quite strongly since the mid-1930s. For instance, in southern France it was known as a regular breeding bird in 1936 only in the Camargue, the Gard and Corsica, although pairs sometimes nested elsewhere in the provinces bordering the Mediterranean and occasionally even further north. Nowadays it is much more widespread as a regular breeder, and even nests as far up the Rhône valley as Valence. Moreover, odd pairs nest even further north in France. In 1955 three pairs nested in Sussex in southern England, and another pair nested on Alderney, in the Channel Islands, in 1956. If this range expansion continues in France, pairs or groups may be expected to nest again in Britain in the not too distant future. Bee-eaters also started nesting in the former Czechoslovakia after 1950, and extended their breeding range northwards in the former Soviet Union.

The almost as colourful Roller has retreated steadily south-eastwards since about 1850 from most of its former haunts in north-west Europe, such as Denmark, south Sweden and western Germany. In Denmark, for instance, it was not an uncommon breeding bird during the first half of the 19th century, but by 1874 the last known breeding case had occurred. Today, the only breeding Rollers in Sweden are to be found on the Baltic island of Fårö, adjacent to Gotland, but before 1925 they were much more widespread and fairly common birds in the south-east part of the country below latitude 60°N, i.e. from the neighbourhood of Stockholm southwards. Figures 33a and 33b (after Durango 1946) show the extent of its breeding distribution in southern Sweden before 1925, and between 1925 and 1946. The chart lines in Figure 33a show the average rainfall for the period May–October for the years between 1880 and 1909.

S. Durango, a leading Swedish ornithologist, made a careful study of the Roller in Sweden in the 1930s and 1940s, and concluded that its decline was due to the north-eastward extension of the Atlantic-type climate as a result of the general amelioration in the North Atlantic. He found that there was a close correlation between the two and believes that the wetter summers brought by the amelioration were unfavourable to the Roller, which depends for its food chiefly upon large, active insects, such as grasshoppers, locusts, crickets and cockchafers, as well as small lizards, which are more easily obtained in warm, dry summers with plenty of sunshine. Nestling mortality became very high because of the inability of the parents to find enough food.

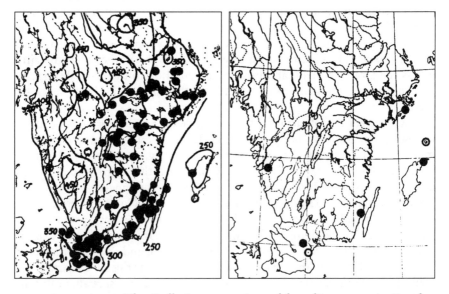

FIGURES 33a & 33b *The Roller's contraction of breeding range in Sweden (white circles = probable breeding only).* 33a *Breeding distribution up to 1925. The lines show the average rainfall in mm during May–Oct for 1880–1909.* 33b *Breeding distribution between 1925 and 1946. (After Durango 1946)*

In contrast to its fortunes in north-west Europe, the Roller has increased to some extent in eastern Europe and Russia, where the climate has tended to become drier this century. It has also tended to increase in its French stronghold of the Rhône delta since 1936 and, like the Bee-eater, has been pushing slowly northwards up that great river valley. Here the general amelioration presumably did not result in wetter summers, being protected from the stronger Atlantic westerlies by the mountain ranges of central France. For the situation since 1960, see Chapter 11.

The southward retreat of three other species—the Hoopoe, Wryneck and Red-backed Shrike—during the climatic amelioration is believed to have been for primarily the same reason as in the case of the Roller, i.e. the unfavourable effect of the generally wetter summers on the activity, and therefore availability, of their insect prey in those parts of north-west Europe most affected by the Atlantic westerlies.

Until about the middle of last century Hoopoes bred quite commonly as far north-west in Europe as the Netherlands, Denmark and southern Sweden. As I have mentioned in an earlier chapter, they were probably regular breeders in southern England during the warm, dry medieval period and perhaps even later, though during the past 150 years only about thirty confirmed instances of nesting are known. Hoopoes ceased to breed in Denmark after 1876, and indeed were already almost extinct there by 1860. Although they seem to have suffered a fair amount from human persecution in the past, particularly in southern Europe, the prime cause of the withdrawal has undoubtedly been the difficulty of finding enough of their favourite prey—grasshoppers, crickets, mole-crickets, bush-crickets, large insect larvae and small lizards. Both the mole-cricket *Gryllotalpa gryllotalpa* and the field-cricket *Gryllus campestris* have likewise declined in north-west Europe, while in England they have all but vanished since 1850, presumably because of climatic change. With the gradual cessation of the 1850–1950 climatic amelioration, and a slow return to more 'continental' summers, there have been a few signs of a return by Hoopoes to the north-west. For instance, a pair nested in Denmark (Jutland) in 1970 for the first time for nearly a hundred years, and breeding records seem to be on the increase in southern England, where there were four reports in 1977, three in 1984 and two in 1990. They have also reappeared in some parts of northern France to breed after an absence of half a century.

The Wryneck also appears to have been affected by the difficulty it has experienced during the wetter summers of the amelioration in finding sufficient of its favourite prey, in this case ants and their larvae. It has been suggested that lack of sunshine may restrict the development of the ants' eggs and larvae, but not enough seems to be known about this yet. So the precise proximate causes of the decline are still rather mysterious, but there is little doubt in my mind that the climatic amelioration is ultimately responsible. In England and Wales, where the Wryneck was once widespread and locally common (especially in central and south-east England), a decline began about 1830. J.F. Monk (1963) and R.E.F. Peal (1968) have traced its subsequent withdrawal south-eastwards. By 1970 it had become a very rare breeding bird confined to the south-eastern counties of England (Figure 34). After 1977 it apparently ceased to nest in England, although birds have been seen in the breeding season most years; but since the late 1960s there has been an extraordinary attempt at colonisation in north Scotland, the cause and origin of which will be discussed in Chapter 11.

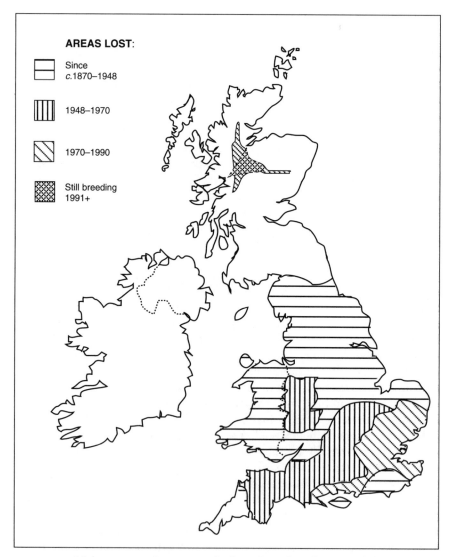

AREAS LOST:

Since
c.1870–1948

1948–1970

1970–1990

Still breeding
1991+

FIGURE 34 *20th century changes in the breeding distribution of the Wryneck in Britain (adapted from Monk 1963; Parslow 1973; Sharrock 1976; Gibbons et al. 1993).*

On the mainland of Europe, the Wryneck similarly contracted its range south-eastwards this century, and it has become a very rare breeding bird in Belgium, the Netherlands and much of the northern maritime provinces of France, where the summers also are less sunny than formerly. It has also declined in Germany, Austria, Switzerland, the Czech Republic and Hungary.

The same may be said of the Red-backed Shrike. (Although Passerines, the shrikes are included in this chapter for convenience.) Since the late 1930s it has, according to Laurent Yeatman (1976), practically disappeared as a breeding bird from France north-west of a line drawn from Mézières, Chalons, Sens and Angers to Vannes, i.e. most of Brittany, Normandy and Picardy. In Belgium it has also become rare, while a similar decrease has been noted since 1930 elsewhere in north-west Europe, for instance in Denmark, Sweden and north-west Germany.

The decline of the Red-backed Shrike in Britain has been disastrous. D.B. Peakall (1962) showed that it commenced more than a hundred years ago and apparently accelerated from 1940 onwards. Before 1850, Red-backed Shrikes bred regularly as far north as the Scottish border, but by 1920 they had disappeared from much of northern England and Cornwall. By 1950 they were largely restricted to the southern part of England and Wales, and from 1971 to the south-eastern counties of England. The total breeding population dropped from more than 300 pairs in 1952 to eighty-one pairs in 1971, to fewer than fifty pairs in 1974 and about six in 1984. By 1988 the last pair known to have nested successfully in England was recorded in Norfolk. The following year only unmated birds were present in three different localities, all in Norfolk, and in 1990 only one possible pair was present out of seven birds reported in five localities in England.

S. Durango in Sweden and D.B. Peakall in Britain both concluded that the pattern of the Red-backed Shrike's contraction of range in north-west Europe was consistent with long-term climatic change. As with the Roller, Hoopoe, Wryneck and other species already discussed, the proximate reason appears to be the difficulty experienced by the shrikes in finding sufficient quantities of their favourite prey—small lizards, grasshoppers, crickets, butterflies, bees, beetles and other sun-loving creatures—in the wetter, duller summers resulting from the 1850–1950 climatic amelioration. Like the Roller, Red-backed Shrikes sit on prominent perches and either pounce on their active prey on the ground below, or take them in flight. Sharrock (1976) thought it unlikely that climate was the sole factor involved in the Red-backed Shrike's decline, and mentioned the accelerating loss of suitable breeding habitat in England, especially the loss of heathland, as an important contributory cause. He also mentioned that egg-collecting 'has had a serious local effect, eliminating the species from small pockets'. Although loss of habitat has undoubtedly aggravated the precarious status of the Red-backed Shrike in England, I agree with Kenneth Williamson's assertion (1975) that 'there can be little doubt that climatic change has been the root cause'. The persistence of the Scandinavian blocking anticyclone in

FIGURE 35 *Red-backed Shrike*

the springs of the late 1960s and 1970s which probably led to returning Scandinavian summer residents such as Wrynecks, Lesser Spotted Woodpeckers, Shore Larks, Lapland Buntings and Bluethroats becoming deflected on the easterly winds to Scotland to nest or attempt to do so, may have been responsible for the extraordinary record of a possible breeding pair of Red-backed Shrikes in Orkney in 1970. This subject will be discussed more fully in Chapter 11.

The same explanation also seems to be applicable in the cases of two other shrikes—the Lesser Grey and the Woodchat. Both have contracted their breeding ranges in central Europe in the course of this century. The Woodchat Shrike's retreat southwards was most evident in Germany and Poland where, up to about the end of last century, it bred as far north as the former East Prussia, as well as in Denmark. It has almost, if not quite, disappeared from the Low Countries. In northern France it steadily evacuated large areas of territory between the surveys made in 1936 and 1970. The Lesser Grey Shrike's contraction in range began earlier—before the end of the 19th century. After 1914 it became very rapid, so that by 1936 the Lesser Grey Shrike had become a very local breeding bird in the east and south-east of France with strongholds only in the Rhône delta area and in the region of Vichy. Like the Red-backed Shrike, but more definitely, both Lesser Grey and Woodchat Shrikes have been making a slight recovery since the amelioration waned after 1950, especially the Lesser Grey, which has been expanding its breeding localities in France and southern Germany since 1960 or thereabouts now that the climate is becoming drier. The growing warmth in south and central Europe since 1975 due to the greenhouse effect is very probably assisting this recovery.

We have already discussed in this chapter the case of a woodpecker—the Wryneck—which withdrew south-eastwards into the heart of Europe in apparent response to the warmer, but wetter conditions brought by the climatic amelioration. However, most of the other European woodpeckers appear to have benefited from this change in the climate; of the other nine

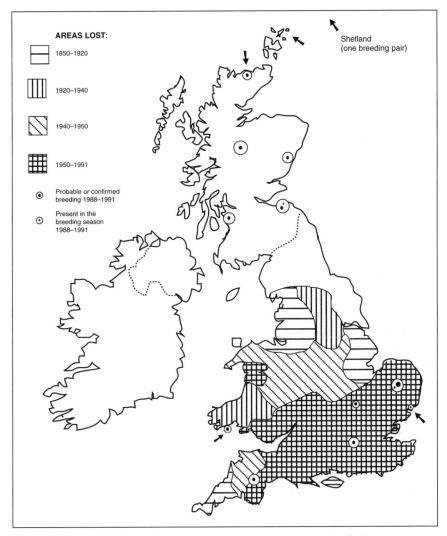

FIGURE 36 *The 20th century decline of the Red-backed Shrike in Britain (adapted from Peakall 1962; Gibbons et al. 1993).*

species, all but two—the Three-toed and White-backed Woodpeckers—have expanded their breeding ranges northwards, north-westwards or westwards at least to some extent.

The largest European species, the Black Woodpecker, which is the size of a rook and of the same colour, except for some red feathers on the head, has been spreading westwards and north-westwards in central Europe since the end of last century from its mountain forest strongholds in southern Germany, Hungary, and the French Alps and Jura mountain ranges. Descending from the mountains, it colonised lowland woods and forests over extensive areas, so that it was breeding in Belgium by 1908, in the Netherlands by 1913 and Luxembourg by 1915. It is now well established in the larger mixed woods of those countries, especially where there are beeches, and is still expanding. The atlas of French breeding birds (Yeatman 1976) showed that it has occupied a vast area of central and northern France, and was then already breeding within reach of the English Channel in Normandy and Picardy. It has now reached the Atlantic coast of Brittany. This invites speculation as to whether or not it will eventually cross the Channel and colonise southern England. So far there are no accepted British records, but Richard Fitter (1959) considered that at least some of the eighty-odd sightings that had been reported up till then were probably genuine. In recent years Black Woodpeckers have also colonised the large Danish island of Zealand where there are currently about eighty breeding pairs or more.

The Black Woodpecker, which once must have had a continuous breeding distribution right across both the northern and temperate forests of Europe, as it has today across the whole of Asia, suffered severely during the Last Glaciation. It must have been wiped out over most of northern Europe, leaving only isolated populations in the mountain forests of the Iberian Peninsula, southern Italy and the Balkans. When the ice retreated and the forests re-established themselves as the climate improved, so the Black Woodpecker spread westward over Fenno-Scandia and central Europe from Russia and the Balkans. But the complete recovery of its lost territories in western Europe was probably halted or even pushed back by the devastating depletion of the forests by man. It is therefore understandable that its recent advances have been explained by some authorities as being due solely to the big coniferous reafforestation programmes of this century, with which they coincided. Reafforestation has obviously been an important and helpful factor in this respect, but I believe that the climatic amelioration has been the underlying stimulus. I admit to a lack of any real proof, apart from the close correlation in dates and the attachment of Black Woodpeckers to old mixed woods, especially those with old beeches, rather than to the extensive conifer plantations as might be expected if that were the chief cause. I am certainly unconvinced that reafforestation has been the primary reason.

Likewise, at first sight, reafforestation alone might seem to be a sufficient explanation of the expansion north-westwards of other woodpeckers in the course of this century. The most extraordinary and spectacular has been the

FIGURE 37 *Syrian Woodpecker*

Syrian Woodpecker, characteristically a woodpecker of lightly wooded country which shuns conifers, and which was originally confined to the Middle East. It seems to have spread into Europe from Istanbul where it was first recorded in 1860, reaching Bulgaria in 1890. By 1931 it had arrived in Romania and by 1939 in Hungary. Since then it has expanded quite rapidly over the former Yugoslavia, Austria and the former Czechoslovakia (Figure 38). After the mid-1950s it seemed to find its climatic tolerance limit on the 70°F (21°C) July isotherm, and its advance slowed down (but see Chapter 12). However, it may also have come up against stiff ecological competition from the very closely related Great Spotted Woodpecker, which is a sparsely distributed species in south-east Europe largely confined to high and dense woodlands which the Syrian Woodpecker avoids; the latter favours instead scattered small broad-leaved woods and clumps of trees in cultivated country. It even occurs in gardens and parks in Bucharest, Budapest and other towns and cities.

The Great Spotted Woodpecker itself, which has a wide distribution throughout the temperate and boreal forest zones of Europe and central Asia to China and Japan, has extended its northern limits in Britain and Fenno-Scandia since about 1870. Prior to this, in the 17th and 18th centuries it was known as an inhabitant of natural and semi-natural wood-

157

land as far north in Britain as Sutherland, and still was so in the early 19th century; but it then underwent a rapid decline and southward contraction of range, so that by 1851, or perhaps even earlier, it became virtually extinct in Scotland and most of England north of Cheshire and Yorkshire. Extensive tree felling, competition for nesting holes with the increasing population of Starlings, and predation by red squirrels were all blamed for the Great Spotted Woodpecker's retreat, but by 1870 it began to recover and push north again. It quickly regained lost territory in northern England and by 1887 was breeding again across the Scottish border in Berwickshire. It had reached the Firth of Forth by 1900 and Argyllshire by 1931, and subsequently spread north to Sutherland to recover all ground lost in the last century. It continued to gain ground in Scotland up to about 1960. In England and Wales it also greatly increased after 1920. The increase did not slow down until about 1960. Then around 1970 a marked increase began again and did not subside until the 1980s. This has been correlated with the abundance at that period of scolytid bark beetles, an important food for Great Spotted Woodpeckers and vectors of Dutch elm disease, which was then rampant in Britain (Marchant *et al.* 1990). The woodpecker population has since stabilised. During the 1850–1950 climatic amelioration a similar increase and extension of range to that in Britain was reported over most of north-west Europe, including Fenno-Scandia.

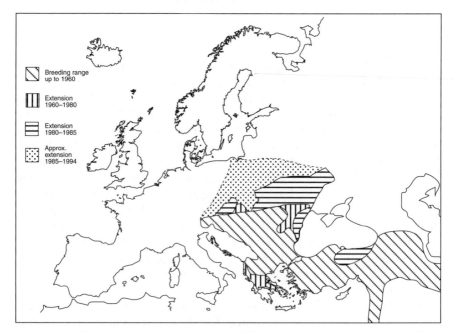

FIGURE 38 *The north-west expansion of the Syrian Woodpecker's breeding range in Europe (adapted from Voous 1960; Harrison 1982; Cramp et al. 1985).*

Although disafforestation may have been a serious influence in the Great Spotted Woodpecker's decline in the early 19th century, if this were the chief cause it is surprising that it did not take effect earlier during the heavy destruction of Scotland's forests in the 17th and 18th centuries. Again, climatic change seems to have been the root cause, the decline having been perhaps due to the excessively wretched run of summers and winters experienced during the earliest decades of the 19th century and the recovery to the progressive amelioration subsequent to 1850 and more particularly after 1890. This view is also held by Marchant *et al.* (1990) who remarked that 'such a climatic warming could be expected to benefit the arboreal invertebrates on which *Dendrocopos* woodpeckers feed'.

Disafforestation—particularly of its favourite deciduous forests and their replacement with conifers—was, however, almost certainly the main cause of the disappearance of the Middle Spotted Woodpecker from many parts of Europe, especially in the west. For instance, it has vanished as a breeding species from the Low Countries, much of north-west Germany, Denmark (last nested in 1959) and Sweden (last successful nest in 1980); elsewhere it became mainly local and scarce. However, recent slight northward and north-westward extensions of range in northern France and increases in southern Germany suggest that climatic change may have played a part as well. Its recent extinction as a breeding species in Sweden may be due to the continuing post-1950 cooling in Scandinavia, which is now out of step with neighbouring parts of the Northern Hemisphere. (See also Chapter 11.)

Although there was little evidence of change during the 1850–1950 amelioration, Lesser Spotted Woodpeckers increased to a marked extent in England and Wales, and perhaps elsewhere in north-west Europe, between 1970 and 1983, following which they declined (Marchant *et al.* 1990). Denmark, for instance, has been colonised since 1964 (Cramp and Simmons 1985). The increase, at least in Britain, has been associated, as with the Great Spotted Woodpecker, with the spread of Dutch elm disease, which provided a more abundant supply of the bark beetles carrying the fungus. The decline of the Lesser Spotted Woodpecker followed the subsidence of the outbreak of Dutch elm disease in the early 1980s. A big decrease has been recently reported from Finland.

The two green woodpeckers of the Old World—the Grey-headed and the Green—have definitely expanded their ranges in Europe during the past hundred years or so. The Green Woodpecker, the only one which occurs in Britain, has much the more restricted distribution, occurring only west of the Caspian Sea, whereas the Grey-headed is found in a broad belt right across temperate Asia to Japan, as well as over much of China and Indo-China. It is thought that the Green Woodpecker evolved from an isolated European population of Grey-headed Woodpeckers which became cut off from the main population by a major advance of the ice sheets at some stage in the Pleistocene glaciations. After the final retreat of the ice sheets of the Last Glaciation, the Grey-headed Woodpecker spread back into Europe

159

from southern Siberia where it encountered competition from the now specifically distinct, but ecologically similar Green Woodpecker. Both species are characteristic of broad-leaved (deciduous) or mixed forests and, where their ranges overlap, the Grey-headed tends to stick to the thicker ones, especially those containing a higher proportion of conifers. It is more likely, therefore, to be seen along forest glades than around the edges and in the marginal habitats which are favoured by the Green. However, in those regions where the Green does not occur, the Grey-headed occupies those habitats too, frequenting parks and large gardens, and feeding on the ground a good deal, just as its competitor does in Europe.

This ecological competition with the Green Woodpecker in Europe seems to have slowed up the Grey-headed's westward advance from Siberia following the Last Glaciation, but nevertheless it has continued to advance westwards when climatic trends allow. For instance, its expansion across central France this century had almost reached Nantes by 1950, and has now reached the Atlantic coast in Brittany and the shores of the English Channel in Normandy, almost within sight of the Channel Islands. Presumably the 1850–1950 climatic amelioration assisted this advance.

The Green Woodpecker tends to suffer very severe losses in bad winters, such as those of 1946–47 and 1962–63, especially near the northern limits of its range. In spite of such occasional set-backs, the generally mild winters of the long-term climatic amelioration enabled it to extend its range considerably northwards in Britain and Scandinavia, presumably because anthills, containing its main prey, were active earlier. In northern Britain Green Woodpeckers seem to have retreated southwards during the rigours of the Little Ice Age, but they were still breeding as far north as southern Northumberland as late as the early years of the 19th century. However, they became increasingly rare so far north and remained thus until about 1925 when they began spreading north again. The English Lake District was colonised soon after 1945 and instances of breeding were reported over the border into Scotland by 1951. Green Woodpeckers now breed as far north in Scotland as Aberdeen, having covered more than 200 km in twenty years (Sharrock 1976). Their spread was more rapid in the more heavily cultivated districts of the east of Scotland than in the wilder areas of the west. They were still spreading northwards in Scotland in the 1980s (Thom 1986). In Denmark there has been a similar story of decline and retreat in the mid-19th century, followed by a recovery since 1930, especially in Jutland. An extension of breeding range also took place in Norway and Sweden from the 1930s.

In North America, the decrease and contraction of range of the Pileated Woodpecker—an ecological counterpart of the Old World Black Woodpecker—and the Red-headed Woodpecker around the turn of the century, followed by recovery and expansion in recent years, may at least in part be connected with the climatic amelioration. The Red-headed, which is susceptible to failures of the acorn and beechmast crops, retreated westward from the eastern seaboard as the climate there became warmer and wetter, but has been returning since the amelioration waned, bringing a return to drier, more 'continental' conditions. Its overall population, however, has decreased to a marked degree over the past seventy years, apparently independently of any climatic effect. Competition for nest sites with the rapidly increasing European Starling has, for example, been blamed for this situation.

Chapter 9

LANDBIRDS: PASSERINES (PART I)

We now move on to discuss the effects of the 1850–1950 amelioration on that large collection of families of perching birds, or Passerines, belonging to the order Passeriformes.

Beginning with the larks, there is some evidence that the northern race of the Shore Lark (or Horned Lark) only began its colonisation of the Scandinavian tundra after 1800 (Voous 1960). David Lack (1954) thought (rightly, in my view) that this westward extension of range, like that of other species shortly to be discussed, 'might perhaps be correlated not with a present change in climate but with the gradual filling of the avifauna of north-western Europe from the east since the last glacial period'. He further pointed out that 'both the birdlife and the plantlife of these northern habitats are richer in species in Siberia than in Europe, a balance which seems [to be] gradually being redressed'. It is only since the Scandinavian population has become established that the Shore Lark has taken to wintering regularly around the shores of the North Sea.

Following an increase between 1950 and 1975 in the numbers wintering along the east coast of Britain, male Shore Larks were discovered in 1972 singing at likely breeding sites in the Scottish Highlands. In 1973 a pair almost certainly reared a brood, while in 1976 a pair may have done so. Finally, in 1977, news was published that at least two males were singing, that a nest with eggs had been found that summer, and a juvenile seen subsequently. So it looked as if the Shore Lark was yet another Scandinavian species endeavouring to extend its breeding range to Scotland. However, none have been reported since 1977. In Finnish Lapland it has declined as a breeding species since the 1930s, when it used to be common, so that only a few pairs still breed today. (See also Chapter 11.)

The explanation of the Shore Lark's very wide distribution in North America compared with the Palaearctic seems to be that, after the Pleistocene glaciations, it colonised that continent from Siberia through the Arctic tundra and, finding a complete absence of competition from other species of larks, it was able to occupy habitats that in Europe are occupied by Skylarks, Crested larks, and the like (Voous 1960).

Professor Voous believes that during the Post-glacial Period the Crested Lark invaded central and western Europe from the Mediterranean region rather than from Asia. An important period of northward expansion probably took place in the very warm early medieval period (Medieval Warm Period), and it was towards the end of this climatic phase that the earliest reference to it in Europe appeared, in a manuscript of 1360 by the monk Claretus of Prague. By the 16th century it had spread over much of central and western Europe, but during the Little Ice Age of the following two centuries, with their frequent severe winters, it receded. However, with the amelioration from the 1850s onwards it spread again to the north and north-west, through Denmark, reaching southern Sweden and Norway by 1900. The expansion continued thereafter, in spite of set-backs caused by the hard winters of the 1940s and the occasional ones subsequently, and by the early 1960s it had become a common bird along the continental coast of the English Channel. An additional factor which favoured the Crested Lark's advance is its attachment to the dry, often rather barren habitats associated with man's activities, such as roads, railways and industrial sites. More recently a contraction of range has occurred in southern Scandinavia and European Russia, with decreases elsewhere in Europe. These will be described in Chapter 11.

During the period of the climatic amelioration, the more familiar (at least in Britain) Skylark greatly increased, and it pushed its breeding range north-

FIGURE 39 *Crested Lark*

163

wards this century to the southern edge of the taiga zone in western Siberia. However, the major influence was almost certainly the northward extension of agriculture in this region. Since 1980 declines have occurred in Scandinavia and in many parts of western and central Europe. The chief reason for these, and similar declines in Britain, has been the change in farming methods (Marchant *et al.* 1990).

It seems that in general, as with the Red-backed Shrike, Wryneck and Roller, the climatic amelioration, with its wetter, more maritime climate, proved unfavourable to the Woodlark on the north-west boundaries of its breeding range in Europe, where it is known to be sensitive to small climatic changes (Harrison 1982). Thus from about 1850, even earlier in Ireland, north-west England and north Wales, it decreased and tended to contract its range south-eastwards throughout north-west Europe (e.g. in northern France, the Low Countries, northern Germany, Denmark, Sweden and Switzerland). However, the run of particularly warm springs and summers between 1920 and 1950 led to a partial recovery in spite of the quite frequent cold winters of the 1940s, including the especially severe one of 1946–47. In England, for example, Woodlarks extended their distribution to breed as far north again as Yorkshire, as well as reappearing in many suitable habitats further south; but they never returned to north-west England or Ireland, apart from isolated breeding records. Judging by his poem "O, Stay, Sweet Warbling Wood-lark" (the words unmistakably reflect that bird's wistful song) and the line "So calls the Woodlark in the grove" in another poem "Here is the Glen", the famous Scottish poet Robert Burns was apparently familiar with this species as far north as south-west Scotland in the 1780s when the British climate was distinctly more continental than it has been since.

Woodlarks appear on the whole to be able to withstand hard, dry winters of the normal continental type, but not wet, rather sunless breeding seasons. The breeding range in Britain and Denmark seems to be mainly confined within those regions with a mean July temperature of 16°C or above (Figure 40). However, since the Woodlark is believed to be a much more insectivorous species than the Skylark (Voous 1960), the influence of temperature may be indirect—exerted through its effects on the Woodlark's favourite insect prey, such as flies, grasshoppers and weevils. Most European grasshoppers and bush-crickets require temperatures in excess of 16°C to become active; they are difficult to detect amongst the vegetation when inactive, so good is their camouflage. Woodlarks may be absent from some apparently suitable habitats because of the absence of some of their specialised insect prey species (e.g. grasshoppers) which are often unexpectedly localised, due sometimes to modern agricultural and forestry practices.

Other factors suggested as responsible for declines in the Woodlark population include loss of habitat through reclamation for agriculture and afforestation, and the invasion of former breeding areas by scrub due to an absence of grazing by rabbits, following their decimation by myxomatosis in

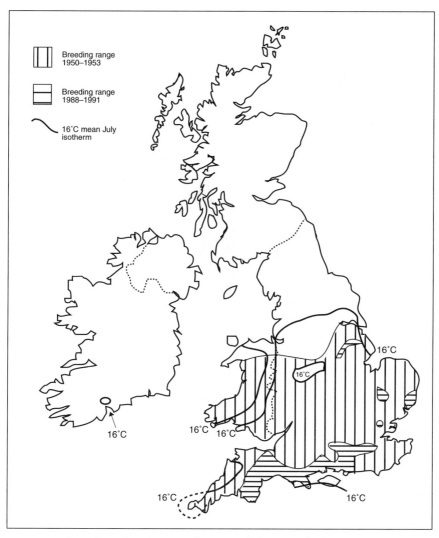

FIGURE 40 *Breeding distribution changes of the Woodlark in Britain and the possible relationship to the 16°C mean July isotherm (adapted from Parslow 1973; Sharrock 1976; Gibbons et al. 1993).*

the mid-1950s. These may well have been important locally, but there is little doubt in my mind that the bird's overall fluctuations in population and changes in range have been due to some basic, long-term cause, most probably climatic. However, H.P. Sitters (1986) considered that climatic change quite possibly played only a minor role, availability of suitable habitat being the key factor.

Although Woodlarks are known to suffer badly in very severe winters, the fact that those of the 1940s failed to check seriously their increase and

165

FIGURE 41 *Red-rumped Swallow*

expansion suggests that they may be only short-term checks; this strengthens the notion that it is spring and summer temperatures and the amount of sunshine, plus availability of the right insect food, which are the most important factors. On the other hand, it must be admitted that Sitters (1986) may be right in this contention that habitat is even more important, since increases have been marked when this suddenly becomes available as a result of forestry activities, even following hard winters, as happened on the Hampshire/Surrey border in 1978 and 1979 (see Clark 1984). Moreover, as Sitters points out, there is growing evidence that many British Woodlarks migrate overseas and may thus escape hard winters.

By 1952 Woodlarks were declining and contracting their range once more, and this trend was apparently accelerated by the severe winter of 1962–63 and the hard one of 1963–64, following which the population was reduced to only about 100 pairs (Parslow 1967). In spite of a succession of mild winters following that, they continued to decrease until 1976, when they began a recovery which will be described in Chapter 11.

Turning now to the swallow family, we find that in spite of marked decreases in the numbers of breeding Swallows (or Barn Swallows) in European towns and villages, due apparently to improved hygiene having reduced the populations of flies (Voous 1960), the species as a whole extended its breeding range northwards and westwards from the early part of this century. It became, for instance, a more regular breeding bird in northern Scandinavia, in the west of Ireland, north-west Scotland, Orkney and Shetland, and also spread north to the Faeroes. It has been attempting to colonise Iceland as well since 1915, when a pair built a nest on a gasworks in Reykjavik, or perhaps even earlier (1911?), but so far has not been very successful. Gudmundsson (1951) suggested that the main reason for the lack of success may be the scarcity of insects. There seems little doubt that the Swallow benefited from the climatic amelioration in northern latitudes. In Norway, for instance, it nested successfully from 1968 as far north as the Hestefoss in east Finnmark. For details of a subsequent widespread decline in northern Europe, see Chapter 12.

The Red-rumped Swallow has also certainly benefited from the amelioration. Since 1950 it has been advancing north-westwards through the Balkans towards Italy, and northwards and north-eastwards through Portugal and Spain to southern France. Before 1950 it was confined to the Rhodope Mountains just north of Bulgaria's frontier with Greece, but by 1968 had spread throughout southern Bulgaria, and is still advancing. It is now colonising central Italy. In south-west Europe it nested for the first time on the French side of the eastern Pyrénées in 1963 and is now well established there. The *Atlas des Oiseaux Nicheurs de France* (Yeatman 1976) even indicates a probable breeding record in the Rhône valley, well north of Avignon. In view of this expansion, it is by no means surprising to learn that Red-rumped Swallows are appearing more often as vagrants further north in Europe, including the British Isles. I myself once saw one, in April 1968 on Romney Marsh in Kent. Up to 1957 only seven had been reliably reported in the British Isles, but between then and 1976 a further thirty-nine were seen.

Moving on to the pipits and their near relatives the wagtails, we find that the Tree Pipit also expanded its breeding range northwards in Fenno-Scandia, notably in Norway, and also in Scotland after 1870. It is now one of the commonest birds of the mature birchwoods of the northern Scottish county of Sutherland, where it was proved to nest for the first time in 1868. The expansion continued through the 1960s into the adjacent county of Caithness, where it is now breeding. There seems little reason to doubt that this spread was prompted by the climatic amelioration, but it is more difficult to explain the slight declines which have occurred at times in southern England and elsewhere in western Europe since 1930. Reclamation for agriculture and building purposes of its habitats does not provide a fully satisfactory explanation. Since about 1980 there has been an overall and noticeable decline in south-east England, hitting an all-time low in the BTO's Common Birds Census sample in 1992. This recent decline is also difficult to explain.

Unlike the Tree Pipit, the warmer, wetter Atlantic-type conditions affecting north-west Europe during the climatic amelioration not surprisingly proved unfavourable to the Tawny Pipit, a bird of rather warm, arid, open countryside. Its centre of distribution in Europe is around the Mediterranean and the interior of the continent. During this century it recoiled towards those regions, and consequently became rarer as a breeding species in the north-west coastal areas, such as northern France, the Low Countries, Germany, Denmark and southern Sweden. However, it has shown some signs of a recovery since 1958, for example in northern France, Denmark and newly afforested areas in the south of Sweden. This may be correlated with the gradual swing back to a drier continental climate as a result of the climatic deterioration which set in after 1950.

There is some evidence that the Red-throated Pipit, which replaces the Meadow Pipit on the tundra within the Arctic Circle, decreased and retreated to the north along the southern limits of its range during the warmer years of this century. It now appears to be recovering lost ground, for example in northern Norway, where it was nesting as far south as 62°20′N by 1975.

Before 1850 the Grey Wagtail was absent from much of central Europe, and it is still curiously absent from a huge area of eastern Europe. Beyond the Urals it is found right across temperate Asia to Manchuria, Kamchatka and Japan. It is thought that it formerly occurred throughout temperate Europe but died out as a result of a period of prolonged and severe winters—perhaps those of the 17th and 18th centuries during the coldest part of the Little Ice Age. Severe winters certainly do cause heavy mortality among Grey Wagtails; that of 1962–63 apparently accounted for about 95% of those in Britain, though recovery was remarkably rapid. The hard winters of 1978–79, 1981–82 and 1984–85 also took a heavy toll. Thus the northward spread of the species in western and central Europe since 1850 probably represented a reoccupation of former breeding territory, encouraged by the improving climate of the mid-19th century and subsequently.

Grey Wagtails were first reported breeding in the Netherlands in 1915, in Sweden in 1916, in the south of Norway in 1919, and in Denmark (Jutland) in 1923. In the British Isles, where they seem to have been well established before 1850 (there are records from 1678), Grey Wagtails increased and spread into the lowlands in the south and east of England from 1900 and especially during the 1950s, and also advanced north-westwards to the extreme north of Scotland. Although the climate was already deteriorating in the north after 1950, the long run of mild winters from 1963–64 to 1976–77 continued to aid their spread in northern Europe, notably in Fenno-Scandia and the Baltic states. In Britain the mild winters of the late 1960s and early 1970s enabled the Grey Wagtail to attain an unusually high level of population (Marchant et al. 1990).

Professor Voous (1960) believes that the White Wagtail, which has an extraordinarily wide range in the Palaearctic, even breeding in Iceland and on the east Greenland coast, may have colonised these last mentioned countries, plus Jan Mayen island and the Faeroes, from Norway within quite recent times, due to the improvement in the North Atlantic climate between 1850 and 1950. The first Greenland records of White Wagtails date from

FIGURE 42 *Fan-tailed Warbler*

1885. The British race of the White Wagtail—the Pied Wagtail—has markedly increased in Ireland since 1900, occupying previously vacant districts in the west. As with the Grey Wagtail, White/Pied Wagtail populations suffer from heavy mortality in severe winters.

Although as a species the Yellow Wagtail has a vast breeding range right across the Palaearctic Region and even into western Alaska, certain well marked southern races have been extending their breeding ranges northwards in recent years, presumably as a result of the climatic amelioration. The Ashy-headed race *cinereocapilla* has, for example, crossed the Alps from Italy and has been breeding in Bavaria since 1968. Similarly, the Black-headed race *feldegg* has pushed north from the Balkans to breed for the first time in Hungary in 1970, and has apparently also begun nesting in Bavaria lately.

The Dunnock (or Hedge Sparrow as some people still call it) extended its breeding range northwards by a considerable amount in Scandinavia during the climatic amelioration, and also in the north of Scotland. For instance, it colonised Orkney and the Outer Hebrides in the latter part of last century, and subsequently increased and continued to spread. The discovery of a nest and eggs on Fair Isle in 1974 suggested that the species was attempting to colonise Shetland.

As might be expected with such a group of predominantly insectivorous birds, the warblers have been especially sensitive to climatic changes. Of

169

forty-one species in Europe, at least twenty-five are known to have responded to the climatic amelioration by expanding their breeding ranges northwards or westwards. The two outstanding cases are Cetti's Warbler and the Fan-tailed Warbler (see also Chapters 11 and 12). At the beginning of the present century both species were almost entirely confined in Europe to the shores of the Mediterranean; only in the southern Balkans, Italy and the Iberian peninsula did they occur very far inland. By 1982 Cetti's Warbler had spread so far north in western Europe that it was breeding in southern England, the Low Countries and north-west Germany, and continuing to increase in spite of the severe winter of 1978–79, while Fan-tailed Warblers were nesting all along the French coast of the English Channel as far north as Dunkirk, and were poised for an invasion of southern England.

Cetti's Warbler, which used to be regarded as a sedentary species, first began to spread north soon after 1920 during a particularly warm phase of the amelioration which continued through the 1930s. This was noticed first of all in the west of France; Cetti's Warblers began breeding in Anjou in 1924, they had reached the Loire basin in the south of Brittany by 1927 and the Seine basin by 1932. In south-east France a similar northward expansion took place up the Rhône valley. Their advance slowed down between 1940

FIGURE 43 *Cetti's Warbler*

and 1952, a period during which there were several hard winters, but it resumed in earnest soon afterwards. By 1960 they had colonised the north coast of Brittany and reached the Channel Islands. Four years later they were breeding in Picardy and south Belgium, and by 1970 were widespread and locally numerous there. The first Cetti's Warbler in the Netherlands was seen near the Belgian frontier in 1968; during the 1970s the species was resident in many localities and probably breeding. By 1975 it was breeding as far north in Germany as Lower Saxony (see Figure 44).

The first Cetti's Warbler recorded in England (excluding three discredited pre-1916 records) occurred at Titchfield Haven on the Hampshire coast in the early spring of 1961, and was followed by another at Eastbourne, Sussex, in autumn 1962. Apart from three more in southern England, the invasion of Britain did not really begin until 1971. Since then there has been a rapidly growing flood of records. Breeding was first proved in England in 1972 (east Kent). By 1975 eight pairs were proved to have nested—in Devon, Kent and Norfolk—and in that year sixty-one singing males were located in east Kent alone, where Cetti's Warbler was considered to be the third commonest warbler in parts of the Stour Valley (Harvey 1977). The following year Kent alone held almost a hundred singing males, and eighty or more pairs bred or were suspected of doing so.

Elsewhere in Europe, Cetti's Warbler had also been on the move. For instance, it had increased in northern Italy and spread eastwards along the River Po and northwards up the River Ticino west of Milan, and in 1977 was engaged in colonising Switzerland. A full account of Cetti's Warbler's range expansion up through north-west Europe has been published by Bonham and Robertson (1975).

The Fan-tailed Warbler is very much a southern species, and in Europe is on the northern edge of its range, which extends across most of Africa, India and south-east Asia. Its period of range expansion began rather later than that of Cetti's Warbler, commencing during the particularly warm years of the 1930s, when it advanced up the Atlantic coast of France as far north as the Vendée. Susceptible as it is, however, to cold winters, it received a severe set-back during the run of cold winters between 1939 and 1947, being effectively eliminated from newly colonised territory. It even became scarce in some of its strongholds along the Mediterranean coast, such as the Camargue, and was virtually wiped out even there by the terrible winter of 1962–63. However, with a subsequent return to generally milder winters Fan-tailed Warblers rapidly made up lost ground. They returned to the Vendée during the 1960s (some apparently as early as 1959), and spread north into Brittany after 1970. By 1977 they had occupied many areas of the north coast of France, and were breeding in coastal marshes and river valleys over much of southern and western France. They also continued their advance into the Low Countries, reaching the Netherlands in 1972, where the first proof of breeding was obtained in 1976, and Belgium by 1975. During 1976 and 1977 at least two birds arrived in southern England.

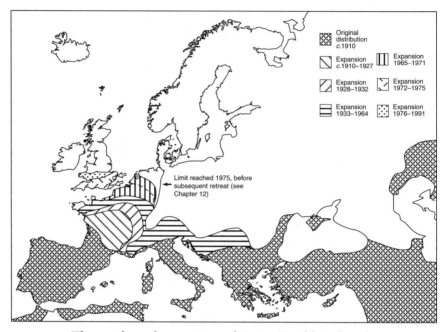

Original distribution c.1910

Expansion c.1910–1927

Expansion 1928–1932

Expansion 1933–1964

Expansion 1965–1971

Expansion 1972–1975

Expansion 1976–1991

← Limit reached 1975, before subsequent retreat (see Chapter 12)

FIGURE 44 *The northward expansion of Cetti's Warbler's breeding range in Europe (adapted from Bonham and Robertson 1975; Harrison 1982; Cramp and Simmons 1992).*

In 1977, twenty-five singing males were located in the Netherlands; meanwhile Fan-tailed Warblers advancing from northern Italy had penetrated Austria (by 1971), Switzerland (1972), southern Germany (1975) and north-west (former) Yugoslavia (1970), where forty-five pairs were reported in 1974 (Simms 1985). To the south, Malta and Crete had both been colonised from 1967.

Runs of mild winters have been considered to be the cause of the extensions of range of both Cetti's and the Fan-tailed Warbler by various authors (e.g. Bonham and Robertson 1975, Ferguson-Lees and Sharrock 1977). Others have suggested genetic changes to account for it, but I see little reason to seek an explanation beyond the influence of the climatic amelioration this century in reducing winter mortality of these largely resident birds (as suggested by the authors already cited), and in improving the availability of their insect prey throughout the year, particularly at critical times. Bonham and Robertson (1975) suggested in the case of Cetti's Warbler that the manner of its northward spread results from the dispersal in autumn of many juveniles, and perhaps some adults, in a basically northern direction to suitable habitats in which they succeed in overwintering. Some of these, at least, remain in the new haunts the following spring, with the result that new populations arise and, in time, breed. Juveniles from these new breeding colonies repeat the northward dispersal tendency, with the result that the range expansion continues in a succession of waves. The Fan-tailed Warbler, and indeed some other expanding but normally sedentary species, appear to spread in the same way. As most people expected, the advances of both species were halted, and at least temporarily reversed, by the series of hard winters of 1977–78, 1978–79, 1981–82, 1984–85 and 1986–87.

Since the particularly warm phase of the 1930s, Savi's Warbler (one of the *Locustella* warblers), a bird of fens and swamps, has also been spreading north-westwards in Europe. Unlike Cetti's, but like most other warblers, it is only a summer resident in its breeding haunts. In the course of this extension of range, Savi's has recovered a good deal of the territory it lost during the latter part of the 19th century, which was blamed upon the large-scale drainage of swamps and marshes during this period. For instance, when it was first described in Britain in 1824, Savi's Warbler was found to be a locally plentiful inhabitant of the then vast fens and marshes of East Anglia and the Thames estuary. As also happened on the European mainland, the subsequent drainage of a very substantial part of these wetlands destroyed many of its former haunts and naturally led to a marked decrease. Nevertheless, I think it unlikely that this alone was the cause of the south-eastward contraction of range which tended to occur in north-west Europe in the 19th century, and which led to its complete disappearance as a breeding species from England after 1856. Sharrock (1976), too, remarked that it is 'tempting to blame the egg-collectors and trophy-hunters of the time, who were always keen to add new species to their hoards, but it seems possible that Savi's Warbler was declining then over much of its peripheral European range'.

As in the case of the Bittern, Black-tailed Godwit and other fenland species which vanished as breeding birds from England in the mid-19th century, Savi's Warbler was probably also already declining due to the generally unfavourable climatic conditions then prevailing, particularly the run of disastrous summers from 1809 to 1820, which must surely have adversely affected breeding success in areas on the edge of a species' range. Moreover, in spite of extensive drainage—which in any case was largely carried out when Savi's was already almost extinct—large areas of suitable habitat still remained in which it could have survived were it not already declining for other reasons. After all, as Jim Flegg (1970) has pointed out, other similar species with equally limited preferences for wet reed- and sedge-beds, such as the Reed Warbler, Great Reed Warbler and Marsh Warbler, did not decline seriously from this cause, except (I might add) here and there on the periphery of their range in western Europe.

Furthermore, the recolonisation by Savi's Warbler of former areas in north-west Europe, including England, has occurred in spite of the fact that there is far less suitable habitat remaining there (except locally in the Zuider Zee area of the Netherlands) than at the time of its extinction. It is true that its initial recolonisation in south-east England occurred in localities providing ideal habitat which have only recently been artificially created through man's activities. In particular, one may mention its stronghold in the marshes of the Stour Valley in east Kent, which was formed subsequent to 1960 on low-lying land flooded through mining subsidence, and the RSPB's marshland reserve at Minsmere, Suffolk, which did not exist before the area was flooded for military reasons in 1940. However, larger areas remained suitable for Savi's Warbler in the Norfolk Broads and elsewhere between 1850 and the time of its return a century later.

Yeatman (1971) has shown that on the continent of Europe the recent extension of the breeding distribution of Savi's Warbler has occurred despite widespread drainage and reclamation of suitable wetland habitat. This expansion seems to have begun during the 1930s from the east and south. In the east, Savi's advanced from Poland across northern Germany. In the 1950s it spread into Denmark, although breeding was apparently not proved until 1964; it now seems to be quite widespread in that country. Sweden was first reached in 1944 and it is now breeding in the south, while in France it has been spreading north since 1940. It also reappeared in Switzerland after 1942.

The recolonisation of south-east England seems to have commenced soon after 1950. In 1954 a male Savi's Warbler appeared in one of the species' former breeding haunts—Wicken Fen, Cambridgeshire—and sang through most of the summer. The first breeding record, however, came from Stodmarsh in the Stour Valley in 1960; there is evidence that a few pairs had bred annually in this area since 1955, and that colonisation may have begun soon after 1950 or even earlier. Apparently, for some years, singing males were understandably passed over as Grasshopper Warblers by unsuspecting ornithologists; the two songs are much alike and difficult to distinguish, even by experienced birdwatchers. Savi's Warblers have since nested in scattered localities elsewhere, as far west as Devon and as far north as Lancashire and Yorkshire.

To sum up, though the reclamation of wetland has undoubtedly had an adverse affect locally in Europe, the main cause of the decline of Savi's Warbler in the first half of the 19th century, and its subsequent recovery since 1930, seems to have been climatic change—most probably the fluctuating incidence of good or bad summers.

The recent marked northward extension of range in northern Europe, including Britain, of the Savi's Warbler's close relative, the Grasshopper Warbler, has also been correlated with an increase in average summer temperatures since 1930. Thus it has been advancing from the south-east in Finland, Sweden and Norway. Between becoming established in southern Sweden in the mid-1950s and 1968, its breeding population increased sevenfold, according to Thiede (1975b). Since reaching Ångermanland in central Sweden in 1964, it has continued to spread north, while in Norway it started to breed in the south in 1967. In the British Isles a marked westward extension of range occurred in Ireland and possibly also in south-west England, while in the north of Scotland it apparently extended its breeding distribution northward. During the 1960s there was a marked increase in Britain, but this was followed by a big population crash in the early 1970s, since when the Grasshopper Warbler has declined almost without interruption. In France, where it is a well distributed breeding species, except in the south, an increase and extension of range was reported from 1936 (Yeatman 1976).

Parallel with this increase and range extension in north-west Europe, Grasshopper Warblers have more frequently been occupying drier habitats than those with which they used to be associated. For example, they have been taking advantage of the tangled cover in young forestry plantations which have been springing up as a result of the big afforestation programmes in the British Isles and elsewhere in Europe. However, the new plantations have probably merely compensated for the loss of marshier habitats, rather than in themselves stimulating the increase of the Grasshopper Warbler. In other words, under the pressure of population growth, initiated by climatic improvement, Grasshopper Warblers may have been more and more obliged to occupy young forestry plantations and other drier habitats in the absence of wetter ones.

Another of the *Locustella* warblers has also expanded its population in Europe. This is the River Warbler, which as far as we know has never bred in Britain in historic times. Its main range lies in eastern Europe and European Russia as far north as Estonia and St. Petersburg. However, since about 1950, perhaps even earlier, River Warblers have been spreading westwards, like several other eastern European birds, across southern Finland, Sweden, Denmark, Germany, Austria and the former Czechoslovakia. Breeding in Sweden was first confirmed in 1987. Though still rare, it is now nesting as far west in Germany as Schleswig-Holstein, and has also moved further west in Bavaria since 1971. By the 1980s singing males had penetrated as far west as southern Norway and the Netherlands, and even to eastern England, where males have been discovered singing in spring on several occasions. Again, higher average summer temperatures seem most likely to have been responsible.

Most of the European *Acrocephalus* warblers have likewise expanded their ranges to minor or major extents, presumably in response to the climatic amelioration. The Reed Warbler, for example, has been spreading northward in southern Fenno-Scandia since the end of the First World War, southern Finland having been colonised from 1922, and southern Norway from 1947. It has also extended its breeding range in Britain since about 1950: slightly northwards in northern England and markedly westwards in south-west England, where pairs have been breeding as far west as the Isles of Scilly since 1962. Apart from an isolated pair in Co. Down in 1935, they did not begin to nest in Ireland until 1981, but now do so regularly. The isolated pairs which have bred or attempted to breed in northern Scotland and Shetland since 1970 are thought, like the Red-backed Shrikes and

Wrynecks, for instance, to have come from the expanding Scandinavian populations.

The very similar Marsh Warbler has also extended its range considerably to the north in Fenno-Scandia since 1930, in which year it bred for the first time in Sweden. However, in Britain, where it is also on the periphery of its range and confined to the south of England, it has decreased to a marked extent since about 1950. Loss of habitat has been blamed for this decline, but it seems likely that the chief cause is climatic—possibly the cooler and wetter summers which were experienced in the 1950s and 1960s. It still remains to be seen if the very recent tendency towards drier, more continental-type summers will arrest this decline (see Chapter 12). In general, the Marsh Warbler's European breeding population is contained within the 16.5°C mean July isotherm.

The Marsh Warbler's sibling species, Blyth's Reed Warbler, is very much its eastern counterpart, their ranges only overlapping in parts of European Russia. From 1934 Blyth's Reed Warbler advanced westwards into southern Finland, but until the 1970s showed little sign of spreading any further. It is one of a number of passerines which have been spreading westwards this century from central Asia.

The Great Reed Warbler, on the other hand, which has a wide distribution across central Europe and central Asia (reaching Japan), has been advancing slowly northwards and north-westwards this century. Indeed it seems that it was doing so as early as the middle of last century, for it first appeared as a breeding species in southern Denmark in 1861, having apparently spread north from Germany. It is nowadays widely distributed over much of Denmark, following a main influx during the particularly warm period of the 1930s, but is still very local. It invaded Sweden in 1917, but in 1960 was still restricted as a breeding bird to the extreme south, although increasing and spreading. Since 1930 it has been regularly observed in southern Finland, where it now breeds along the shores of the Gulf of Finland. The Finnish ornithologists O. Leivo and Olavi Kalela have demonstrated quite clearly that the range expansion of this and other species is related to climatic improvement, particularly with regard to the marked increase since 1920 in the mean spring and summer temperatures of northern Europe. Great Reed Warblers have been appearing in southern England more frequently since about 1950, where singing males have quite often held territories in possible breeding sites in recent years, such as in extensive reed beds in Devon, Dorset, Kent, Surrey, Hampshire, Norfolk and Lincolnshire. It seems only a matter of time before breeding is proved in England; it may indeed have occurred already.

In northern Britain, the Sedge Warbler has colonised Orkney since the mid-19th century and the Outer Hebrides since 1930, presumably due to the climatic amelioration; I have little data from elsewhere in Europe, where it breeds as far north as the extreme north of Fenno-Scandia, apart from an increase and northward expansion in southern Finland up to the mid 1970s, and reports of widespread decreases since then in Finland, Sweden, Estonia, the Netherlands, Germany and central Europe generally.

Apart from some minor improvements, the British Sedge Warbler population also went into a serious decline after 1968 until an all-time low was reached in 1985, since when there has been a sustained recovery, at least up to the time of writing. As pointed out by Marchant *et al.* (1990), 'there is no doubt that the cause of the Sedge Warbler's post-1968 decline has been the persistent rainfall deficit in the species' wintering range in the Sahel and adjacent savanna zones of Africa. The regional improvements in West African rainfall in the middle to late 1970s, the failure of the rains in 1983, and the recent easing of drought conditions in the Sahel, are all evident in the Sedge Warbler indices'. These indices are the British Trust for Ornithology's Common Birds Census and the Waterways Bird Survey. The cause of the decline elsewhere in Europe is undoubtedly the same (see Berthold *et al.* 1986).

Of the six species of warblers belonging to the genus *Hippolais*, all but one of which breed in Europe, four have undergone extensions of range during the present century. One species, the Icterine Warbler, decreased along the north-western edge of its range and retreated eastwards from about 1930. For instance, it disappeared from the Geneva district at about the time as that region of Switzerland was colonised by its sibling and ecological competitor, the Melodious Warbler. Its retreat in those parts of Europe where its breeding range borders that of the Melodious Warbler is therefore readily explained in terms of ecological competition from its expanding sibling, but the reason is not so obvious in northern Europe where the Melodious Warbler is absent (Durango 1948). It may be that, apart from ecological pressure from a related species, the warmer but wetter summers produced since 1930 by the climatic amelioration have not suited it. There has, however, been some evidence of a recovery by Icterine Warblers in northern Europe recently, possibly due to the more frequent hot, dry summers since about 1975.

A north-eastward expansion of range by the Melodious Warbler, which completely replaces the Icterine Warbler in south-west Europe and extreme north-west Africa, has been taking place since the mid-1930s, presumably in response to the climatic amelioration. Thus, it has colonised not only southern Switzerland, but also the extreme western part of the former Yugoslavia, and has continued to gain ground from the late 1970s onwards.

The Olivaceous Warbler prefers more arid climatic conditions than the Icterine and Melodious Warblers, and is largely confined to the neighbourhood of the Mediterranean and the Middle East south of the Black, Caspian and Aral Seas. Apart from southern and eastern Spain, its main stronghold in Europe lies in the Balkans. In this region it has been spreading steadily north-westwards since about 1935, probably because of the increasing aridity in the Middle East which, as we have already seen, has affected a number of species, especially marsh and water birds. Olivaceous Warblers have, for example, extended their breeding range through Hungary to eastern Austria. Much the same explanation may also be offered in the case of the Olive-tree Warbler, which has spread north-westwards through Bulgaria and the former Yugoslavia during the same period. The Olive-tree Warbler's world distribution, incidentally, is entirely confined to the eastern end of the Mediterranean.

Turning to the typical warblers of the genus *Sylvia*, the Whitethroat extended its summer breeding range a little to the north in northern Europe during the recent climatic amelioration and presumably because of it. For example, from 1880 it spread farther north in the extreme north of Scotland, and also colonised the Outer Hebrides. However, in 1969 a big population crash occurred, resulting from the drought in its winter quarters, the Sahel (which also caused the big losses of Sedge Warblers). (See Chapters 11 and 12.) The Lesser Whitethroat, which probably spread into north-west Europe from the south-east following the Last Glaciation, made slight gains on the edge of its range between about 1890 and 1950. In Britain, for instance, it extended its breeding range westwards and northwards slightly during this period, and continued it subsequently through Northumberland and Cumbria, reaching south-east Scotland by the mid-1970s.

Another *Sylvia* warbler with an eastern-orientated breeding range in Europe also spread during the amelioration. This was the Barred Warbler, which advanced steadily west and north-west between 1930 and about 1960, for instance recolonising Denmark where it used to be quite common before 1900. Since about 1960 it has retreated again, so that by 1975 only some fifteen to thirty breeding pairs were still left in Denmark (Dybbro 1976). However, in southern Finland and the southern tip of Scandinavia the range extension continued until at least the early 1970s, with pairs breeding as far north in Sweden as Södermanland (60°N) and, from 1972, in south-east Norway (Christie 1975).

The Barred Warbler belongs to that group of birds, which includes the Red-backed Shrike, Thrush-Nightingale and Wryneck, which fare best in a dry, continental-type climate with relatively high summer temperatures. Thus, although the generally wetter summers of the climatic amelioration as a whole have not suited it, the run of mainly dry, sunny ones between 1920 and 1950 apparently did so. Like the Red-backed Shrike, with which, incidentally, it apparently has a symbiotic-type association,[1] it preys a good deal upon adult insects, such as ants, grasshoppers and small chafers, which are most active in warm sunshine, and it is probably through its food, therefore, that it is influenced by the climate. In whatever precise way the climatic factor operates on the Barred Warbler, however, it is clear from its breeding distribution that it cannot tolerate the maritime climate of Europe's western seaboard.

The British Isles are on the periphery of the Blackcap's breeding distribution in north-west Europe and, although common in England and Wales, it is much more local in Scotland and Ireland. It has, however, been increasing and spreading westwards and northwards in Ireland since the latter part of the 19th century in spite of erratic fluctuations in population; today, although still absent from large parts of western Ireland, it appears to be attempting to penetrate those areas too. Similar fluctuations occur every now and then in Scotland, especially along the northern limits of its distribution, but it seems to have attempted to extend its range further north there, too, as a result of the climatic amelioration, even breeding as far north as Orkney and Shetland in the late 1940s. In 1985, a pair bred on the Faeroe Islands for the first time. The marked increase since the mid-1950s over much of Europe, including Britain, is discussed in Chapter 12. Another effect of the amelioration, with its long runs of mild winters, has been the increased tendency of Blackcaps to spend the winter in southern England and Ireland, especially since about 1950. Some birds even winter in northern England and a few as far north as Caithness in Scotland (Stafford 1956). Though it is possible that some of these are British breeding birds, most seem to come from further north and east in Europe; for example, one wintering bird in Ireland was found to have been ringed in Austria in August.

The Orphean Warbler, a large species of the Mediterranean region and the Middle East, has not shared the expanding fortunes of most of the other warblers during the amelioration; instead it has contracted southwards during this century. In the 19th century it apparently nested as far north as southern Germany, Luxembourg and north-central France, but now its northern limit has retreated to south of a line running approximately along the Massif Central in France, northern Italy, southern (former) Yugoslavia and the Bulgarian frontier with Greece and Turkey. The cause of this retreat is not properly understood, but it seems reasonable to assume that the wetter summers in the north-west European mainland resulting from the amelioration were in some way unfavourable to it. It is worth noting that its present-day breeding distribution is almost entirely confined within the 20°C mean July isotherm; and chiefly within the 22°C July isotherm. Perhaps, being more arboreal, it depends more upon active adult insects caught in flight than do other *Sylvia* warblers.

The Dartford Warbler is resident throughout its limited breeding range in south-west Europe and north-west Africa, which in fact represents its entire world distribution. Unlike most warblers of the Mediterranean region, its breeding range extends to southern England, where it was first discovered near Dartford, Kent, in 1773—hence its English vernacular name. At that time it was much more widely distributed in England than at present, breeding on the then extensive gorse and heather heathlands of southern England, from Kent in the east to Cornwall in the west, as far north as East Anglia and the south Midlands, and even locally in Shropshire and Staffordshire. Unfortunately, not only has the Dartford Warbler suffered from the steady reclamation and fragmentation of its specialised habitats, but its vulnerability to heavy mortality in the periodic severe winters has impaired its ability to recolonise those more northerly and westerly fragments of its habitat which remain. This vulnerability to very cold winters has serious effects not only in those areas on the edge of its range, but in its strongholds as well, such as south-west France and Spain.

In spite of what are at times extremely heavy losses, recovery is usually remarkably rapid. For example, following two hard winters in succession—1961–62 and the especially severe one of 1962–63—the British population of Dartford Warblers was reduced from 460 pairs in 1961 to only eleven known pairs in 1963, yet by 1966 the population had doubled and, as a result of a long run of mild winters, had risen to about 560 pairs by 1974.

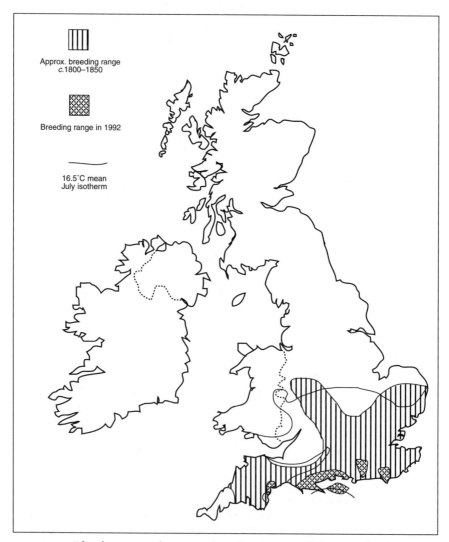

FIGURE 45 *The former and present breeding range of the Dartford Warbler in Britain and its possible relationship to the 16.5°C mean July isotherm (1992 distribution adapted from Gibbons et al. 1993).*

180

From the main strongholds in Dorset and south-west Hampshire, recolonisation of traditional sites in neighbouring counties was gradually effected. For more recent developments the reader is referred to Chapter 11.

It is quite possible, as with the Orphean Warbler, that the tendency towards wetter summers during the 1850–1950 climatic amelioration was unfavourable to the Dartford Warbler. It is perhaps significant that the Suffolk population died out early this century following some thirty years of increasing summer wetness, and at a time when the Dartford Warbler's habitat had not been too seriously diminished by reclamation or afforestation. Moreover, the distribution of the Dartford Warbler in the early 19th century seems to have been confined within the 16.5°C mean July temperature isotherm (Figure 45), and this may still be the limit today. At that period this isotherm encompassed central England, as it does nowadays. In this connection, Professor Lamb (1966) has shown that the average temperatures for the peak summer months of July and August in central England were only 15.5–15.6°C at the beginning of this century. It is little wonder therefore that the Dartford Warbler contracted its range at about that time.

Moving on to a consideration of the *Phylloscopus* or leaf warblers, the Chiffchaff's present-day distribution lies mainly between the July isotherms of 10°C in the north and 25.5°C in the south. Although occurring in a more or less continuous broad band right across Europe and Asia, except the extreme north and east, there is a curious gap in southern Scandinavia which also seems to extend to Scotland. Voous (1960) suggested that this gap probably resulted from the way in which northern Europe was repopulated by this species following the Last Glaciation. One group, he believes, invaded from the south, while another approached from the east via Finland, and they have not yet closed the gap and joined up. This process is believed to be still in progress, as evidenced by the slow colonisation of Denmark and the south-west tip of Sweden from northern Germany since 1860. I would suggest that the process was held up or even reversed during the Little Ice Age, and only got going again when the climate ameliorated once more after 1850.

Significantly, Chiffchaffs also greatly increased and spread north and west in Ireland from 1850, occupying most of the country by 1900. More recently—since about 1950—they spread northwards in Scotland to Caithness and Sutherland (Simms 1985), colonising the Inner Hebrides from 1959, and began the process of colonising the Outer Hebrides as well. Afforestation may have helped the spread of the Chiffchaff in these northern areas, but the improved climate was probably the underlying cause. Chiffchaffs continued to increase in Britain through the 1960s, as shown by the BTO's Common Birds Census, peaking in 1968 and 1970. Thereafter, they decreased sharply, reaching a low point in the hot summer of 1976. A partial recovery then ensued until the early 1980s, when their population again dropped steeply. Since 1984 they have recovered rapidly and are continuing to increase. As with other warblers which make the Sahara crossing to their winter quarters, the Chiffchaff's population fluctuations are broadly in line with the drought-affected rainfall patterns in the Sahel region of West Africa. However, they have not been quite so serious, because many Chiffchaffs winter north of the Sahara. The picture elsewhere in Europe has been much the same as in Britain (Marchant *et al.* 1990). Like the Blackcap, the Chiffchaff has tended to winter on a larger scale than before in southern England and Ireland because of the generally milder winters resulting from the amelioration, and now, perhaps, because of the growing greenhouse effect in central Europe.

The Willow Warbler, whose distribution extends further east in the Palaearctic Region than the Chiffchaff, has also extended its breeding range slightly westwards in Ireland and in northern Britain since 1850, but how much this is due to afforestation or to climatic improvement it is difficult to judge. Perhaps it is a bit of both. Since 1950 the British population has been remarkably stable, but fluctuations have been more marked on the European mainland, especially in Fenno-Scandia.

Another leaf warbler which extended its breeding range considerably to the north in Scotland from 1850 is the Wood Warbler. Its predilection for sessile oak woodlands, which are most prevalent in the west and north of the British Isles, is well known, but the correlation between its distribution, as shown in the BTO Atlas (Sharrock 1976), and those areas of the British Isles with an average annual rainfall of more than 625 mm (25 inches), is most striking. A comparison of the BTO Atlas distribution map of the Wood Warbler with the annual rainfall map in Manley (1952) suggests that it was the generally wetter summers of the 1850–1950 amelioration that lay behind the Wood Warbler's northward spread in Scotland. The correlation is impressive for Norway also. In Scotland, this warbler reached the north of Sutherland by the 1940s, and it now breeds in suitable localities throughout the country. However, since about 1960, by which date the climatic amelioration had drawn to a close, this expansion of range in Scotland has ceased (Thom 1986; Marchant *et al.* 1990).

In Ireland the Wood Warbler is almost entirely absent, though sporadic breeding occurred in 1932, 1938 and 1968. However, in 1968 it was discovered that the species had established a foothold in oakwoods in Co. Wicklow, where the average annual rainfall amounts to more than 750 mm (30 inches). In England, Wood Warblers, much subject to population fluctuations, declined and deserted many areas after 1940 which they formerly occupied, especially in the east. Most of these areas lie within the region which experiences less than 625 mm annual rainfall, and it may perhaps be that this region has become generally even drier since 1940. There seem, however, to have been subsequent signs of slight recoveries in the Wood Warbler population here at times, and this may be due to temporary periods of higher rainfall. In passing, it is interesting to note that the Wood Warbler is more specialised in its choice of habitat in the maritime areas of north-west Europe than it is elsewhere in its continental range.

Bonelli's Warbler is another species whose range is almost entirely confined to Europe, though it also breeds in north-west Africa and at the eastern end of the Mediterranean. Its centre of distribution is more south-westerly than the other leaf warblers, but since the 1930s it has been pushing north (e.g. France and Germany), and breeding was confirmed in the Netherlands in 1976. It has occurred increasingly frequently in England since the first one was reported in 1948, and some spring arrivals have even stayed for a while and sung in likely breeding sites. This has led to confident predictions that Bonelli's Warbler would soon attempt to colonise southern England as well as the Netherlands. However, so far this has not materialised. It may have been frustrated by the post-1950 climatic deterioration in the Arctic zone, although the present growing influence of the greenhouse effect may revive such hopes.

In addition to the leaf warblers already discussed, two others—the Greenish Warbler and the more northerly Arctic Warbler—have been steadily expanding their breeding ranges westwards from Russia since about 1900. The Arctic Warbler now breeds as far west as central Finland and northern Norway, and is becoming more frequent as a wanderer to the British Isles. The westward spread of the Greenish Warbler has been more impressive. It reached what was formerly East Prussia in 1905 and was breeding there by 1910; during the early 1930s it spread through Poland into north-east Germany, and between 1935 and 1940 it swept north-west-wards through Estonia and Latvia into south-east Finland. By 1953 it was breeding on the Swedish island of Gotland, and it is now well established along the eastern side of Sweden as well as in the southern half of Finland. With the arrival of singing males in the 1970s, the colonisation of Denmark began. As with the Arctic Warbler, Greenish Warblers are appearing more frequently on migration in the British Isles as well as in neighbouring countries. Before 1940 only one had been recorded in Britain (in 1896), but since then more than 200 have been reported, mostly in early autumn, though a few have occurred in late spring. The westward expansion of these two species has been attributed to the warmer springs and summers of north-central Europe brought about by the 1850–1950 climatic amelioration.

Since the 1920s, when the climatic amelioration was really under way, the Firecrest has been increasing in central Europe and expanding its range north-westwards to include Belgium, the Netherlands (1928), Denmark (1961) and south-east England (1961–62), all of which were formerly outside its breeding limits (Figure 47). During the 1960s Firecrests began to appear in southern Fenno-Scandia as well. In the Netherlands the Firecrest was first found breeding at two localities in Noord Brabant, and a little later was proved to breed in Overijssel and southern Limburg. At about this time there were signs that it was extending its range northwards into south Jutland (Denmark) from northern Germany, but breeding was not proved in Denmark until a pair was found nesting on the island of Amager, south of Copenhagen, in 1961. This was followed by other breeding records on

Zealand and Falster during the 1960s. This invasion is believed to have taken place across the Baltic from the former East Germany or Poland (Dybbro 1976). Meanwhile, breeding was finally proved in North Schleswig and Jutland during the 1960s and the species is now well established, if local, in Denmark.

The first satisfactory evidence of Firecrests breeding in England occurred in 1961, since when they have been shown to breed, as a summer visitor, in a number of widely scattered localities in south-east England. Although it is possible that breeding Firecrests have hitherto been overlooked in Britain, because ornithologists did not expect to find them there until the chance discovery of a small population in the New Forest in Hampshire, the pattern of discovery fits in well with a colonisation from the south-east, and with the pattern of spread on the European mainland. By 1975, moreover, more than a hundred pairs had been reported from thirty-one localities, including singing males as far west as Gwent (Wales) and Worcestershire (where breeding was proved), and as far north as Yorkshire (Sharrock 1976; Ferguson-Lees *et al.* 1977). Since then, the British population has fluctuated a good deal; see Chapter 11.

The Firecrest's sibling, the Goldcrest, which in periods of mild winters breeds commonly over most of the British Isles, has been increasing and spreading in Ireland and Scotland ever since the climatic amelioration began

FIGURE 46 *Firecrest*

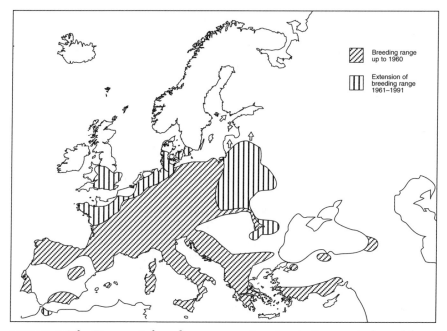

FIGURE 47 *The Firecrest's breeding range expansion in Europe (adapted from Voous 1960; Cramp and Simmons 1992; Gibbons et al. 1993).*

to take effect after 1850. Nevertheless, this expansion has generally been attributed largely to increased afforestation with conifers in formerly tree-less areas. This has undoubtedly been very important, but the Goldcrest could hardly have taken full advantage of the new forests, particularly in northern areas, without the benefit of the increased proportion of mild winters associated with the long-term amelioration. Unlike the Firecrest, which is a summer visitor over much of its northern range, the Goldcrest remains throughout the winter in the British Isles and is partially resident elsewhere in Europe also, and it suffers very heavy mortality in severe winters such as those of 1962–63 and 1985–86 in particular. However, as with the Dartford Warbler and some other species, recovery is usually remarkably quick. It has, for example, recovered very well from the 1985–86 winter population crash, aided by recent mild winters. Whatever the precise causes, the Goldcrest has colonised many islands in the Inner and Outer Hebrides, Orkney, Shetland and also the Isles of Scilly in the course of this century, in some cases very recently.

Among the flycatchers inhabiting the Old World, a pair of sibling species, the Collared and Pied Flycatchers, both extended their ranges in Europe to a considerable extent during the climatic amelioration. Since both are highly insectivorous, as is characteristic of the family, it is not surprising to find that they benefited from a general rise in spring and summer temperatures in northern Europe. It is thought that the two species evolved from two populations of a single species which became separated by the maximum advances of the ice sheets in the Pleistocene, and that their breeding ranges have begun to overlap only in quite recent times.

The more southerly Collared Flycatcher has extended its range in central Europe very slowly northwards during this century. Since 1965, for instance, it has consolidated its breeding population in east Saxony and Hesse (Thiede 1975b). In the southern Baltic area it has established temporary breeding populations during periods of mild springs that caused returning migrants to overshoot the normal northern limits. As a result of such overshooting during the middle of the last century, a permanent breeding population became established on the large Swedish island of Gotland, where nowadays it is much more common than the Pied Flycatcher. Recently, it has also established itself on the neighbouring island of Öland.

The northward advance of the Pied Flycatcher has been more spectacular than that of the Collared, especially in northern Finland since 1947. It has also spread westwards in France since 1936 and has recolonised the Netherlands from the east since 1940, being now locally common in the eastern provinces. A similar spread and increase has also been recorded in Denmark since about 1920. In Britain the Pied Flycatcher spread slowly northwards into Scotland during the latter part of the 19th century and the early part of the 20th century, then it put on a pronounced spurt between 1940 and 1952, when parts of the Scottish Highlands were colonised, as well as previously unoccupied areas further south in England and Wales. This marked period of expansion in Britain of course coincided well with events on the European mainland. Since 1960 it has slowed down and even been reversed locally in Britain, and also in adjacent parts of the continental mainland, though not in Sweden (Marchant *et al.* 1990). Pied Flycatchers began to nest in Ireland for the first time in 1985. Finnish birds have recovered from the population crash of the early 1980s, which followed hard winters (Väisänen *et al.* 1989).

The provision of nest boxes since the end of the Second World War has undoubtedly greatly aided the Pied Flycatcher's increase and spread, especially in those areas where natural holes are scarce; but it is unlikely that it would have done so, at least to anything like the same extent, had not basic natural factors such as climatic improvement favoured it. As a result of his long-term study of the Pied Flycatcher in Britain, Bruce Campbell (1954–55, 1965) concluded that climatic factors might be important.

Another factor (mentioned by Sharrock 1976) which may have furthered its expansion in northern Scotland (which, incidentally, has continued since 1965, albeit more slowly) is the provision of additional natural nest sites by the northward spread of Great Spotted and Green Woodpeckers. Like the Wood Warbler, the Pied Flycatcher is fond of nesting in sessile oakwoods in Britain, and it is in such woods that it is most common. However, as with that species, it seems to me that rainfall is the important factor limiting its distribution there. It appears to require not less than 625 mm (25 inches) per annum, and preferably over 750 mm (30 inches). Thus the rather wetter summers of the recent climatic amelioration seem to have been distinctly to the Pied Flycatcher's advantage.

Europe's smallest flycatcher, the Red-breasted, has been spreading north and west since about 1930 in southern Finland and also in northern Germany where, however, it becomes progressively scarcer and less regular as a breeding species the further north one travels in Schleswig-Holstein. Between 1940 and 1989 Red-breasted Flycatchers nested at least ten times in Denmark, mostly in the east of the country. They have also been breeding on the islands of Öland and Gotland in south-east Sweden since 1944. As a result of this expansion, they have appeared more frequently on migration in the British Isles. As all the Red-breasted Flycatchers of the Palaearctic Region—the species breeds right across temperate Asia to Sakhalin and Kamchatka—winter in India and south-east Asia, it has been suggested (Ferguson-Lees 1970) that the entire European population represents a comparatively recent westward extension of what was originally a central and eastern Asiatic range. Perhaps the greatest part of this extension took place in the very warm Little Climatic Optimum of early medieval times.

The breeding range of the Spotted Flycatcher is centred on the western Palaearctic, and it breeds far to the north in Fenno-Scandia, though not in Iceland. In the British Isles it has bred in the Hebrides and Orkneys since 1940, but has not yet bred in Shetland. It also appears to have increased in Ireland during this period. Since 1964, however, Spotted Flycatcher breeding populations have generally declined in the British Isles and also in Denmark, allowing for short-term fluctuations. Marchant *et al.* (1990) thought that climatic factors might be operating through their influence either on breeding success or on mortality levels in their overwintering territories (perhaps linked with the West African Sahel droughts). They suggested (rightly, in my opinion) that, as Spotted Flycatchers feed almost entirely on flying insects, fine, warm weather in May and June is likely to enhance breeding success. Therefore a combination of unfavourable climatic factors from the 1960s to the 1990s may be responsible for the decline.

Of the many species of North American flycatchers, the Great Crested Flycatcher has apparently spread slightly north-eastwards since about the middle of this century, colonising part of Nova Scotia. Another North American bird, the Northern Mockingbird, has spread slowly northwards into Canada and now breeds in discontinuous pockets in southern Alberta, southern Ontario, south-west Quebec, Nova Scotia and Newfoundland (Godfrey 1966). Presumably the extensions of range of both species may be correlated with climatic improvement.

At least ten species of Palaearctic thrushes and chats advanced north, north-west or west as a result of the 1850–1950 climatic amelioration. However, the Rock Thrush is an exception. In the Little Ice Age it was far more widely distributed in the highland areas of central Europe than at present. For instance, it formerly occurred in central Germany, several parts of north-east France, and also along the northern fringe of the Carpathian Mountains; but it disappeared by the end of the 19th century, except for a few strongholds in north-east France, such as the Vosges Mountains, where it clung on until the First World War period. Since 1950, however, it has shown signs of increasing in its present range (e.g. in the Ukraine and Switzerland) and of spreading once more. Thus it seems that the recent climatic amelioration proved unfavourable to the Rock Thrush, but that its fortunes may have revived following the waning of the amelioration. A drier summer climate seems to suit it better.

The Blue Rock Thrush also bred further north in central Europe up to the middle of last century than it does today, but there are signs that it too may be attempting to extend its range once more.

Two other southern species, the Black-eared Wheatear and the Black Wheatear, have withdrawn further south, probably because of the more oceanic climate which the amelioration brought to western Europe. According to Yeatman (1976), the Black Wheatear has all but disappeared from its former localities in the Mediterranean zone of France, while the Black-eared Wheatear is now almost entirely restricted to this zone; at one time (e.g. 1885) it nested as far north as Bourgogne. On the other hand, the Black-eared Wheatear has spread northwards in the Balkans since 1895 (Matvejev 1985).

The Wheatear (or Common Wheatear) has a very wide distribution in the Northern Hemisphere, breeding not only throughout Europe and north-central Asia as far as the extreme north of Fenno-Scandia and Siberia, but also on Jan Mayen, Spitsbergen, in Iceland, Greenland, north-east Canada and Alaska. It thus exhibits a notable toleration of climatic conditions, though of course it is only a summer resident in these northern latitudes. Since all these northern populations, American and Eurasian, share common wintering grounds in Africa, it is believed that North America was only colonised comparatively recently, perhaps as recently as the Medieval Warm Period (Little Climatic Optimum). Moreover, the 1850–1950 climatic amelioration in the North Atlantic area seems to have allowed it to extend its colonisation of Greenland and the Canadian Arctic still further, and (apart from a check due to the post-1950 Arctic cooling phase) the growing influence of the anthropogenic greenhouse effect may enable it to continue this trend.

The widespread decline of the Wheatear in lowland England and Ireland since about 1900, and especially since 1930, is not yet fully understood. Comparatively recent losses of habitat to agriculture and forestry, the effects on habitat of the lack of grazing through the decline of sheep rearing caused by the agricultural recession before 1940, and the temporary disappearance of rabbits due to myxomatosis in the mid-1950s, have obviously all been important contributory factors to this long-term decline; however, I agree with John Parslow (1973) that it is unlikely that they alone are the root causes. Yeatman (1976) mentions a similar decrease in the number of

FIGURE 48 *Black Redstart*

Wheatears in France and many other European countries that has been attributed to the loss of habitat through agriculture and urbanisation. For instance, he stated that it has disappeared as a breeding bird from many parts of the lowlands of northern France, as has also happened in the Low Countries, Germany, Finland and European Russia. Although chronologically the decline fits in with the climatic amelioration, it is nevertheless difficult to find a real connection between the two, unless it is due to possibly drier conditions than formerly in south-east England. It is interesting to note that, judging by the distribution map in the BTO Atlas (Sharrock 1976), the Wheatear is mostly absent as a breeding species from those areas with an annual rainfall of less than 750 mm (30 inches). The decline in Britain, especially lowland England, continued through the 1960s and early 1970s; but since 1976 there has been a more or less strong recovery which coincides rather well with the counteracting by the greenhouse effect warming of the post-1950 Arctic cooling.

The Black Redstart, which made news when it began in a big way to occupy the bombed sites of London during the Second World War, has been spreading north-westwards and northwards in Europe since the middle of the last century. Typically, it is a bird of rocky, mountainous areas, and its expansion to the lowlands has been facilitated by its gradual adaptation to breeding on human habitations and other buildings. Professor Karel Voous (1960) summarised its natural habitats as follows: 'rocky, apparently warm slopes with numerous broken rocks and stones and deep clefts and gullies in the rocks; also steep cliff faces, from sea level up to near the snow line in the

189

high mountains; ...'. Thus the modern extension of the Black Redstart's breeding habitat—towns, villages and industrial sites, with their steep cliff- and rock-like structures containing numerous ledges—is not so different from its natural habitat. Although this interesting adaptation has enabled the Black Redstart to spread through lowland areas, it seems that the real stimulus to do so has come from climatic change—from the gradual warming up of the climate of western Europe which commenced, as we have seen, about 1850 and reached a climax in the 1930s.

Black Redstarts first appeared in the northern plains of France and Germany in about 1870, and began to spread into the Jutland peninsula of Denmark from Germany towards the end of the century. The first Danish breeding record occurred at Ribe in south Jutland in 1872; thereafter occasional pairs nested until colonisation began in earnest after 1900, though breeding birds remained few in number until 1930. Subsequently they spread to many new places throughout Denmark, especially in the east, where nowadays some 200 pairs breed (Dybbro 1976). From Denmark, and perhaps other parts of the south Baltic shore, other pairs colonised the extreme south of Sweden from 1910 onwards. The Baltic states of Estonia, Latvia and Lithuania were colonised in the 1920s. Quite recently Black Redstarts have begun colonising southern Finland and Norway.

Meanwhile the continental coasts of the English Channel and the North Sea were occupied all the way from Normandy to Friesland by the 1940s, so it is not surprising that southern England was colonised during this period. Apart from an isolated breeding record from Durham in 1845, Black Redstarts were not proved to breed in England until 1923, when a pair nested (and also in the following year) on the sea cliffs at Fairlight, near Hastings, Sussex. However, strong evidence has since come to light that a pair bred at Pett Level, near Fairlight, as early as 1908, and that perhaps other pairs nested in previous years this century as well. Whether or not they bred earlier, 1923 seems to have seen the start of the invasion of Britain by Black Redstarts. Following a slow start, the main influx occurred in 1942, when they not only bred in the bomb-devastated areas of central London, but also in several other towns, while singing males occupied territories in many other localities scattered throughout England from Plymouth to Sheffield. With the gradual disappearance of the bombed sites, industrial sites such as power stations seem to have become the preferred habitat in urban areas, and the English breeding population, which is still largely restricted to the south-east, nowadays fluctuates between about thirty and 120 pairs. In perhaps the best year so far, 1986, 119 territories were held by singing males, with breeding proved or suspected in eighty-one cases (Spencer *et al.* 1991). Incidentally, in France, according to Yeatman (1976), central Normandy was only colonised in 1958.

Although climatic amelioration seems to be primarily responsible for the expansion in range of the Black Redstart, Kenneth Williamson (1975) suggested that the 'heat-island' phenomenon may have enabled it to establish itself firmly as a British breeding bird. As he explained, this phenomenon is produced by the sum total of countless domestic and industrial fires and heating systems, by warm effluents poured into rivers by industry, 'and by dust and smoke particles which encourage condensation of water vapour and so create a layer-cloud which obstructs the radiation of heat from land to space'. In Greater London, a difference in minimum night temperature of 7°C has been recorded between urban and rural areas during May, while with smaller towns such as Bath and Reading the difference can still amount to at least 2–3°C. Williamson points out that although the Black Redstart began a determined colonisation of England in the 1920s, 'it was not until birds took to haunting London's buildings and bombed sites that its numbers grew rapidly', and that, moreover, the species is now becoming associated with power stations. It seems to me that this tendency may well enable the Black Redstart to cling on to its outposts in north-west Europe during unfavourable climatic phases. In support of Williamson's suggestion, I may add from my own observations as well as those of others, that in the Low Countries and elsewhere in the extreme north-west of Europe, Black Redstarts are also associated with large towns, industrial complexes, and power stations, whereas in the warmer parts of central Europe they commonly nest on buildings in small villages.

The Red-flanked Bluetail is one of a group of Asiatic passerines which have been extending their ranges steadily westwards into Europe, on and off, for many years. As its name implies it is a beautiful bird—at least the

FIGURE 49 *Red-flanked Bluetail*

male is; his upperparts, including the tail, are bright blue while the under-parts are creamy white with bright orange flanks. However, except at close range, these bright colours apparently do not show up well in the dark, silent, coniferous forests of the Siberian taiga zone which are its home. It seems that during the Last Glaciation, when the great ice sheets advanced further south over northern Europe and western Asia than they did further east, pushing the taiga zone southwards before them, the Red-flanked Bluetail likewise retreated into east-central Asia and vanished from its former haunts in the west. Since the final retreat of the ice, it has presum-ably been gradually recovering lost territory in western Asia and northern Europe, expanding more quickly during periods of climatic amelioration and slowing down or halting altogether in periods of climatic deterioration. For example, it is probable that during the long medieval Little Climatic Optimum it made a considerable advance to the west, perhaps as much as 2,000 km (1,500 miles) to reach the Ural Mountains, but then more or less halted during the subsequent Little Ice Age.

Up to the 1930s, the western limit of regular breeding was the River Pechora, a little to the west of the Urals (Stegmann 1938), but, with the marked warming of the climate around this period, the Red-flanked Bluetail began to move again. By 1937 it was breeding as far west as the Kola Peninsula in extreme north-west Russia. Since 1949, when the first bird was reported in Finland, some 1,500 km west of the Pechora, a small but regu-lar breeding population has become established in eastern Finland (breeding first proved 1971), and it is thought that the species may be more numerous than actual observations suggest (Mikkola 1973). In recent years, as might be expected, Red-flanked Bluetails have begun to be reported not only in northern Norway, but on migration elsewhere in western Europe, such as Germany, Italy and the British Isles. Up to and including 1976, seven were recorded in Britain since the first one in Shetland in October, 1947.

Thus, as concluded already by Sovinen (1952) and Mikkola (1973), the recent westward expansion of the Red-flanked Bluetail and other eastern birds may be explained as part of the gradual re-establishment during the Post-glacial Period of former pre-glacial distributions. The recent climatic amelioration, with its warm springs, has produced a continuation of the expansion, which is still proceeding under the growing, favourable influence of the greenhouse effect.

The advance of the ice sheets during the Ice Age is also thought to have been responsible for the existence in the western Palaearctic Region of two distinct, but extremely similar, species of nightingales—a good example of a pair of 'sibling species'. The theory is that there was once a single species of nightingale, but the glaciations split the population into two, isolating one section in the south and south-west, and the other in the east. Isolated as they were over a long period of time, they diverged sufficiently to become distinct species, the Nightingale evolving from the south-western population and the Thrush Nightingale from the eastern. When the ice retreated they both expanded their ranges to regain lost ground, and eventually met again sometime in the Post-glacial Period, but no longer interbred, even where their ranges overlapped in central Europe. Ecological segregation appears to have occurred in the areas of overlap, the Thrush Nightingale preferring damper habitats and thus avoiding direct competition with its sibling.

Extensive periods of climatic deterioration, like the Little Ice Age, must have slowed up or even reversed the range expansion of both species. Indeed, we know that the Thrush Nightingale had already extended its breeding range much further west in former times, as in the 18th and early 19th centuries it was still known to inhabit eastern Hungary, the Vienna district of Austria, the middle reaches of the River Elbe in Germany, and south-east Sweden as far north as Stockholm. During the height of the Little Climatic Optimum its distribution was probably even more extensive, and it had most likely been on the decline throughout the ensuing Little Ice Age, especially in the very cold 17th and 18th centuries.

By the middle of the 19th century, when the Little Ice Age came to an end, the Thrush Nightingale had vanished from all of the regions just mentioned. However, with the onset of a new amelioration of the climate its retreat halted, and by the 1930s, when the climate of western Europe was at its warmest, it started to advance again. Since then, it has spread northwards in southern Finland, recolonised southern Sweden, and increased and spread further west in northern Germany and Denmark. Since 1964 the south-eastern coastal area of Norway has been colonised as well, and the species is also spreading south-westwards in northern Germany, breeding nowadays in Brandenburg (Thiede 1970), Schleswig-Holstein, and also around Hamburg (Rheinwald 1977, 1993).

Now that the Thrush Nightingale is becoming established in southern Norway, it is not beyond the bounds of possibility that it could attempt to nest in eastern England and Scotland, just as some other Scandinavian birds have done recently. Already there have been about a hundred occurrences in Britain since 1957, compared with only two before that, and some of these have been singing males in May. Many of these have been adults 'overshooting' on the spring migration from their winter quarters in tropical east Africa.

There is evidence that the Nightingale also decreased a good deal during the Little Ice Age and contracted its breeding range in western Europe. However, loss of breeding habitat during this period, as well as the influence of climate, was probably an important cause. It is not impossible that during the very warm medieval period Nightingales nested in Britain as far north as south-east Scotland. After all, they still bred as far north as northern Yorkshire at the beginning of the present century.

With the climate improving up to a peak in the 1930s, there was also some improvement in the numbers of Nightingales, which increased noticeably in England, particularly in the south-west, from the mid-1930s until about 1950, and they even colonised some areas on the periphery of their range. Subsequently, they withdrew from these areas and decreased in many other places, especially from 1957. At the present time, Nightingales rarely breed further north than northern Lincolnshire or as far west as Wales. Likewise, they have withdrawn since 1950 from a large part of north-west France, and have become distinctly scarcer in the Low Countries and north-west Germany. Again, this recent decline has been largely blamed upon alteration or destruction of the habitat, such as the neglect of coppices-with-standards in England and the replacement of broad-leaved woods with conifers. There is little doubt that these have been important factors locally, but climatic change is almost certainly the prime cause. As Laurent Yeatman (1976) wrote: the nightingale shows 'une préférence pour les terrains jouis-sant d'étès chauds'. In the post-1950 period, as we have seen, the springs often proved to be unseasonably cool and late (Marchant *et al.* 1990), but the generally warmer springs and summers since about 1975, apparently due to the growing influence of the anthropogenic greenhouse effect, show signs of improving the Nightingale's fortunes on the north-western edge of its European breeding range.

There is little doubt that the Blackbird benefited to a tremendous extent from the climatic amelioration in north-west Europe. As Kenneth Williamson (1975) asserted: 'Among the passerines, no bird's status has improved more dramatically than the Blackbird's'. Before 1850 it was considered a shy and solitary bird which rarely ventured outside its wood-land haunts. Thomas Bewick described it in 1797 as 'restless and timorous', 'easily alarmed' and 'difficult of access'. In north-west Europe, the climatic improvement, with its more frequent mild winters resulting in lower mortal-ity, led to a gradual increase in Blackbird numbers, so that by 1850 the species was expanding into adjacent habitats, such as farmland, parks and gardens. Lack (1954) pointed out that worm-eating species, such as the Blackbird and Song Thrush, find food for their young more easily in wet summers than dry ones; so the wetter summers of the amelioration favoured them as well.

From about 1870 Blackbirds also expanded the northern and north-west-ern limits of their European range, pushing gradually further and further north in Scandinavia and southern Finland (Figure 50), and colonising (after 1900) many of the outlying islands of western and northern Scotland, including Shetland, as well as (in about 1950) the Faeroe Islands. Since about 1930 they have even taken to wintering regularly in southern Iceland, where once they were rare and usually perished in the hard winters (Gudmundsson 1951). These wintering birds probably migrate to Iceland from the more northerly populations now breeding in Scandinavia, particu-larly from Norway.

An analysis of recoveries of ringed Blackbirds by Robert Spencer (1975) showed that the expanding Finnish population has taken to wintering as far west as eastern Britain, along with the Swedish breeding population from which it apparently originated. During the present century Blackbirds also

FIGURE 50 *The Blackbird's colonisation of Finland (updated from Kalela 1949; Cramp and Simmons 1988). (Redrawn from Spencer 1975)*

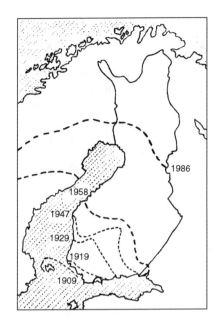

spread westwards in Ireland to colonise all the western areas from which they were formerly absent. Elsewhere also their numbers have continued to rise in remarkable fashion. By the 1930s, they had spread from rural gardens into suburban gardens and city parks in Britain, the Netherlands, Denmark, Germany and other European countries, while since then they have even successfully occupied gardens and small open spaces in city centres. Helped by the 'heat-island' effect, and the food provided by the human population, winter mortality is much lower in urban than in rural areas. The waning of the long climatic amelioration after 1950 saw a slowing of the Blackbird's increase and a period of stabilisation, relatively little interrupted by the severe winter of 1962–63. However, after 1975 the British population began a slight but steady decrease, which was apparently due to increased mortality levels in the series of more or less hard winters between that year and 1987. The mild winters since then may presage a recovery.

In recent years Blackbirds have expanded more and more successfully into open habitats such as moorland, and at the higher altitudes more usually associated with the Ring Ouzel. Indeed, there is evidence that interspecific competition occurs where Blackbirds and Ring Ouzels overlap. The Ring Ouzel is in fact a mountain relative of the Blackbird, but is nevertheless also found breeding, to some extent, in moors and tundra down to sea level. Its present discontinuous breeding distribution in Europe and the Caucasus, which represents its entire world range, is thought to have arisen from the way in which the species was driven southwards by the advancing ice sheets of the Last Glaciation, and also down from the mountains on to low moorland and tundra by the expanding alpine glaciers. When the ice and glaciers retreated in the Post-glacial, the Ring Ouzel populations became isolated, as some recolonised the mountain ranges, while others moved northwards along with the northward retreat of the tundra and its associated moorlands. The long intervening period of isolation of these populations led to the formation of the present-day subspecies.

During the periods of significant climatic improvement, such as the Little Climatic Optimum (Medieval Warm Period), the Ring Ouzel presumably retreated further northwards and also higher up the mountains, returning during colder phases such as the Little Ice Age. It certainly decreased in central Europe and the British Isles during the 1850–1950 climatic amelioration, contracting its breeding range to higher altitudes, especially where faced with competition from colonising Blackbirds. By about 1950 the Ring Ouzel had disappeared as a breeding bird from Cornwall, the Isle of Man and many parts of Ireland, but from the mid-1960s to about 1980 there were some signs of a recovery in some areas which could be linked with the climatic deterioration in northern latitudes between 1950 and about 1980. For example, the BTO Atlas survey revealed evidence of a spread into the Outer Hebrides, Orkney and Shetland, and possibly Cornwall (Sharrock 1976). The north European race was also fairly recently discovered nesting (several pairs) in the Arrée Mountains of Brittany in north-west France (Yeatman 1976), and also in Belgium. It remains to be seen what effect the present greenhouse warming will have in due course.

The Song Thrush is known to suffer heavy losses in severe winters such as those of the 1940s, 1962–63, 1978–79 and 1981–82, which may have been at least partly responsible for the marked decline observed in the British Isles since about 1940. Apart from minor, short-term recoveries, this decline was still continuing in 1992. In Shetland, a small population, which became established between about 1900 and 1946, died out following the severe winter of 1946–47. Decreases were reported from urban areas of Poland up to the 1980s (Cramp and Simmons 1988), but in some parts of north-west and northern Europe, including Denmark (apart from a marked temporary decrease from 1975–79), Finland and the Netherlands, Song Thrushes have increased (Marchant *et al.* 1990), as also in Germany.

Before 1800 the Mistle Thrush seems to have been a relatively scarce bird over much of its European breeding range, including Britain, and was largely confined to woodlands, especially coniferous ones; in some regions it was a distinctly mountain species, frequenting subalpine forests. In Britain it was chiefly confined to southern England and Wales, being rare or absent further north, and in Ireland it was unknown. Then in 1807 (or perhaps a year or two earlier) it nested for the first time in north-east Ireland, and within about forty years had spread over the whole country. Meanwhile, by about 1850 it had also become much more common in northern England and southern Scotland, and was spreading further north. By 1950 it was breeding almost everywhere in the British Isles except Shetland, though very rarely and sporadically in the Outer Hebrides and Orkney. On the European mainland it also increased, especially from about 1870, and colonised many new areas, especially in the lowlands such as those of the Netherlands and north-west Germany.

A notable feature of the Mistle Thrush's expansion has been its adaptation, like the Blackbird, to living on cultivated land, including the large gardens and parks of towns and suburbs, in many parts of Europe including Britain. Afforestation, especially of conifers, has assisted its spread in some regions, such as Scotland, where suitable habitat did not previously exist. However, these factors do not by themselves explain the whole story satisfactorily, and one cannot help associating the Mistle Thrush's population explosion and spread, like that of the Blackbird's, with the climatic amelioration from 1850–1950 which broadly corresponds with it, at least in mainland Europe. The climatic improvement began to be felt in the British Isles rather earlier than on the continent, and this probably explains its early expansion there.

From the end of the 1970s until the early 1990s, the British breeding population of the Mistle Thrush decreased, at least partly due to a series of set-backs caused by the frequent hard winters of that decade, but very recently, following the very mild winters of 1988–89 and subsequently, it is showing an upturn in fortunes. Marchant *et al.* (1990) considered that recent major changes in farming practices, particularly the switch from sowing cereals in spring to autumn, was probably a more important influence in the decline of the 1980s than the series of hard winters of that period. Time will tell. Incidentally, a big decrease was observed in Finland during the 1980s, and this is more likely to have been due to hard winters than to changes there in farming practices.

The Fieldfare is another species which appears to have survived the Last Glaciation in the heart of southern Siberia, and which has been steadily extending its range westward ever since the ice sheets retreated north. A characteristic bird of the taiga zone, it had reached as far west in northern Europe as Fenno-Scandia by about 1750, but much less far west in central Europe. However, thereafter it extended its range in that region too. By the beginning of the 19th century it began to breed in Poland, from whence it spread in the course of the next hundred years into Germany, the former Czechoslovakia and Switzerland. The first nests were discovered in eastern Germany around 1850, in Switzerland from 1923, eastern France from 1953 and Denmark since 1960. Fieldfares started nesting in the Belgian Ardennes from 1967, and in the Netherlands from 1972, where sporadic breeding had occurred since 1903. By 1989 it was reported that at least 700 pairs were breeding in south Limburg, and that this population was still expanding. Fieldfares were not known to breed in the British Isles until 1967, in which year a pair nested successfully in Orkney, while a second pair appeared to have done so in County Durham, northern England, as well. After 1967, breeding became regular in Shetland for several years, and more or less regular in the Pennine Hills of northern England. In the Scottish

Highlands, where the first nest was discovered in 1970 (Inverness-shire), breeding has also become regular, if sparse, over a wide area.

It seems likely that the long-term, pre-1950 climatic amelioration facilitated the Fieldfare's westward expansion across central Europe. Indeed, Finn Salomonsen (1951) described a remarkable instance of the way in which the amelioration enabled the Fieldfare to establish an outpost in the New World during the very warm 1930s. The Fieldfare winters in some abundance in southern Norway, but in a rather capricious manner: in some winters they are more abundant than in others, and on occasions they may occur in enormous flocks totalling many thousands. With the onset of severe weather they usually move south-west to the British Isles. On 19 January 1937, a sudden drop in the air temperature, due to an influx of cold air from the mountains of southern Norway, apparently caused huge flocks of Fieldfares to leave the south-west Norwegian coast that afternoon and head for Britain. Unfortunately for them, they were caught up in a severe south-easterly gale and blown north-westwards over the north Atlantic at about 100 km per hour. In the course of a pitch-black night they apparently lost all sense of direction and were carried by the gales to Jan Mayen and the coast of north-east Greenland, arriving there on 20 January. Once ashore and rested, they re-orientated and flew in their normal south-westward migratory direction, and appeared a week later in south-west Greenland, having crossed the Greenland ice-cap. Here they settled in the mixed birch and willow forests in the interior of the Julianehåb district; this fact was not discovered until a decade later, by which time they were a locally common breeding species.

Professor Salomonsen considered that the reason the Fieldfare was so successful in establishing itself in south-west Greenland, following its fortuitous arrival, was primarily the climatic amelioration. The mean July temperature in the Julianehåb district had increased from 9 or 10°C to 11°C, which corresponded with the mean July temperature in the Fieldfare's birch forest breeding habitat in northern Scandinavia. Moreover, the winter temperatures were more than 5°C higher than they were fifty years previously, thus making it possible for the Fieldfare to survive the Greenland winter. Since 1937, natural selection seems to have eliminated the migratory urge in this population which otherwise might have led to insupportable losses at sea. As Kenneth Williamson (1975) remarked, it is 'not difficult to imagine how a similar event in a warm epoch many centuries ago could have deposited Redwings in Iceland, there to form resident colonies'. Incidentally, although single Redwings have arrived in Greenland on a number of occasions from both Iceland and Scandinavia, they have never yet arrived in sufficiently large numbers to establish a breeding colony as has the Fieldfare. I should perhaps mention, in passing, that the milder winters have resulted in both Fieldfares and Redwings wintering in Iceland in greater numbers than formerly. The Fieldfare did not become established as a breeding bird in Iceland until after 1950.

FIGURE 51 *In central Europe Fieldfares often breed in suburban gardens and parks, as here at Eppelheim, near Heidelberg, Germany. (Photograph: John F. Burton)*

Sharrock (1976) considered that the pattern of the British breeding records of the Fieldfare over the period 1968–72 pointed to a link with Scandinavian rather than central European stock. To me the pattern then and since suggests that the birds attempting to colonise northern England have come from Denmark or/and perhaps north-west Germany—an extension of the central European expansion. In this I am influenced by the coincidence of the big build-up of the Fieldfare colonies in west Jutland subsequent to 1967, the year in which a pair were thought to have bred for the first time in northern England. Fieldfares attempting to breed in southern England may come from the recently established population in Belgium and the Netherlands. The Scottish population, on the other hand, probably originated from Norway, perhaps in response to a retreat by northern populations there in the face of the post-1950 climatic deterioration in the Arctic, as seems to have happened to some other species, such as the Redwing, Lapland Bunting and Wood Sandpiper. No doubt the question will be answered in due course by ringing recoveries.

In discussing the possible successful colonisation of Britain by the Fieldfare, Batten *et al.* (1990) mention human disturbance as a threat. This is of course a possibility, if excessive; but since 1989, when I took up residence in south-west Germany, the tolerance shown by Fieldfares of a high level of human activity in the suburban gardens, squares and parks where many of them breed (Figure 51), often in loose colonies, has greatly

199

impressed me. They are intolerant of crows, Magpies, cats and other such potential predators, but like the urban Blackbird take little notice of human beings going about their normal business. I have often watched them at quite close quarters at their nests and collecting invertebrates as food for their young from lawns. In Heidelberg and Mannheim I have watched them feeding on 'green' islands in the midst of busy, congested roads. This leads me to believe that if Fieldfares succeed in establishing themselves in England they may well occupy similar suburban habitats.

[1] There is, curiously, a superficial similarity in appearance between the Barred Warbler and the adult female and juvenile plumages of the Red-backed Shrike, notably the barring of the under-parts.

Chapter 10

LANDBIRDS: PASSERINES (PART II)

The Bearded Tit (or, as some people would have us call it, Bearded Reedling, since it is not really a member of the tit family) has both contracted and expanded its breeding range in western Europe since 1800. It is not easy, however, to sort out how much of this has been due to climatic changes and how much to alterations in its habitat, or to other factors such as collecting, particularly in the 19th century. Because of its specialised breeding habitat—large, dense reedbeds, which have been eradicated in many former haunts by drainage—the Bearded Tit has today a highly discontinuous distribution in Europe. Moreover, since it suffers severe mortality in hard, prolonged winters, small isolated populations in fragments of the original habitat have often been unable to survive in sufficient strength to make a subsequent recovery. For instance, Bearded Tits were once to be found breeding in extensive reedbeds over a wide area of south-east England, but owing to land reclamation and direct persecution by man they declined throughout the 19th century, so that by 1900 their range had contracted almost entirely to Norfolk. Fortunately, the enforcement of improved protective legislation at about that time led to a gradual recovery, and Suffolk was recolonised, but they continued to be confined to these two counties until 1960. During this period they were nearly wiped out by the severe winters of 1916–17, 1939–40 and 1946–47. Apparently, prolonged heavy snowfalls or the formation of thick ice over the reed litter, which prevent the Bearded Tits finding their staple winter food of reed seeds, are the critical factors, not the severity of the frosts. Thus the bad winter of 1946–47, with its heavy snow-falls persisting through to March, was much more catastrophic than that of 1962–63 — the worst winter of the century so far. After the former winter, only a few pairs remained in England and the Netherlands, while the entire population of western Europe was estimated to have fallen to about 100 pairs or less. In 1962–63, on the other hand, the breeding population in East Anglia, which by then had risen to nearly 300 pairs, was little more than halved.

Between 1947 and 1962 the British population of Bearded Tits had grown steadily (though not reaching 100 pairs again until 1957) and had even begun to expand, as those on the continent of Europe had also done. The population expansion in the Netherlands was greatly helped by the creation, between 1948 and the late 1960s, of vast areas of reeds as a result of the energetic efforts of the Dutch since the Second World War to reclaim much of the former Zuider Zee. By the autumn of 1965 the reedbeds there were estimated to contain about 20,000 birds. Large-scale ringing of these birds revealed that many of them were erupting and dispersing to England and Wales, and even Ireland, and as far afield on the European mainland as south-west France. It was not surprising to find, therefore, that the Bearded Tit rapidly extended its breeding range, not only in the Low Countries and northern Germany, but elsewhere in north-west Europe, and even to Switzerland. By 1968 more than 100 pairs were nesting in Denmark; southern Sweden had been colonised already in 1966, and in 1974 they were nesting in large numbers as far north as Lake Tåkern in central Sweden, with outposts even further north-east (e.g. Lake Roxen). There are indications (see Mead and Pearson 1974) that the recent population build-up in the Netherlands may have originated from earlier influxes of the hardier east European race *russicus* in the 1940s. Moreover, this possibility is supported by the discovery that the Dutch and British Bearded Tits appear to be a hybrid between the nominate race, *biarmicus*, and *russicus*.

Following a particularly successful breeding season in East Anglia in the fine summer of 1959, the British population exploded during the equally fine autumn of that year and dispersed all over England and Wales, augmented as we have seen by Dutch birds. Except for the years immediately following the hard winter of 1962–63, such eruptions have become a feature of autumns ever since, especially in the period of mild winters which continued in an unbroken run until the severe winter of 1978–79. Many of these birds settled down in suitable locations in various parts of southern and eastern England, with the result that breeding colonies are now to be found from Lancashire and Yorkshire in the north to Kent in the south-east, and Dorset in the west. The hard winter of 1978–79 and subsequent ones caused only temporary set-backs (Marchant *et al.* 1990).

From all of this, it very much looks as if climate is the basic factor controlling the numbers and distribution of the Bearded Tit. Most likely it was much more widespread in the early Middle Ages, on account of both the very warm climate and the much greater availability of suitable habitat; but it presumably suffered a great deal during the many severe winters of the Little Ice Age with their frequent heavy and prolonged snowfalls. Thus, it must already have become a far more localised bird by the time of its discovery in England in the late 17th century. Kenneth Williamson (1975) considered that the 1850–1950 climatic amelioration was chiefly responsible for its increase during this century, with which conclusion I concur. As he pointed out, owing to the long run of mild winters from 1963 to 1978 (and again since 1986–87), climatically the Bearded Tit has had everything in its favour. With the aid of the growing influence of the greenhouse warming, it

FIGURE 52 *Penduline Tit*

should be able to hold on to its territorial gains in the north, where the winters are gradually becoming milder, and could even extend them wherever suitable large reedbeds (two hectares plus) exist. By 1974, at least 590 pairs were reported to be breeding in Britain (O'Sullivan 1976), and the present breeding population is considered to be slightly higher than this (Batten *et al.* 1990; Marchant *et al.* 1990).

Like the Bearded Tit, the Penduline Tit, which has an extensive range across central Eurasia, tends to extend its distribution in an irruptive manner. Following a southern contraction of its northern distribution limits (e.g. Poland) between 1870 and 1890, these areas were subsequently recolonised in the 1920s and 1930s, and it has been expanding northwards and westwards quite spectacularly since 1940. Thus, it has penetrated as far west as north-west Switzerland and in Germany as far west as the Rhineland, and has made attempts to breed in Denmark. Isolated nests were even discovered as far west as the Netherlands by 1962, while individual birds have occurred in north-west France, the Channel Islands and on the east coast of England from 1963. In south-west Europe, during the same period, Penduline Tits have also been spreading outwards from their marshland strongholds to wherever suitable habitat is available. This has happened especially in south-east Spain, while in the south of France a remarkable increase has occurred since 1974 following a marked decline after 1936. The latest information of the spread is given in Chapter 12.

The reasons for these population fluctuations of the Penduline Tit are not known for certain, but it seems likely that the north-westward expansion in central Europe may, at any rate, have been initiated by the warmest period of the climatic amelioration—that is, between 1920 and 1940. The relatively few (seven) hard winters since 1950 have presumably helped the Penduline Tit's spread up to the present, even though the amelioration drew to an end about 1950. Moreover, the current greenhouse effect warming may now also be playing an increasing part. The temporary decline in northern Poland and elsewhere between 1870 and 1890 can be linked with a short-term cold phase during this period, when harsh winters occurred frequently until 1896.

The Crested Tit, an inhabitant of coniferous forests, advanced a considerable distance to the north in central Finland between 1929 and 1949, apparently at the expense of the more northerly Siberian Tit which retreated a similar distance. The Finnish ornithologist, Einari Merikallio, who described this change in their distribution in 1951, attributed it to the long-term climatic amelioration. If this is correct, as seems most likely, then it seems certain that it is the climate that finally determines the boundary between these ecologically competitive species, the Siberian Tit being the better adapted to life in the colder winters of the more northerly pine forests. The Crested Tit has also spread slightly northwards in northern Scotland, although most people have attributed this expansion to the maturing of the extensive conifer plantations. The same reason has also been advanced in Denmark where a remarkable colonisation took place subsequent to 1890, almost certainly from northern Germany. However, in both these instances the climatic improvement may have facilitated the expansion. Incidentally, this may also be true of the big increase and northward spread of the Coal Tit in northern Scotland since 1844, where everyone agrees reafforestation has undoubtedly been the main factor involved.

A small northward extension of the Marsh Tit's range into south-east Scotland occurred after 1920 and was still continuing fitfully in the late 1980s (Thom 1986). However, since the late 1960s, and particularly in the mid-1970s, a shallow long-term decline has affected this species in Britain (Marchant *et al.* 1990), and was still continuing in 1992 (BTO Common Birds Census). The reasons for this decline are unknown, although relatively minor fluctuations have been recorded in relation to hard winters. On the European mainland there has been no clear evidence this century of any notable changes in distribution, although population fluctuations have occurred. This is also true of its sibling, the Willow Tit, which, however, increased in Britain between 1963 and 1977, when there was a long run of generally mild winters, though it has since declined somewhat. There has, moreover, been a southward contraction of its range in Scotland since 1950.

The influence of the 1850–1950 climatic amelioration seems more obvious in the cases of the Blue Tit and the Great Tit, both of which spread significantly to the north up to 1950 in northern Scotland and Fenno-

Scandia. Again, reafforestation is regarded by many as the chief cause, but Pennie (1962) pointed out that there was no obvious connection between the expansion of the Great Tit and the spread of plantations in either Sutherland (Scotland) or Norway, while Haftorn (1958) correlated its advance in central Norway since 1930 with the improved winter climate. In addition to the marked colonisation of the extreme north of Scotland since the mid-1920s and a strengthening of its position in the Inner Hebrides, the Great Tit has been attempting to colonise the Outer Hebrides since 1966 in the wake of the Blue Tit, which started to do so a few years earlier. Both species, incidentally, have colonised the Isles of Scilly off the south-west coast of England since 1920.

The Azure Tit, central Asian counterpart of the Blue Tit, also extended its breeding range towards the end of last century (e.g. in the 1860s and 1870s), when it reached as far west as the Baltic coast. During this period it over-lapped the Blue Tit's range to a far greater extent than today, and the two species hybridised more frequently. It had been steadily retreating to the east during the course of this century, but has lately shown signs of advancing west again in Belarus (Byelorussia). Again, like the Penduline Tit, this may at least be partly due to the greenhouse effect warming since about 1975.

The milder winters in eastern North America too have enabled the Tufted Tit to extend its breeding range slowly northwards into south-east Canada since about 1920. On the other side of that continent, the common Bushtit has spread further north in south-west Canada. The first nest on Vancouver Island was discovered in 1945, since when the species has become common there.

Increasing atmospheric pollution was blamed for the decline of the Nuthatch in London and the neighbourhood of other industrial conurbations during the second part of the 19th century. This may well have been so. I can remember the Nuthatches which frequented Greenwich Park in east London during the 1940s and early 1950s being very sooty; they vanished from this locality in 1953, but returned some years later after the introduction of London's smokeless zones. However, atmospheric pollution does not appear to explain the retreat southwards of the species in northern England and North Wales about the middle of last century, or the recovery which took place subsequent to 1927. These events in fact seem to be compatible with the climatic changes which have occurred since 1800. It seems to me most probable that the Nuthatch declined and retreated southwards because of the wetter, generally unfavourable climate of the first half of the 19th century, but recovered when the climate improved to a marked degree after 1920. The Nuthatch's northward extension of breeding range (and expansion in Wales) became most pronounced after 1940. It now breeds as far north in Northumberland as the Scottish border, and from the 1960s was recorded more frequently across it; by 1989 it was proved to breed for the first time in south-east Scotland (Marchant *et al.* 1990).

The discovery in the 1960s of breeding Nuthatches in three or more separate locations in central Scotland, well north of the most northerly English breeding birds, is fascinating. Although it is possible that these Nuthatches leap-frogged from Northumberland, the more likely explanation is that they were of southern Scandinavian origin, like the Wrynecks, Red-backed Shrikes and certain other species which nest further north in Scandinavia than in Britain, but have lately colonised or attempted to nest in Scotland. I have not yet, incidentally, come across any reports of Nuthatches having spread northwards in northern Europe as a result of the climatic ameliora-tion, and neither had Marchant *et al.* (1990) in 1989.

The Short-toed Treecreeper, a bird of the lowland broad-leaved forests of Europe and north-west Africa, is believed to have become separated from its sibling, the Treecreeper (or Common Treecreeper), so familiar to British birdwatchers, as a result of the Last Ice Age. It probably evolved from a retreating population of Common Treecreepers which became isolated in the broad-leaved woodlands of the Mediterranean region by the advance of the ice sheets, and consequently had to alter their predilection for coniferous forests and adapt to their new haunts. Meanwhile, the main population of Common Treecreepers survived in the retreating belt of coniferous forests further east in Eurasia. When the ice of the Last Glaciation finally retreated, these coniferous forests advanced northwards and westwards at a much faster rate than the similar expansion of the broad-leaved forests further south. Thus, the Common Treecreepers themselves also spread west and north at a faster rate than did the Short-toed Treecreepers, which spread northwards with the broad-leaved forests. It is thought that the Short-toed Treecreepers advanced so slowly that although several species of broad-leaved trees managed to colonise the British Isles before they became sepa-rated from the main continent by the rise in sea level, the birds failed to do so. Common Treecreepers, however, colonised the whole of the British Isles and, in the absence of competition, occupied not only the coniferous wood-lands but the broad-leaved as well—on high ground and low ground alike. Where the two species eventually met and overlapped in central Europe they apparently competed ecologically with each other, and the Common Treecreeper became more or less restricted to the coniferous forests, espe-cially those of the uplands and mountains.

Since the very warm period of the 1920s and 1930s the Short-toed Treecreeper seems to have been endeavouring to extend its breeding range further north in Europe. For instance, although breeding was not proved in Denmark until 1946 (in south Jutland), it was suspected of doing so much earlier. Since 1950 it has become well established as a breeding species in south-east Jutland, and nests or possible nesting pairs have been reported elsewhere in southern Denmark in recent years. Lately it has shown signs of trying to colonise southern England. Birds have been reported since 1969 at several mainly coastal localities, and there were suggestions, subsequently withdrawn, of breeding in Dorset. It is possible that the long-term climatic amelioration was responsible for recent events, but it remains to be seen whether the spread in Denmark and any attempted colonisation of England will prosper in the face of competition from the Common Treecreeper. If the amelioration encouraged the Short-toed Treecreeper's recent expansion of range, one is prompted to ask if the much longer and warmer Little Climatic Optimum of the Middle Ages did so as well? Quite possibly it did, but the Short-toed Treecreeper failed to establish itself permanently.

Apart from a slight north-westward extension of range in northern Scotland to colonise Stornoway in the Outer Hebrides, there has been little evidence of any attempt by the Common Treecreeper to extend its range further north or north-west in Europe. This species is, of course, widespread in North America, where it is known as the Brown Creeper, and where it occurs (being the only species) in both coniferous and broad-leaved woodlands. It is generally believed to have colonised North America from eastern Asia during one of the warm interglacials of the Pleistocene Ice Age.

Although, likewise, there seems to be little or no evidence of any significant extensions of range of the Wren (called Winter Wren in North America) in either the Old World or the New which can be correlated with climatic change, it does exist for two North American species, Bewick's Wren and the Carolina Wren. The former only colonised southern Ontario during the course of this century, being recorded for the first time in 1898, with the discovery of the first nest not being made until 1950. Since about 1970, however, Bewick's Wren has seriously declined east of the River Mississippi and has virtually disappeared from the eastern part of its range in the Appalachian Mountains (Robbins 1985) for reasons as yet unclear, although severe winters in that decade may have been implicated. Following a spread north through the north-east United States, the Carolina Wren colonised southern Ontario during the present century, and advanced slowly northwards after it was first recorded nesting at Point Pelee in 1905. The milder winters were almost certainly responsible in both cases. Although the Long-billed Marsh Wren does not appear to have extended its range significantly in North America, Finn Salomonsen (1948) stated that it bred in western Greenland during the exceptionally warm period of the amelioration between 1920 and 1941.

Although most recent changes in distribution recorded among American birds which can be correlated with the 1850–1950 climatic amelioration have taken the form of extensions of range to the north or east, the Red-eyed Vireo has been exceptional in spreading to the south and west. A bird of mature woodlands with plenty of secondary growth, it has to some extent been extending its range southwards down the Pacific coast of the United States and westwards into the Rocky Mountains. It has not, apparently, retreated in the northern part of its range, however. Although climatic change has been suggested by American ornithologists as the primary cause, the precise way in which it has operated does not appear to be known.

Another exception to the general rule is the highly migratory Bobolink, which has slowly extended its breeding range to the west, almost to the Pacific coast. The reason for this does not seem to have been ascertained, but

207

its contemporaneous decline in the eastern part of North America has been blamed on changes in agricultural practice, especially the mechanisation and early mowing of hayfields. The Bobolink is particularly associated with flowery hay meadows where grass is allowed to grow high. The decline of hay production in the course of this century is believed to have led to its virtual disappearance in New England, for instance, where it used to be numerous.

Turning now to a consideration of those species which have advanced northwards or eastwards, we find that several relatives of the Bobolink in that purely American family of birds, the Icteridae, are involved. They include Brewer's Blackbird, the Yellow-headed Blackbird, the Brown-headed Cowbird and the Western Meadowlark. These are all birds of open country, such as the prairies, and it has been suggested that their eastward extensions of range have been made possible by the felling of woods and forests in the east to create more pastureland. However, since the expansion of all these species began during the very warm years of the 1920s and 1930s, it may be that the climatic amelioration was the underlying cause. The Brown-headed Cowbird, for instance, was first proved to breed in Nova Scotia in 1929 and subsequently colonised Prince Edward Island and Newfoundland.

Some of the American finches and their allies are also among those which have expanded northwards or eastwards. The chiefly non-migratory Northern Cardinal, a brilliantly all red grosbeak with (in the male) a black face, has staged a remarkable advance northwards from its former breeding range in the southern United States. Since the early part of last century it has spread far to the north to colonise not only the whole of the eastern part of the United States, but much of south-east Canada as well, where it first nested in 1901 at Point Pelee in Ontario. A characteristic bird of woodland edge and tangled thickets, it is very adaptable and is just as at home in semi-deserts and swamps, and in city parks and gardens. This and its exploitation of bird tables have assisted its remarkable range extension, but the gradual improvement in the climate until quite recently, especially the higher winter temperatures, is probably the underlying cause.

This may also have been the chief cause of the even more remarkable and quite sudden expansion of range to the east by the Evening Grosbeak from its headquarters in the western part of North America. This began about 1920 and carried on in a great swathe right across the continent to the Great Lakes, and beyond to the Atlantic coast. In Canada, for instance, it first bred in Ontario in 1920 and now breeds as far north-east as Cape Breton Island, Nova Scotia.

The Dickcissel, a bunting-like finch, which has also extended its breeding range considerably further east following the Second World War, is a more complicated case. A typical bird of the prairies and similar open country, it used to be common in the 19th century over a wide area of the United States, but by the beginning of the present century it underwent a marked decline, and disappeared altogether from the eastern part of its former range by about 1914. However, soon after 1945 it began to be seen more frequently in the east. Since 1950 it has recolonised a large part of its breeding range and is continuing to push eastwards towards the Atlantic coast. All sorts of attempts have been made to explain the decline and subsequent recovery, such as changes in agricultural practice, but these are not wholly convincing. The fact that Dickcissels are notoriously subject to short-term fluctuations in population suggests a climatic factor, perhaps operating through their chief summer prey, grasshoppers. It is noticeable that their decline broadly corresponded with the mildest part of the climatic amelioration, i.e. from 1890 to 1940, when the climate became wetter in summer in the eastern coastal areas of North America—a trend unfavourable to the activity and development of grasshoppers. Since 1950 the climate has generally become hotter and drier in this region.

Of the American sparrows—closer allies of the buntings than they are of the Eurasian sparrows—the Clay-colored and Henslow's Sparrows have both extended their breeding ranges in quite recent years, probably in response to the amelioration. The breeding range of the Clay-colored, a sparrow of scrubland, lies principally in west-central North America, but it has been spreading eastwards. In 1966 a nest was found by the Canadian ornithologist, W. Earl Godfrey, as far east as Ottawa. Henslow's Sparrow, a local bird of damp, weedy meadows, on the other hand, is distributed in the breeding season in a broad band across the eastern half of North America in the latitudes of the Great Lakes. In the 1960s it extended its breeding range as far north as Ottawa, and it has also been seen on several occasions in Quebec (Godfrey 1966).

Of the many Eurasian buntings and finches, thirty-five species to my knowledge have shown alterations in their breeding distributions which I think may be correlated with the 1850–1950 climatic amelioration. Of these, twenty-two expanded their ranges, mostly to the north, north-west or west, while the rest contracted, mostly to the north.

Starting with the buntings, I discussed the Cirl Bunting's apparent colonisation of England in the 18th century at some length in Chapter 4. Having established itself in south-west England during a warm phase of the 18th century, this typically Mediterranean species of woodland edge spread northwards during the first part of 1850–1950 climatic amelioration, but began to decline and contract its range to the south early in the present century. The frequent severe winters of the 1940s very likely accelerated the decline, and the particularly severe winters of 1962–63, 1977–78, 1978–79, 1981–82, 1984–85 and 1986–87 did not help matters. By 1989, as a result of a thorough survey by Andy Evans on behalf of the RSPB, the breeding population was put at a maximum of 119 pairs. The continuation of the survey in 1990 led to the discovery of up to 133 possible breeding pairs—a welcome increase (Spencer *et al.* 1993). On the continent of Europe, following a period of expansion in the second half of the 19th century, a similar withdrawal occurred on the north-west limits of its range; for example, in

Belgium, Luxembourg and northern France (Yeatman 1976), and in south-west Germany, from where it has now almost disappeared. In 1980 the German breeding population had been estimated at between 100 and 120 pairs. Maybe the greenhouse effect warming will aid a recovery.

Another Mediterranean-based species, the warmth-loving Rock Bunting, also retreated a little to the south from 1930 in western Europe (e.g. central France), possibly because, as with the Cirl Bunting, the summers of the amelioration were too wet. Voous (1960) pointed out that its distribution limits 'follow mountain zones and slopes which in July are probably on the average not colder than 68°F [20°C] and not warmer than about 86°F [30°C]'. Yet another Mediterranean species, the Black-headed Bunting, which is found in the Middle East as well as in the south-east quarter of Mediterranean Europe, has been extending its breeding range northwards through Bulgaria and into Romania.

The Ortolan Bunting has a much wider distribution in Europe than the Rock Bunting, and, as well as breeding in the countries along the northern edge of the Mediterranean region, nests within the Arctic Circle in Fenno-Scandia. Voous (1960) sets its distribution limits as 59°F (15°C) in the north and 86°F (30°C) in the south, while the Swedish ornithologist Durango asserts that it is only found breeding in those parts of Sweden where the total annual rainfall is 23 inches (58.4 cm) or less. In spite of heavy annual losses to bird-catchers when on migration, it extended its breeding range during this century, especially after 1940; but since about 1960 it has been reported to be declining in parts of Belgium, the Netherlands (only thirty-three breeding pairs left in 1989), France, Germany and especially Norway, where it used to be common in the south-east, but has decreased to a great extent and become much more localised. Its previous northward advance in Fenno-Scandia was attributed to the spread of cultivation, but this itself has chiefly been made possible by the climatic amelioration, so it seems highly likely that the warmer climate has been the prime cause of both. The present regression in the Ortolan's range in north-west Europe may therefore be due in some way to the gradual climatic cooling since 1950, even though there have been some especially hot and dry summers since the late 1970s.

The history of the Corn Bunting in north-west Europe this century is also rather mysterious, but is probably likewise connected with the climatic amelioration. For instance, since 1900, and especially since 1930, it has disappeared or become very local and rare over most of Ireland and Wales where it used to be common, especially in coastal districts; the same is true of south-west England, the Hebrides, Orkney and Shetland, though more or less stable populations have been maintained in such disjunct areas as Aberdeenshire, parts of the Western Isles of Scotland, Lancashire, the West Midlands, the Fens around the Wash, and Kent (Thompson *et al.* 1992). Corn Buntings are highly dependent upon agriculture in Europe, and profound changes in farming practice have been blamed for their decrease in Britain and elsewhere. Although in some areas such changes have probably been a contributory factor, John Parslow (1973) has pointed out, rightly in my opinion, that they have not been sufficiently marked everywhere in the Corn Bunting's former range to account fully for the bird's decrease. More recently, Paul Donald of the BTO, which is currently carrying out a survey of the Corn Bunting's population and precise habitats in the British Isles, wrote, 'It would be wrong, though, to blame the decline of this and other farmland birds solely on farming methods. Studies by BTO staff suggest that the decline began earliest and has been most severe in Ireland and western Britain, the areas where traditional methods have survived best. Indeed, the main stronghold is now the intensively farmed Fens of Cambridgeshire and Lincolnshire. There must be other factors at play' (Donald 1993*a*). Evans and Flower (1967), discussing the situation in the Scottish islands, thought that the decline might be correlated with an increase in rainfall associated with the climatic amelioration. Certainly, a study of the map in the BTO Atlas (Sharrock 1976) reveals quite clearly that the current distribution of the Corn Bunting is concentrated in those parts of the British Isles with an annual rainfall of less than 1,000 mm (40 inches). Thus in the western parts of the British Isles, if this is truly a controlling factor, one would expect to find the Corn Bunting confined to coastal districts, as indeed it is, since these are the only areas with less than 1,000 mm per year on average. This can be seen from a more detailed rainfall map than the overlay supplied with the above-mentioned atlas. A similar decline of the Corn Bunting occurred during the same period in north-west France, and its present-day distribution hugs the coast in western Normandy and Brittany just as it does in western Britain (see Yeatman 1976).

Three northern species of buntings which do not breed in Britain extended their ranges westwards in Fenno-Scandia from about 1900. They are the Little Bunting, the Rustic Bunting and the Yellow-breasted Bunting (Figure 53). All have breeding distributions which extend in a broad band right across European Russia and northern Asia. The Rustic Bunting has now spread as far west as Sweden and is extending its range there, while the Little Bunting has reached Swedish Lapland. Before 1900 the Little Bunting was not known to breed west of Archangel and the river Dvina in northern Russia. The Yellow-breasted Bunting has been on the move since as early as 1825; it advanced as far west as Lake Onega by 1875 and eastern Karelia and Lake Ladoga by 1900. By 1925 it had penetrated deep into south-east Finland and had almost reached the frontier of Estonia. Although its colonisation of Finland was slow, about 100 pairs were believed to be nesting there by 1958. The species actually reached the west coast at Oulu on the Gulf of Bothnia as early as 1920. As one would expect, all three species have now become much more frequent visitors on migration to the British Isles and elsewhere in western Europe. The climatic amelioration is generally considered to be responsible for the westerly advance of all three species.

In extreme north-west Britain, the very adaptable Reed Bunting, which enjoys an extremely wide distribution in the Palaearctic region between lati-

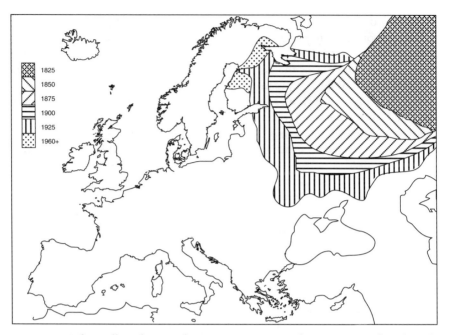

FIGURE 53 *The Yellow-breasted Bunting's westward expansion of its breeding range in Europe (adapted and updated from Lack 1954; Voous 1960; Cramp and Simmons 1994).*

tudes 35°N and 70°N, increased and extended its breeding distribution in the Hebrides and colonised Shetland in about 1949. It did not succeed in colonising Iceland, although it very occasionally reaches the Faeroe Islands. It is not known for certain whether or not this slight extension of breeding range in northern Britain was connected with the long-term climatic amelioration, but from the 1950s, and especially between 1965 and 1976, the Reed Bunting greatly increased in much of Britain, and expanded its choice of habitats from wet into dry situations, such as barley and oil-seed rape fields, young conifer plantations and even chalk downland scrub. The same ecological expansion has also been noticed in the northern provinces of France (Yeatman 1976). This ecological change and increase is considered, I feel sure correctly, by Marchant *et al.* (1990) to have been an overspill from the primary reed habitat, as a consequence of the high population levels reached following the run of generally mild winters in the 1950s and again from 1963 to 1975. Reed Buntings normally recover well from the set-backs caused by hard winters, but they have so far failed to recover the high population levels of the 1970s following the cold winters of 1977–78 and 1978–79, and the frequent severe ones from 1981–82 to 1986–87. Moreover, they have tended to desert the drier habitats for the primary wetland ones. Although the British population has now stabilised at a low level, there is still a tendency to decline rather than increase. Changes in agricultural practice have been suggested as a possible reason, making it more difficult than previously for Reed Buntings to overspill into farmland.

Finally, there is some evidence that the two most northerly breeding buntings in the Northern Hemisphere, the Lapland and Snow Buntings, retreated northwards in at least some parts of their breeding range during the recent climatic amelioration. For instance, the Snow Bunting all but disappeared as a breeding species from the Scottish Highlands and the Faeroes during the warmest part of the amelioration. Both showed clear signs of returning south with the onset of the climatic deterioration in the high Arctic from about 1950. The Lapland Bunting even began what appeared, at the time, to have been an attempt to colonise the Scottish Highlands for the first time in ornithological history, but this seems to have come to naught. Readers will find a fuller treatment of these two species in Chapter 11.

Moving on now to close relatives of the buntings, several northern finches also retreated northwards to a certain extent during the 1850–1950 climatic amelioration. These included the Brambling, Arctic Redpoll and Twite, and again I will postpone a full discussion of their cases until Chapter 11. Of the more southerly species, at least twelve expanded their breeding ranges northwards or westwards, probably in response to the climatic amelioration. The most spectacular case is the Serin, whose rate of spread at one time almost matched that of the Collared Dove. The course of its expansion in Europe has been well documented at various times by a number of

FIGURE 54 *Serin*

authors, including Ernst Mayr (1926), Viking Olsson (1971), I.J. Ferguson-Lees (1971) and Ian Newton (1972). The following account is based upon them; likewise the map (Figure 55), which I have endeavoured to bring up to date at the time of writing.

Before 1750 the Serin was confined to sunny open woodland and woodland edge in the western and central part of the Mediterranean region, breeding in north-west Africa and from the Iberian peninsula through Italy to the southern Balkans. Sometime soon after that date, however, it began to expand its breeding range northwards, and by 1875 had occupied Switzerland, southern Germany, Austria, western Hungary and the former Czechoslovakia. Fifty years later it had colonised most of France, except the north, and much of Germany and western Poland, while odd pairs had also nested in the south-east Netherlands and eastwards as far as Romania. For a while, between about 1940 and 1960, the rate of expansion slowed down almost everywhere in the north, though less so in the north-east and east, where throughout this century it had been faster than in the north-west. Nevertheless, by 1960 the Serin had consolidated earlier gains and extended its regular breeding limits to include all continental Europe (apart from the coastal strip from Brittany to Jutland) as far east as a line running from central Estonia south to the Black Sea at a point not far west of Odessa (Ukraine). In Turkey it had spread even further east by 1960, almost to the Caucasus.

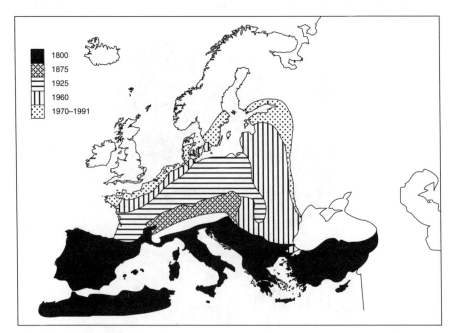

FIGURE 55 *The Serin's northward expansion of its breeding range in Europe from 1800 (updated from Olsson 1969).*

After 1960, the Serin accelerated its advance again, so that by the 1970s it was breeding regularly almost throughout the northern coastal districts of France and the Low Countries, Jersey (Channel Islands), southern Denmark, southern Sweden, the southern coastal districts of Finland, and from St. Petersburg (Russia) south to Odessa (Ukraine). However, from the 1970s the pace has slowed down. Although Serins began nesting in Jutland from 1961, the anticipated colonisation of this part of Denmark has not really taken place, and it is still scarce there; nor has it yet happened in southern England, though pairs definitely nested in Dorset in 1967, Sussex in 1969 and Devon in 1978. Since 1981 up to nine pairs have been recorded annually in southern England, and one or two pairs have been proved to have bred in some of these years. Clearly, the species is having difficulty in establishing itself here. Personally, I doubt that it will succeed, in spite of the increasing numbers of vagrant occurrences, unless the global warming due to the greenhouse effect eventually exerts a stronger influence. With the colder climate in the Arctic region since 1950 affecting much of north-west Europe at present, I think the Serin has now reached the furthest limits of its expansion, at least in the north.

Olsson (1971) remarked that 'the steady advance of the Serin from country to country defies adequate explanation other than that of a genetic change in migratory habits'. He went on to say that climatic factors have been suggested, but asks: 'If this were so, surely many other species would have reacted in the same way?' Of course, as related in this book, many other species have in fact advanced primarily in response to climatic change, particularly the climatic amelioration between about 1850 and 1950. The expansions of most species have not been so spectacular as that of the Serin, though some such as Cetti's Warbler, the Fan-tailed Warbler, the Firecrest and the Black Redstart have been almost as spectacular, while the Black-headed Gull and Collared Dove have surpassed it. One apparent difficulty in attributing the Serin's expansion to climatic amelioration is that its advance seems to have begun well before 1850, perhaps as early as 1750 or thereabouts. However, as Hubert Lamb (1966) pointed out, even in the Little Ice Age 'striking' short-term ameliorations occurred in Europe, and those 'around the 1630s, 1730s, 1770s and 1840s must all have been quite impressive'. He also stated that climatic improvement was 'fairly continuous from 1700 to the early 1900s, though there were setbacks in the mid–late 1700s and late 1800s'. Thus it is quite likely that these brief ameliorations were sufficient to stimulate early advances of the Serin, with static periods in between. Otherwise, the main period of its expansion fits in well with the major period of climatic amelioration from about 1850 to about 1950. As we have seen, it slowed up during the colder decade of the 1950s and renewed its momentum after the return of milder conditions in central and eastern Europe in the 1960s and 1970s. Olsson's further objection (1971) that 'the climatic range of the Serin today extends from mild, humid, maritime conditions in the west to dry, continental weather in the east, from the summer heat of north Africa and Asia Minor to the cold of Sweden and Finland' does not seem to me to be a real one. So do the breeding ranges of many European birds. Surely, it is the improvement in summer temperatures in the north and north-west, plus the ecological niche apparently available in these regions for another exclusively seed-eating small finch, which have enabled it to expand from the warm, sunny habitats of its original range. By the way, although resident in the latter, it is a summer visitor only to its annexed territories in the north. Its spread east through the north-east Balkans, the Ukraine and Turkey may be related to slightly milder conditions there between October and May.

It is well known that, more or less contemporaneous with its range expansion, the Serin became adapted to a variety of other habitats besides woodland edge, mostly in association with man, such as parks and gardens, vineyards and orchards, and even industrial wastelands. Sharrock (1976) commented that 'it is not known whether the adaptation to new situations actually triggered the expansion, or whether the spread compelled birds to adopt a more catholic choice of site'. Since I believe in climatic change as the principal cause, I favour the latter idea. Moreover, I am encouraged in this belief by the fact that Olsson and others have noted the way in which during the early stages of colonisation of a new area, Serins seem choosy in their selection of habitat; the vanguard occupy the best sites, however scattered or isolated these may be, while the later arrivals infill the less desirable places.

A montane relative of the Serin, the Citril Finch, which chiefly inhabits coniferous woods near the tree-line in the mountain ranges of central and south-western Europe, spread northwards a little in the 1950s and 1960s, for example in the Harz Mountains of Germany. It is not clear, however, if this was related in some way to the 1850–1950 climatic amelioration, or, on the contrary, to the post-1950 deterioration which, by 1980, seems to have been counteracted by the warming in southern Europe due to the greenhouse effect.

As pointed out by Kenneth Williamson (1975): 'the progressive amelioration of climate brought about the overlap and interbreeding of what had been held to be two distinct species of Redpoll, the high-Arctic *Acanthis hornemanni* and low-Arctic and boreal *A. flammea*'. Like Finn Salomonsen, Ken Williamson never accepted the separation of these two as distinct species, and the effects of the 1850–1950 climatic amelioration proved them right. Thus, in Greenland, the low-Arctic race *A. flammea rostrata* (Greenland Redpoll) advanced northwards and invaded the breeding range of the high-Arctic *A. h. hornemanni* (Hornemann's Redpoll), and interbred with it here and there. Similarly, in the northern Palaearctic (e.g. northern Norway) the more southerly *A. f. flammea* (Common or Mealy Redpoll) spread north into the territory of the high-Arctic *A. hornemanni exilipes* (variously known as the Arctic, Hoary or Coues's Redpoll) and produced hybrid populations. The Redpolls breeding in Iceland also seem to be a hybrid population which Williamson believed was derived from an inter-mingling of the Greenland Redpoll (*rostrata*) and Hornemann's Redpoll (*hornemanni*) 'during some relatively warm period of the past, perhaps the Viking era, when the latter was in retreat'.

The small dark race *A. flammea cabaret* inhabiting the British Isles and the alpine regions of central Europe, and often known as the Lesser Redpoll, has also undergone fluctuations in range during the present century which are probably primarily associated with climatic changes, although afforesta-tion with conifers seems to have played an important part, too. For instance, between 1900 and 1910 there was a marked increase and spread into lowland, and particularly southern, areas of Britain where they chiefly colonised large hawthorn hedges and damp woodlands of alder and sallow. That was followed by a widespread decrease and withdrawal northwards between 1910 and 1950, especially during the first two decades. This, it may be noted, corresponds well with the warmest phase of the 1850–1950 climatic amelioration. From 1950, when the amelioration began to wane, until 1980 the Lesser Redpoll steadily increased and spread once more into southern, lowland regions, becoming common in districts where before it was very rare or absent. It also spread northwards around 1930 into the birchwoods and new forest plantations in the northernmost counties of Scotland, and, by 1950, the Scottish islands off the west coast (Thom 1986; Marchant *et al.* 1990). This expansion undoubtedly received much encour-agement from the big conifer reafforestation programme following the end of the Second World War. It also led to a colonisation by British Lesser Redpolls of the coastal dune conifer plantations of the Netherlands, Belgium and Denmark (west Jutland) from the early 1960s (attempts were made even earlier), and also on the French coast of the English Channel. They spread eastwards into the south of Sweden and northern Germany, and south-east-wards from France into western Germany, by 1981, and they are now breed-ing in south-west Germany.

An expansion has also been taking place of the Lesser Redpoll breeding population of the montane region of central Europe, in this case to the north. Thus, since 1970 Lesser Redpolls have been breeding in the High Tatras and colonising Czech Bohemia, including the neighbourhood of Prague, and also nearby Saxony in eastern Germany (Thiede 1975*b*). As with the spread of this bird in Britain, the expansion of the central European population subsequent to 1950 may also have been associated with the trend towards a deterioration in the climate between 1950 and 1980, but has been dealt with more conveniently here rather than in Chapter 11. In Britain, however, the breeding population began a marked decline after 1980 which was still continuing in 1992. Nevertheless, in 1988 it was still higher than it was when the BTO began their Common Birds Census in the 1960s (Marchant *et al.* 1990). The reasons for this are at present unclear, although they may be bound up with the climatic trend in Britain since 1980, particularly the series of cold winters from 1977–78 to 1986–87 and their possible effects on birch and spruce seed-crops, and Redpoll winter survival. Yet winters have been mild since then, but Redpolls continue to decrease.

The Siskin has had, in some respects, a somewhat similar history in Britain to the Lesser Redpoll. Until about 1850 it was chiefly confined to the native pine forests of the Scottish Highlands. Thereafter it spread slowly to new areas and perhaps more rapidly from about 1900. During this period it also spread to Ireland due, it is believed, to the widespread and large-scale establishment of Scots pines, but, as with the Lesser Redpoll, climatic change may also have been involved. In the course of this century it has expanded its breeding range to cover most parts of Ireland. Since about 1940 the increase and expansion of its range in England and Wales has been especially marked, so that today it breeds in coniferous woods in many areas as far south as the English Channel coast. The great increase in the planting of conifers, particularly spruce, since the First World War, and especially since the Second World War, has undoubtedly facilitated this bird's recent expansion, but an underlying cause may be the generally drier conditions which the British Isles have experienced since the 1850–1950 amelioration waned. The former may have favoured better and more regular seed production in the spruce, larch and other conifers, thus providing more ample food supplies for Siskins. It is noticeable that the Siskin spread relatively little during the warmest phase of the amelioration. Continental European breeding populations fluctuate to a marked extent and are subject to much eruptive behaviour, but, according to Marchant *et al.* (1990), no long-term trends are known.

To return to a consideration of those finches which spread northwards or westwards, apparently because of the climatic amelioration, it would seem that the Goldfinch, Greenfinch, Bullfinch, Hawfinch, Linnet, Trumpeter Finch, Scarlet Rosefinch, House Sparrow and Tree Sparrow qualify for inclusion in this category.

Since about 1920 the Goldfinch has expanded its range northwards in southern Finland and in Scandinavia in cultivated areas. It bred for the first time in Norway in 1955, but has since colonised much of the southern portion of that country. After reaching a peak in the 1970s, the British population of Goldfinches decreased markedly between 1977 and 1986. This has been attributed to the use of herbicides and other agricultural practices (Lack 1986), but as the Goldfinch is also adversely affected by hard winters, it is difficult to exclude the effects of the winters of 1977–78 to 1986–87. Since 1987 Goldfinch numbers have been recovering again. The Greenfinch, too, spread northwards to a marked extent in Fenno-Scandia and northern Scotland in cultivated and urban areas during the same period, while the Bullfinch did so in Britain, Ireland and Denmark, and perhaps elsewhere in northern Europe after 1930, and especially after 1955. Associated with this increase was a spread into a wider variety of habitats, including urban parks and gardens. In Britain, Bullfinches reached a peak population in the mid-

1970s, but thereafter began to decrease and continued to do so up to 1991, since when there have been signs of a recovery.

Although there appears to be little or no published evidence of a northerly advance of the Hawfinch elsewhere in its extensive European and Asian breeding range, it certainly did so in Britain before 1950. Indeed, there is some reason for believing that it is a relatively recent colonist, as the first breeding records were not authenticated until about 1830 (Newton 1972). However, they may have bred much earlier, as in the 17th century Sir Thomas Browne wrote that in Norfolk they were 'chiefly seen in summertime about cherrietime'. Of course, it is possible that in the 17th century, the coldest century of the Little Ice Age, the species may have been in decline following a colonisation during the medieval Little Climatic Optimum; and perhaps it became extinct or nearly so in the almost equally cold succeeding century, when both Thomas Bewick and Gilbert White described it as only an occasional visitor, generally in winter. Thus, in Sir Thomas Browne's time, Norfolk may have been one of the last strongholds of a then declining Hawfinch population.

A perusal of Gilbert White's 18th century records of Hawfinches in the Selborne area in north-east Hampshire is interesting. On 20 March 1791 he wrote in his *Naturalist's Journal* (Greenoak 1986–89): 'Mr. Burbey shot a cock Gross-beak [White's name for this species], which he had observed to haunt his garden for more than a fortnight: Dr Chandler had also seen it in his garden.' Morris (1990) believes that this is the male which is still preserved in the Gilbert White Museum at Selborne, X-rays of which he examined. These prove that the bird was preserved using 18th century techniques and was unlikely to have been mounted subsequently. It may well have been preserved by White himself. Continuing his entry for 20 March 1791, White stated: 'Birds of this sort are rarely seen in England, & only in the winter. About 50 years ago I discovered three of these gross-beaks in my outlet, one of which I shot.' He also mentioned this earlier record in his *Natural History of Selborne* (1789), adding 'since that, now and then one is occasionally seen in the same dead season.' In his next entry in his *Journal* for 21 March 1791 he wrote: 'A hen gross-beak was found almost dead in my outlet: it had nothing in it's craw.' The only other definite record in his *Journal* is for 31 December 1776, which reads: 'A gross-beak was shot near the village. They sometimes come to us in the winter.'

Whenever precisely Hawfinches recolonised England, there seems little doubt that before 1835 they were more or less confined to the south-east and the Midland counties. Subsequently they spread quite rapidly northwards to northern England and, by around 1900, to southern Scotland, and westwards into Wales and Somerset. After 1915 the pace slowed down, and since about 1950 decreases have occurred in several areas, though increases have continued to be reported elsewhere, especially in northern England (Parslow 1973). There seems to my mind little doubt that climatic change has been the prime cause of the main fluctuations in the Hawfinch population. The big expansion between 1835 and 1950 corresponds well with the overall climatic amelioration of that period.

The northward spread of the quaintly named Trumpeter Finch—its incessant flight call reminds one of a toy trumpet—from north-west Africa into south-east Spain has been summarised by Wallace *et al.* (1977), upon which the following account is based. It has been extending its breeding range northwards in Morocco with the result that, from 1961, wanderers have been appearing more frequently in southern Spain than previously this century. By 1969 Trumpeter Finches were becoming very numerous in the province of Almeria where, unfortunately, they were caught in large quantities by the local bird trappers. Although breeding was thought to have occurred as early as 1968, it was not confirmed until 1971 when three nests were discovered. In that year the first vagrants occurred in Britain—in Suffolk in late May and in Sutherland in early June—followed by one on Alderney in the Channel Islands in late October 1973. However, the next one did not arrive until May 1981, in Orkney. As with the similar northward expansion of the Little Swift and the White-rumped Swift in north-west Africa, the climatic amelioration is the probable cause. It will be interesting to see how much further these species will continue to spread to the north.

In the course of this century the Linnet retreated southwards in southern Finland, but it is not clear whether this was due to climatic change or to other factors. Kalela (1949) thought that, as well as the influence of cooler summers, it might at have been at least partly caused by the intensification of agriculture; this has also been advanced as a reason for declines in the British Isles since 1950. It is believed that the loss of marginal land in modern farming and the greater control, especially with herbicides, of weed species, upon whose seeds the Linnet depends to a large extent, is the key factor. In the 19th century, Linnet populations were undoubtedly much reduced by the popularity of the species as a cage bird and the consequent trapping of it on a large scale. With the banning of this trade, Linnets staged a rapid recovery in the first half of the present century and colonised Shetland in 1934. However, since about 1950 they have disappeared from Shetland, and almost so from the Outer Hebrides and some of the western Scottish islands. It seems possible that the colonisation of the Shetland Isles could have been prompted by the climatic amelioration, while the subsequent retreat from these and other Scottish islands may be correlated with the post-1950 climatic deterioration in northern latitudes.

A partial recovery of the Linnet in mainland Britain, following the severe winters of 1961–62 and 1962–63, which reached its zenith in 1966 and 1967, was only temporary. Since 1977 there has been a steep decline (Marchant *et al.* 1990), although after 1987 there was some evidence of a hesitant recovery. On the European mainland also, there have been marked decreases since about 1977. In Finland, the breeding population fell disastrously following the bad winters of 1984–85 and 1986–87, with the result that Linnets virtually disappeared from the northern parts of their range (Hildén 1989).

On high, mountainous ground in a restricted area of the British Isles and Scandinavia the Linnet is, to some extent at least, replaced by the rather similar Twite. A study of the Twite's European distribution suggests a strong predilection for an oceanic rather than a continental climate. There is in fact a strikingly good correlation between its breeding range and areas north of about latitude 53°N with an annual rainfall above 750 mm (30 inches) in

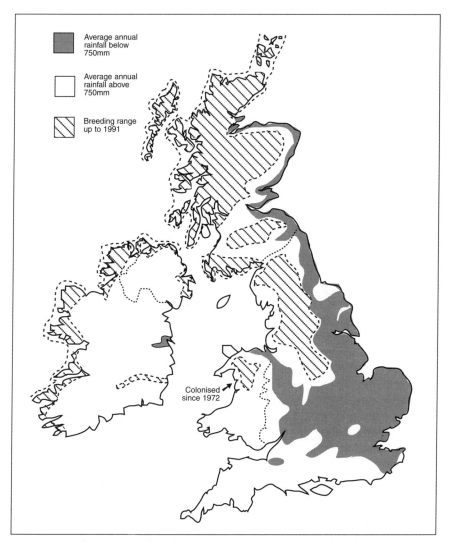

FIGURE 56 *The Twite's breeding range in Britain correlated with average annual rainfall above 750 mm; plus recent range expansion (adapted from Sharrock 1976; Gibbons et al. 1993).*

both the British Isles and Scandinavia. Surprisingly, therefore, Orford (1973) considered that 'there is only a very general correlation with rainfall'. In the Pennines of northern England, where the Twite is at the southern limit of its breeding range, the correlation with upland areas experiencing an annual rainfall of more than 1,000 mm (*c.*40 inches) is remarkably close. Thus it seems that it requires an even wetter climate in the south than in the north; or, put another way, it tolerates a drier climate in the north than the south.

During the 1850–1950 climatic amelioration, the Twite contracted northwards somewhat in the southern parts of its breeding range in the British Isles, possibly in the face of growing competition from the Linnet, which increased during this period. For instance, it disappeared from Devon, parts of Yorkshire, the Cheviots, the Border Fells, the Pentland Hills, the Isle of Man and Ireland, and appeared to decrease in the Pennine Hills (Newton 1972; Orford 1973; Williamson 1975). This may also have been due to a locally drier climate there resulting from the amelioration. The long-term effects on the Twite's distribution in north-west Europe of the growing warming due to the greenhouse effect will be interesting to observe (see Chapter 11).

During the climatic amelioration, the Brambling, an ecological replacement of the Chaffinch in the birchwoods and willow scrub of the northern Palaearctic from Scandinavia to Kamchatka, withdrew the southern limit of its breeding distribution northwards, especially in Fenno-Scandia. In Finland, for instance, between 1910 and 1949 it retreated 110 km in the west of the country and 260 km in the east (Merikallio 1951); at the same time the Chaffinch advanced northwards by similar distances, and according to Williamson (1975) ousted the Brambling from an estimated 45% of its former range. These two species overlap to a relatively limited extent, and are evidently in ecological competition. The Chaffinch's breeding distribution does not extend anything like as far east in Asia as that of the Brambling, but it has spread northwards in Siberia as well, although this has been attributed chiefly to the spread of cultivation. Nevertheless, climatic improvement has probably played an important role in the Chaffinch's advance here too. As will be discussed in Chapter 11, the climatic deterioration in northern latitudes since about 1950 seems to have brought about a reversal of this trend, and since 1960 there are even signs that bramblings are attempting to colonise countries as far south as Denmark and Scotland.

In northern Britain, the Chaffinch extended its range slightly northwards earlier this century to colonise parts of the Hebrides and to strengthen its hold in Orkney, presumably in response to the amelioration; but it is difficult if not impossible to sort out the part played by the extension of tree planting in these areas. Significantly, perhaps, it has not yet managed to colonise Shetland.

The Scarlet Rosefinch (or Common Rosefinch) is another member of the group of northern Asiatic species which advanced westwards during the 1850–1950 amelioration and, so far, have continued to do so. It has apparently been advancing in a series of waves since at least 1850. For example, in Finland it became quite common at that time in the south-west of the country, but subsequently retreated eastwards and virtually disappeared by the end of the 19th century (Merikallio 1958). Soon afterwards, however, it reappeared and spread west again, especially from 1930. At the present time, Scarlet Rosefinches nest in considerable numbers throughout southern Finland as far north as latitude 66°30′N, with about a thirtyfold increase by the mid-1970s on the numbers present in the mid-1940s (Stjernberg 1985). Since the 1940s they have also extended their breeding range across the Baltic Sea into Sweden and have been breeding in rapidly growing numbers since 1954, especially since 1970. In that year they began to colonise Norway; they are now breeding commonly in the south and are still increasing and spreading, especially along the river valleys. This spread along river and mountain valleys is a feature of their mode of progression (Stjernberg 1985). Meanwhile, in the southern Baltic Sea region they began to nest in Denmark and northern Germany in the 1970s, following a westerly advance along the north coast of Poland. Even as far south as central Europe the Scarlet Rosefinch was already extending its breeding range in the 1960s, 1970s and 1980s—for instance, in waterside willow thickets as far south-west as Slovakia. (See Chapter 12 for more recent information on the continuing expansion.) In the Karelian peninsula of north-west Russia, Scarlet Rosefinches also increased rapidly from the 1960s, extending their breeding range northwards to the shores of the White Sea (Stjernberg 1985).

Breeding as they were so close to British shores by 1987, Scarlet Rosefinches (which, incidentally, spend the winter in southern Asia, including the Middle East) began appearing much more frequently on spring migration in Britain, and local ornithologists were soon eagerly anticipating the first breeding records. The first pair to do so nested in 1982 in Scotland. Subsequently, the English Channel and the North Sea proved to be an unexpectedly intimidating obstacle, and only a meagre sprinkling of pairs bred in Scotland until 1992.

Although I believe the spectacular westward advance of the Scarlet Rosefinch was to a large extent assisted by the 1850–1950 climatic amelioration, in common with several other Siberian species, such as the Yellow-breasted Bunting and the Red-flanked Bluetail, I accept that other factors must be at work. Torsten Stjernberg (1985) appears to follow Haartman (1973) in dismissing the amelioration of the North European climate since the 1930s as a main cause, in the belief that it was probably not of major importance in the range expansions of various species. Stjernberg himself suggested that the most important feature of that climatically favourable period was that in the 1930s almost all the summers were good for breeding success, and hence the number of colonists rose considerably. He suggested that changes in the choice of habitat were the main reason for the range expansion of the Scarlet Rosefinch. Among other factors he thought had contributed to its success were habitat changes due to man, its success in exploiting new, mainly open habitats, especially at the expense of sharply declining Linnet populations with which he considered the Scarlet Rosefinch to be in competition, and the relative longevity of the species.

The tiny Common Waxbill, which was introduced into open country in central Portugal, near Obidos, earlier this century, and is resident there, has been extending its breeding range into Spain since at least the 1970s, presumably helped by the warming climate since that decade.

The Rock Sparrow is another of the small number of southern European species which recoiled under the influence of the climatic amelioration. During the earlier part of the 19th century it nested in many of the warm, sunny valleys of south-west Germany, as far north as Thuringia. Towards the end of the century it disappeared from that country and from Austria and Switzerland (Voous 1960), and it has also vanished in the course of this century from Alsace and Burgundy in France (Yeatman 1976). The tendency for the summers in these regions to have been, on average, less warm and more humid during the amelioration is thought to have been the reason for this southward contraction of the Rock Sparrow's breeding distribution.

To some extent the Tree Sparrow has been affected in much the same way. I believe the generally wetter summers of the amelioration were probably responsible for its virtual disappearance from Ireland between about 1890 and 1959, and also from many parts of western Scotland and Wales, south-west England and north-west France. C.A. Norris (1960) correlated the Tree Sparrow's breeding range in Britain with areas of low rainfall. Indeed, it appears to avoid areas with more than 1000 mm (40 inches) annually, which seems to explain very well its distinctly coastal and valley distribution in Ireland, Wales and Scotland at the present time. It must be said, however, that the 1930s and 1940s, when the decline and range contraction of the Tree Sparrow were most marked, was a relatively dry period when summers were mainly hot, and winters cold. The Tree Sparrow then staged a quite dramatic comeback in western Britain and Ireland from about 1960 or a little earlier; a situation which could be correlated with the waning of the amelioration around 1950 and the consequent return to a rather drier climate.

Unfortunately for this theory (at least at first sight), the BTO's Common Birds Census showed that an equally dramatic decline in numbers and contraction of distribution again set in around 1976–77, just when another series of mainly hot, dry summers and very cold winters began. Since the hard winter of 1986–87, the winters have become mild, but the summers have continued to be largely hot and dry. However, the Tree Sparrow's decline goes on and its numbers have reached a very low ebb (Marchant *et al.* 1990; Marchant and Balmer 1993). So it is all very perplexing! Marchant *et al.* (1990) mention possible links between the long period of agricultural recession in the first forty or so years of this century (but consider, rightly in my view, that this was coincidental), and the recent widespread use of herbicides in weed control on farmland, which has adversely affected other seed-eaters. They dismiss the 1850–1950 climatic amelioration as a factor on the grounds that it was long past its peak before Tree Sparrows began their last phase of increase. They also point out that there is no evidence that they were seriously affected by either the use of organochlorine pesticides in the 1950s and 1960s or the loss of elm trees due to Dutch elm disease, as Tree Sparrows do not depend upon elms for nest holes. Summers-Smith (1989) concluded from his investigations of available data that upsurges of the British breeding population are due to irruptions from the continent of Europe following high population levels there, with subsequent gradual declines when immigration ceases. Population levels in north-west Europe have fluctuated a good deal, with declines in northern Germany since the 1970s and in Sweden in the 1980s, but recent increases in Denmark. Clearly, the reasons for the long-term fluctuations in Tree Sparrow breeding populations are difficult to understand; they are almost certainly complex. My own view remains that the more maritime regime of the 1850–1950 climatic amelioration was in fact primarily responsible for the widespread decline and contraction of range eastwards up to the late 1950s, at which date the amelioration waned and the Tree Sparrow populations recovered; and that this recovery would have continued beyond the mid-1970s had it not been for the increasingly widespread use of herbicides to control weeds from that time on. Thus, the latter may have overridden the benefits of the drier climate since the late 1950s.

As pointed out by Kenneth Williamson (1975), the northward spread of agriculture by up to a hundred miles in northern Europe and Siberia, during the 1850–1950 climatic amelioration, enabled such species as the Lapwing, Rook, Starling and House Sparrow 'to extend their spheres of influence'. The same is true of the House Sparrow and Starling in North America, whose spread, following their artificial introduction towards the end of last century, has been extremely rapid. In northern latitudes House Sparrows suffer heavy losses in hard winters, but the increased frequency of milder winters in Fenno-Scandia, for example, during the amelioration (especially in the 1930s) allowed it to achieve rapid gains of territory. It managed to colonise the Faeroe Islands during the 1930s, and by 1940 had occupied the most northern and north-western districts of Scotland.

Similarly, the Starling expanded its breeding range northwards and north-westwards very rapidly during the climatic amelioration earlier this century. Towards the end of the Little Ice Age it had withdrawn south-eastwards, hanging on, to quote Williamson (1975), only in the 'strongly maritime climatic zones of the Faeroes, the northern isles and the Hebrides', where the present local races had long become adapted to the local climate in past periods of isolation. In the British Isles, it had thus more or less disappeared from the greater part of Ireland and Scotland, west Wales and south-west England, and, moreover, had declined considerably elsewhere. A similar decline during this period (i.e. between about 1790 and 1830) apparently occurred in neighbouring parts of north-west Europe.

However, after 1830 Starlings began to recolonise all the areas they had vacated and beyond, and have continued to expand their breeding distribution to the present day. By 1880 they had reoccupied much of northern England and southern Scotland, and had also spread as far west as Cornwall and west Wales. Since then they have colonised the whole of the British Isles with the exception of parts of the west Scottish Highlands, and expansion was still continuing there and in west Wales up to the 1960s (Parslow 1973). Ireland, which had been almost abandoned during the early 19th century, was recolonised towards the end of that century, though the westernmost parts of the country were not reached until the 1930s or even later.

Elsewhere in Europe the Starling also extended its breeding range northwards and north-westwards to a marked extent, for example in Finland and Scandinavia. Starlings even began to nest on Bear Island in the Arctic Ocean between northern Norway and Spitsbergen from 1952, and on Spitsbergen itself from about 1954. A pair attempted unsuccessfully to nest in Iceland, where the Starling has long been a winter visitor, as early as 1935. A nest was also said to have been discovered in a natural cliff site in the mid-1930s (Boswall 1974); but it was not until 1941 that it became established as a breeding species in the south-east corner of the country. After that it expanded its breeding range slowly; Starlings became widespread in the south and increased enormously in Reykjavik and the surrounding area (Waag 1975).

This northward spread of the Starling has been generally attributed to the amelioration of the climate through the increase of the mean annual temperature (Voous 1960), plus the extension of agriculture and a reduction in the number of raptorial birds (Yeatman 1971). As mentioned previously, Williamson (1975) pointed out that, to a significant degree, the growth and spread of agriculture was made possible by the climatic amelioration. Since about 1970, however, the expansion trend has been reversed: Starlings have declined noticeably right across northern Europe, especially in Fenno-Scandia and in Britain. The BTO's Common Birds Census has shown that, since peaking around 1964, the British Starling population has been declining, with some minor fluctuations, ever since, and has been dropping particularly steeply since the early 1980s. O'Connor and Shrubb (1986) suggested that the widespread decline in rural areas could have been caused by the switch to the autumn sowing of cereals, and consequent reduction of spring ploughing, lessening the availability of soil invertebrates as food for Starlings; the use of fungicides may also have had this effect. These factors have also been advanced as reasons by continental authors, together with the effect of the post-1950 climatic deterioration in the north European sector. Marchant et al. (1990) also believe that the widespread European decline may have a climatic basis, but that changes in farming practice 'have surely contributed to the trend'. With this assessment I entirely agree.

The Golden Oriole is another species which benefits from climatic ameliorations, providing that they bring warmer and rather sunnier summers. As I noted in an earlier chapter, there is some evidence that the

Golden Oriole was a familiar bird in England in medieval times; that is, assuming that W.H. Hudson (in Nicholson 1948) was right in suggesting that this was the identity of the 'woodweele' in the following verse by an unknown poet:

'The Woodweele sang and wolde not cease
sitting upon the spray
So lowde he wakened Robin Hood
In the greenwood where he lay'

Chaucer in the *Romaunt of the Rose* wrote of the 'woodweele' or 'Wodewele' in terms that suggested it was widely known and therefore quite common: 'in many places were nyghtyngales, Alpes [Bullfinches], Finches and Wodeweles, that in hir swete song delyten'.

E.M. Nicholson (1948), however, was disinclined to agree with Hudson on the grounds that 'although not impossible it seems unlikely that the Oriole not only extended as far north as Sherwood Forest in the Middle Ages but was sufficiently well known to have a local name', and himself suggested that it may have been the medieval name for the Greenfinch. The Green Woodpecker is another possibility since at a later period it became known as the 'woodwall', but Nicholson considered it 'most unlikely that this was the bird intended either in the ballad or by Chaucer who actually mentions the Green Woodpecker also in the *Romaunt of the Rose* under the name "Papingay"'. I have wondered about the Willow Warbler or even the Wood Warbler as other possibilities, but I must confess that I incline towards Hudson's conjecture, since 'woodweele' is so suggestive of the unmistakable flute-like song of the Golden Oriole, which is variously described as 'weela-weeo' or 'aweela-wilee'. Moreover, the onomatopoeic Dutch name 'Wielewaal' is strongly suggestive of a common origin with the old English name.

E.M. Nicholson's objections to the translation of 'woodweele' as the Golden Oriole seem less tenable when set against the fact that Chaucer and other writers of the early Middle Ages lived towards the latter part of a long and very warm climatic phase, the Little Climatic Optimum, which at its maximum was much warmer than today. It is highly probable that Golden Orioles were much commoner in southern England than they are at the present day; and, moreover, that they nested not only as far north as Sherwood Forest, but perhaps even well beyond. As we shall see, this view is strengthened by the fact that the Golden Oriole spread considerably northwards in north-west Europe as a result of the 1850–1950 climatic amelioration, although it was much shorter and generally less warm. It has gradually re-established itself since 1950 in the southern half of England (notably in East Anglia since 1965) with more than twenty pairs breeding by 1976. It has even nested as far north as southern Scotland and possibly as far west as Wales.

To sum up, I believe that the Golden Oriole was a widespread and fairly common breeding bird in England, especially in the south, during the Little Climatic Optimum, but declined and retreated south-eastwards during the subsequent Little Ice Age, reaching a low point during the mainly very cold 17th, 18th and early 19th centuries. With the gradual improvement of the climate (especially the increase in mean spring temperatures) from 1850 onwards, the species began to recover, and since about 1930 has been extending its breeding range steadily northwards. Denmark was colonised from northern Germany as early as the 1850s, and it subsequently spread over most of the southern and especially the eastern part of that country. In 1976 it was estimated that between 200 and 400 pairs of Golden Orioles were breeding in Denmark, but unfortunately the population has declined since about 1980. There have also been marked decreases in Germany, but the population there does fluctuate a lot as I have seen for myself in Baden-Württemberg. The southern tip of Sweden was first colonised in 1944, and by the mid-1970s there were between ten and twenty pairs breeding there (Dybbro 1976). In Finland, an estimated 1,800 pairs of Golden Orioles (Merikallio 1958) bred in a large area of the south-east, and they have apparently increased and extended their range a little since about 1930. Here the Golden Oriole is at the northernmost limit of its range in the Palaearctic, and a glance at a map of its distribution there will reveal at once how attached it is to drier climates with warm summers. Thus it is well able to breed as far north as south-east Finland with its short, hot summers, but tends to avoid the cooler, maritime summer climates of north-west France (Brittany and Normandy), Norway and the British Isles, except when the climate alters, as it has done recently, towards a drier continental type. It will be very interesting to see if this trend continues now that the global warming due to the anthropogenic greenhouse effect has been gradually exerting an increasing influence on Europe's weather since about 1980.

This detailed examination of the effects of the 1850–1950 climatic amelioration now concludes with an assessment of the Corvidae (crows). Of the twelve species and subspecies breeding in Europe, six (Jay, Nutcracker, Magpie, Rook, Carrion Crow and Jackdaw) expanded their ranges northward or westward while one species (Siberian Jay) and one subspecies (Hooded Crow) retreated northwards. A further species, the Chough, contracted its range south-westwards.

To take the Siberian Jay first, this Boreal species of the coniferous forests of the taiga has been ousted from its former haunts in the Baltic states since about 1925, and also from many parts of southern Finland (still decreasing in the 1980s) and southern Scandinavia, by the northward advance of the Jay (Voous 1960; Williamson 1975), in the same way that the Brambling retreated before the advance of the Chaffinch. The Jay spread northwards in northern Scotland during the same period, no doubt also aided by afforestation, and now breeds regularly as far north as 57°N, but its advance now seems to have halted. As Sharrock (1976) has pointed out, it is strangely absent from several of the southern Scottish counties, and there is good reason to suspect that the rather greyer-backed birds in the more northerly population may have originated quite recently from the race breeding in Scandinavia. As we have seen, a number of other species (e.g. Wryneck and Red-backed Shrike) are thought to have colonised or attempted to colonise Scotland from this source. In Ireland the local race *hibernicus* has also greatly increased and expanded its breeding range north and west this century, especially since 1936, and it continues to do so (Hutchinson 1989).

In central Europe, the Nutcracker, a bird of montane coniferous forests, also spread northwards and westwards after 1930; it is believed that this was in response to the increase of the mean temperature in April when the young leave the nest (Voous 1960). For instance, this happened in Germany where it had spread as far north as the neighbourhood of Hanover by the early 1970s. In Fenno-Scandia the Nutcracker appears to have extended its range northwards slightly during the amelioration; Merikallio (1958) stated that it did not start nesting in south-west Finland until 1873, but by 1956 some 100 pairs were estimated to be present.

The Magpie (or Black-billed Magpie) is believed to have invaded North America from eastern Asia in comparatively recent times, perhaps during the present Post-glacial (Voous 1960). It expanded its breeding range quite rapidly to the east and north this century, and was also reported more regularly in winter north and east of its breeding range in Eurasia and North America (Voous 1960; Godfrey 1966). It was still increasing in Fenno-Scandia in the 1980s. This expansion may well have been helped by the climatic amelioration, as is probably the case in northern Scotland where the Magpie has been pushing northwards since about 1920. A marked reduction in persecution by gamekeepers since the First World War makes it difficult, however, to determine the precise causes of the increase and spread. Magpies were apparently unknown in Ireland before 1676, but subsequently spread from a colonisation of Co. Wexford, and they now breed commonly through-out the country. The cause was probably accidental and not connected with climatic change, but it is perhaps surprising that it did not occur previously.

In northern Europe generally, the Rook expanded its breeding range considerably to the north during the 1850–1950 amelioration. It was able to do this partly through its exploitation of the growth of agriculture; this in turn was extended northwards as a result (at least in part) of the increased length of the growing season resulting from the improved climate. Another consequence of this improvement which operated in the Rook's favour was the lower mortality in the milder winters. R.K. Murton suggested that the availability of grain from early ripening corn, or possibly the insects which live in the growing corn, may be making it easier for Rooks to survive the period of summer droughts as well. He further pointed out that very hot, dry summers are as bad for the Rook as cold winters, since they cause the soil to bake hard and thus prevent the bird from probing deep enough to obtain its staple diet of earthworms and other soil animals. Much the same happens in the winter in the heart of the continental land mass, where long, cold winters are usual and the ground becomes frozen. At this time, Rooks from such regions migrate to the more maritime parts of Europe where, under the influence of the Gulf Stream and the Atlantic westerlies, the winters are generally milder and wet. The fact that the areas of north-west Europe enjoying such milder winters have extended means that continental Rooks no longer need to migrate so far; this has enabled them to push their breeding range further north in the ensuing warmer springs.

In Finland, for instance, the Rook was unknown as a breeding bird until 1877, though well known as a spring migrant. Following subsequent scattered nesting records, the invasion of Finland really began in 1885 when Kirkkosaari in Köyliö was colonised. From about thirty or more pairs nesting in 1896 the numbers breeding in Finland grew to about 3,000 by 1956 (Merikallio 1958). In northern Scotland, Rooks spread north-westwards between about 1880 and 1960, colonising Orkney in the 1870s, the Outer Hebrides in 1895 and, as a result of an abnormal overshooting of spring migrants (Williamson 1975), Shetland in 1952. Elsewhere in the British Isles the Rook population increased rapidly this century, especially from 1930, but began to decline in most areas around 1960. A national Rook census in 1975 showed quite clearly that this decline, which amounted to 43% since the previous census of 1944–46, was then continuing on a large scale (Sage and Vernon 1978). Although various reasons were advanced to explain this decline, which was widespread throughout northern Europe from the Low Countries and Scandinavia to Russia, prominent amongst which were changes in modern agricultural practice, it seems likely that the climatic deterioration in northern Europe since 1950 may also have been involved. A subsequent sample census in 1980 in Britain (Sage and Whittington 1985) revealed signs of a recovery, and the BTO's Common Birds Census has shown that this has, with local exceptions in south-east England, been maintained. This seems to fit in with the growing warming influence of the greenhouse effect since about 1975.

Interestingly, in the drier southern half of France, where the Rook is generally rare or absent as a breeding species, the southern limit of its range gradually extended southwards after 1870 (Yeatman 1976). Presumably these new areas were opened up to the Rook by the wetter, maritime-type climate of the amelioration. The Mediterranean zone remains unsuitable because, as noted above, the soil tends to become too hard in the summer heat, and thus prevents the Rook from probing for its staple foods. It seems unlikely that the Rook will penetrate much further south as the climate returns to the drier continental-type that prevailed before the 1850–1950 amelioration. Incidentally, though, it has established several scattered rookeries in north-west Switzerland since 1963, before which date it was unknown as a breeding species in that country (Schifferli *et al.* 1980).

For much the same reasons as the Rook, the Jackdaw also expanded its breeding range considerably to the north in Fenno-Scandia earlier this

230

century, though it does not probe deeply in the soil for its food like that species (with which it associates), but takes it from the surface. A northward extension of breeding range took place in north-west Scotland during the same period, which resulted in the colonisation of a few places in the Outer Hebrides, Orkney, and by 1943 Shetland, but it gradually came to a standstill in the 1950s. Moreover, the populations of Jackdaws in several parts of Europe have greatly increased in the course of the present century, including not only Britain and Ireland but Spain, France (where there was a marked expansion of range to the south-east from the 1920s), the Low Countries and Denmark. Apart from occasional contrary fluctuations of a relatively minor sort, they are still increasing generally.

The Carrion Crow and the Hooded Crow are two races of the same species; they are believed to have evolved from a common ancestor whose European breeding population became split in two by the advance of the ice sheets during the Last Glaciation; one population isolated in the Iberian Peninsula gave rise to the modern Carrion Crow, while the other, isolated in south-east Europe and south-west Asia, became the Hooded race. Similarly, another population isolated in south-east Asia produced another black crow of the Carrion type. Voous (1960) believed that, following the retreat of the ice sheets with the climatic improvement, the Hooded Crow spread west from its central Asian headquarters to colonise Scandinavia, north-west Scotland and Ireland before the Carrion Crow, advancing from the south, reached these areas.

Where the two forms meet and overlap, which is for some 750 miles (1,207 km) in Europe and 2,000 miles (3,000 km) in central Asia, they inter-

FIGURE 57 *The Carrion/Hooded Crow range overlap in Scotland showing the shifts in the central axis of the hybrid zone between 1928, 1974 and 1991, correlated with an annual rainfall above 1,500 mm (adapted and updated from Cook 1975).*

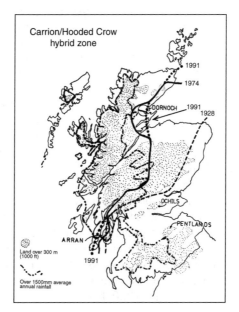

231

breed, and in consequence a hybrid zone has developed, varying in width from 15 miles (24 km) to 100 miles (161 km). During the 1850–1950 climatic amelioration this hybrid zone shifted considerably to the north in western Europe. For instance, in Scotland, Cook (1975) showed that its axis moved significantly north-westwards between 1928 and 1974 (see Figure 57); the difference is much more marked in the north-east than in the extreme south-west, where little change has occurred. I have added to Cook's map the slight changes which have occurred since 1974. In linking this shift of the hybrid zone with the climatic amelioration, he correlated it particularly with the influence of altitude for the following reasons: (a) the change has been most dramatic in the areas of eastern Scotland where the land is under 1,000 feet (300 m) and where the amelioration has allowed more intensive arable farming; (b) the zone is broader where the transition from low to high ground is more gradual; (c) the continued existence of small, possibly relict, populations of hybrids on the high Ochil and Pentland Hills to the south of the present zone; and (d) the tendency for the Hooded Crow to increase in frequency relative to the Carrion with increasing altitude, and to predominate on the highest ground.

This evidence seems undeniable; but there also appears to be a good correlation between the current axis of the hybrid zone, the north-western limit of the Carrion Crow's range, and the eastern limit of areas experiencing an annual rainfall of more than 1,500 mm (60 inches). The Hooded Crow's present breeding range in Scotland lies very largely within this region of higher rainfall, except in the extreme north-east where the Carrion Crow has only recently penetrated. Although in Ireland and the Isle of Man the Hooded Crow breeds throughout the drier as well as the wettest areas, here too the Carrion Crow either does not occur or has only recently arrived. It seems likely, therefore, that the Carrion Crow is better adapted to, and thus more successful in, the drier areas than the Hooded Crow, and has benefited during the amelioration from the increasing dryness of north-east Scotland and eastern Ireland, which latter country it is currently invading. In the absence of Hooded Crows, however, Carrion Crows are perfectly capable, as in Wales for example, of successfully inhabiting regions of high rainfall, especially in the valleys. The present north-western limit of the main range of the Carrion Crow in Scotland is just outside those areas averaging an annual rainfall of less than 1,000 mm (40 inches). In consequence of recent climatic trends, then, the Hooded Crow has withdrawn further and further on to the wetter, higher ground in Scotland. We may yet see this happen in eastern (especially north-eastern) Ireland too, unless the climate becomes wetter again. In the meantime, Hutchinson (1989) reported that Hooded Crows have increased to a great extent in Ireland since 1924. Both races have benefited from much reduced human persecution since 1914.

The retreat of the Hooded Crow as a breeding species in the face of the northward advance of the Carrion Crow in Denmark and elsewhere in Fenno-Scandia since about 1920 has led to a marked reduction in the numbers of migrant Hooded Crows from these areas wintering in south-east England. They are, for example, now rare in winter along the Thames estuary where they used to be common up to the First World War.

Another species of crow, the Chough, unlike most of the tribe, contracted its range in Europe during the amelioration. However, it had been declining since well before the 1850–1950 amelioration, for some 200 years at least, and there is little clear evidence that this climatic change was the main reason, or an important contributory one. In Europe, the Chough seems to have a Lusitanian–Mediterranean-type range (Burton 1974a), with its main centre of distribution in the Iberian peninsula, where it is found far inland as well as on the coast. The British Isles form its most northerly outpost and here it has a chiefly coastal distribution characteristic of a past glacial refuge, though it occurred well inland in Scotland up to and including much of the 18th century, and still does so to a limited extent in parts of Wales. It also used to breed on coastal cliffs in northern England including Yorkshire, and from Kent to Cornwall.

A sedentary species, the Chough probably colonised north-west Europe during the First Interstadial Phase (Aurignacian) of the Last Glaciation, when the climate was warmer than it is today. Subsequently, as the ice sheets advanced again—though not as far as in the First Glacial Phase—the Chough, along with other Mediterranean elements of the fauna and flora, retreated southwards and found refuge in the Celtic Land. As explained in Chapter 2, this was a large area of land which became exposed when the sea level dropped and retreated during the glacial phases, thus forming a direct land connection between the ice-free regions in and to the south of the British Isles, and France and Spain. At this time the Chough may have enjoyed a wide distribution in the Celtic Land, as in Spain nowadays, but became especially adapted to coastal cliffs in the north. When the sea level rose with the melting of the ice at the end of the Last Glaciation and submerged the Celtic Land, the Chough, like several other animals and plants, became isolated on the rocky coasts and cliffs of southern Ireland, South Wales, southern England and Brittany. Later it spread northwards along the coasts in the wake of the retreating ice to colonise northern Ireland, northern England and Scotland. It may indeed have colonised suitable cliffs all round the English coast, including East Anglia, by the Medieval Warm Period. There are, for example, Choughs pictured in early 14th century manuscripts written in East Anglia (Yapp 1981).

A possible alternative explanation of the Chough's distribution in north-west Europe is that it spread north from Spain during the Post-glacial Holocene (Flandrian) climatic optimum and has been declining slowly ever since, apart from a possible resurgence in the Medieval Warm Period (Little Climatic Optimum) when it may have become temporarily more common. It appears to have been a fairly widespread, popular and well-known bird in Britain in the early Middle Ages (Yapp 1981). However, its partiality for coastal habitats in the north suggests an adaptation which has enabled it to withstand severer climatic phases than those which have occurred in the Post-glacial.

The extent to which Choughs have withdrawn since the last century to the westernmost parts of northern France, Britain and especially Ireland, where the climate is the most maritime and the winters therefore are very mild, has been very noticeable. Choughs are known to suffer heavy losses in hard winters, such as that of 1962–63, and I consider it probable that they have been declining in Europe ever since the end of the very warm Little Climatic Optimum. The frequent bad winters of the prolonged Little Ice Age most likely had a particularly adverse effect on them, from which they have never recovered, apart from minor local increases in their strongholds. In addition, they have apparently suffered from some competition for nest sites and food (e.g. in Spain) from the highly successful Jackdaw, which greatly expanded its range and numbers in Europe during the 1850–1950 climatic amelioration. However, the fact that Choughs have decreased even where Jackdaws are absent shows that this is definitely not the chief cause of the decline, though it may contribute to it locally. It is pleasing to note that since the mid-1970s Choughs have been increasing, particularly in Scotland, and this may be connected with the mainly hotter, drier summers since then, which in turn can be linked to the greenhouse effect warming.

Susan Cowdy (1973) showed that ants and their larvae, among other soil invertebrates, form a staple diet of the Chough, at least in Britain; its long, decurved bill is adapted for probing deeply into their nest galleries. It may be that a reduction in the availability of these insects, possibly due to climatic factors or changes in vegetation structure, or both, could also have contributed to the waning fortunes of this elegant and enchanting crow. Certainly it is tempting to make comparisons with the long-term decreases in the British Isles of such ant-dependent species as the Wryneck and the large blue butterfly *Maculinea arion*; but more about this in the next chapter.

Section Three

NEW ICE AGE OR GREENHOUSE WARMING?

Chapter 11

EUROPEAN CLIMATE DETERIORATION 1950–1980

The apparent effect on birds

The reader will by now be well aware that the climate of parts of the Northern Hemisphere deteriorated between about 1950 and 1980. This natural, not man-induced (anthropogenic), deterioration did not become evident in all parts of the Northern Hemisphere at the same time. Indeed, in the lower latitudes, especially in certain sectors such as the British Isles and the adjacent coastal countries of the European mainland, signs of a deterioration did not at first become obvious, and they did not occur at all in the south-west. As one would expect, it was most noticeable in the Arctic regions, more especially in those parts where the amelioration had been most marked and rapid, such as around Franz Josef Land, Iceland and Spitsbergen. In Iceland and Franz Josef Land, for example, where the mean surface air temperature rose between 1910 and 1929 by an average of 1°C and more than 2.5°C respectively, the average temperature had fallen by the 1960s by 0.6°C in Iceland and 2.4°C in Franz Josef Land from the peak level attained during the 1940s. Even more dramatically, an area centred over Franz Josef Land became 5°C colder than in several of the preceding decades. Moreover, the Arctic sea ice had advanced by May 1968 to its most forward position for fifty years (Lamb 1975).

The influence of this cooling of the Arctic slowly spread more widely, so that by the late 1970s it was having a more noticeable effect in latitudes as far south as those in which lie the British Isles. Thus Britain, north-central Europe and eastern Europe began to experience their worst winters since 1962–63, that of 1978–79 being the most severe. The same trend was observable in northern North America. However, by the late 1980s this run of cold winters was succeeded by milder ones, especially that of 1988–89, which was one of the mildest on record.

In contrast to the European Arctic, however, the cooling of the Canadian Arctic began comparatively late. Even in the late 1960s, as Professor Lamb has pointed out, it had hardly started, and some areas had become temporarily even warmer than before. As a direct consequence of this tardy cooling of the northern regions of North America, such a steep increase arose in the gradient between the winter temperatures there and those of the tropical parts of the Atlantic Ocean, that it greatly invigorated the atmospheric circulation over the Atlantic during the 1950s, 1960s and early 1970s. Mild, moist air was driven by the then stronger and more persistent winds far into the heart of Europe and produced the remarkably long, if short-term, run of mild winters in these latitudes during the period when the overall climate of the Northern Hemisphere was actually cooling; thus the state of climatic amelioration was maintained there in spite of the general trend. Little of this warm air, however, penetrated northwards into the Arctic and the winters there became progressively colder.

Therefore, we can now understand how the apparently paradoxical situation arose whereby western Europe experienced some unusually mild winters at a time when the natural cooling of the climate was actually well under way. At the same time the man-made (anthropogenic) greenhouse effect was just beginning to exert an influence. By 1980 it was strongly counteracting the natural cooling trend, as predicted by the American climatologist Wallace Broecker (1975), and, boosted by a probably short-term, natural warming trend in the current decade, is now continuing to cause a rapid escalation in global temperatures. It is not surprising, therefore, that many European species of birds, insects and other animals have continued their range expansions (with in some cases minor checks) still in response to the earlier climatic amelioration, even though the deterioration in the Arctic has already forced northern species to retreat south. The six cold winters of 1977–78, 1978–79, 1981–82, 1984–85, 1985–86 and 1986–87 suggested that this short-term regime of mild winters had come to an end, and that we could expect cold winters to occur more frequently in western Europe in the foreseeable future. The generally mild winters that we have instead experienced in Europe since 1987 (those of 1988–89 and 1989–90 being especially so), are apparently due to the fact that the greenhouse warming is now really taking effect. Incidentally, a fairly narrow zone extending across eastern Siberia, Manchuria and northern Japan (Hokkaido) also enjoyed a similar run of unusually mild winters during that period (see Figure 58).

Meanwhile, much of the Arctic was becoming progressively colder. Between 1961 and 1975, for example, most months of every year were colder there than the average for the corresponding months of the period 1931–60. And even in England, in one of the mild winter zones, none of the winters between 1961 and 1975 had a mean temperature which exceeded the previous thirty-year average, with the minor exception of 1971, which exceeded it by only 0.1°C (Lamb 1975). Taking the world as a whole, G. Kukla of the Lamont-Doherty Geological Observatory, New York, demonstrated by means of a careful survey, using satellite pictures, that the global amount of snow and ice increased by 12% between 1967 and 1972, with the largest increase in 1971 when the Arctic cooling quite suddenly extended its influence to Canada (Lamb 1975). Most of this increase in the Northern Hemisphere occurred in spring and autumn, producing a lengthening of the winter of which most of us were all too well aware at the time. The tanta-

FIGURE 58 *Regional temperature trends for recent decades show that the world has not been warming evenly. The numbered contours indicate changes in average annual temperatures in units of tenths of a degree per decade. Britain and a large part of Europe actually cooled during the late 1970s and 1980s; parts of the former Soviet Union warmed dramatically (source: Climatic Research Unit, University of East Anglia, from Gribbin 1990). (Courtesy of Dr John Gribbin)*

lisingly late, cold springs were indeed quite a feature of the period 1960 to 1985, particularly in the late 1970s and early 1980s. The late 1970s also saw some severe winters in eastern North America, with very heavy snow-falls, as also happened late in the winter of 1992–93.

Professor Lamb (1975) believed that the intensification of the contrasts in temperature, around the perimeter of those regions of the Arctic which cooled during the two decades or so prior to 1970, was the reason for the anomalies in the distribution of the prevailing atmospheric pressure and winds. The result was an abnormal frequency of anticyclones in the zone between 40°N and 70°N—the so-called 'blocking' anticyclones. The increased persistence of one over Scandinavia at that period had the effect of blocking or keeping the Atlantic depressions and the prevailing mild west-erly winds at a distance from the coasts of the British Isles and adjacent parts of western Europe, thereby greatly reducing their frequency, especially between the mid-1960s and 1987. One consequence had been an increase in the incidence of bitterly cold winds from the east and north-east in spring which had delayed the onset of summer in those years; another had been prolonged periods of drought at various seasons of the year, seen at its most extreme in the severe summer drought of 1976—the worst of this century so far. A third effect of Scandinavian blocking anticyclones has been a tendency to hold up depressions in the North Sea area, especially in winter and spring, with the result that the British Isles is subjected to icy polar winds blowing round the depressions from the north and north-west.

Another unusual blocking anticyclone became established over Greenland in the early 1950s, and brought about a progressive increase in the frequency of incursions of cold northerly air over the Norwegian–Greenland seas, especially in the winter months. This development reached its peak during the years from 1966 to 1970 and has since declined somewhat; but it has had, and apparently continues to have, together with the colder conditions in the Arctic zone, a marked effect on winter visitors to Europe from the Arctic avifauna.

RESPONSE OF NORTHERN SPECIES OF BIRDS TO THE NATURAL POST-1950 CLIMATIC DETERIORATION (including the influence of the Scandinavian blocking anticyclone on southern species migrating north in spring)

In response to this colder Arctic climate, there has been a greater incidence of reports of Arctic birds south of their normal range in winter (especially in late winter), with a more southerly scatter in Britain and elsewhere in north-central Europe than formerly, and in addition a later stay than usual into the spring; this has involved particularly such species as White-billed Divers, King Eiders, Gyrfalcons (with winter influxes in 1972, 1973 and 1988) and Snowy Owls. Up to and including 1992 that beautiful and graceful Arctic rarity, Ross's Gull (Figure 59), had been reported fifty-three times in Britain since 1958, out of a total of only fifty-four accepted British occurrences since records have been kept. It now seems to be an annual, if still rare, visitor. The Ivory Gull, another attractive Arctic species, is also becoming a more frequent winter visitor to British and neighbouring European waters. Seven individuals of the extremely rare and beautiful Steller's Eider were also recorded in British waters between 1970 and 1976; there were none in the previous decade and only five in the previous 140 years.

Apart from the greater incidence of such formerly very rare Arctic visitors, a number of other Arctic and subarctic birds, for instance the Lapland Bunting, Shore Lark and Temminck's Stint, have begun to breed further south than had been previously recorded in the past hundred years or more; presumably they are being forced to contract the northern limits of their breeding ranges, and retreat south, in the face of the natural climatic cool-

FIGURE 59 *Ross's Gull, a high Arctic breeding species now appearing farther south in Europe in winter than hitherto this century. (Photograph: Dr John Sparks)*

ing trend in the high Arctic, and in spite of the incipient influence of the global greenhouse warming. Similarly, a number of Boreal and Temperate zone species have also withdrawn the northern limits of their breeding ranges southwards during the same period. In the rest of this chapter I propose to discuss systematically the most striking and interesting of these.

The White-billed Diver, which breeds on lakes and in river estuaries in the tundra of Arctic Russia and Siberia, has been found summering in Finland, and wintering further south and in greater numbers than was normally the case before 1950. For example, in Sweden the annual total of birds recorded in winter rose from a previous highest ever total of twenty-two in 1983 to thirty-seven in 1985, and eighty-eight in 1988. Vagrant White-billed Divers have also appeared as far south in winter as north-west Spain, France, Switzerland, Bavaria and Austria, as well as in the British Isles, Denmark, northern Germany and Poland. Up to the end of 1992, 132 had been seen in British waters since 1958, compared with only eighteen in the years before that.

The relatives of the White-billed Diver, the Great Northern Diver and Black-throated Diver, are not only also appearing further south in winter (e.g. Great Northerns as far as the Mediterranean), but are showing clear signs of extending their breeding ranges southwards too. For example, Great Northern Divers have been showing signs of breeding in Fenno-Scandia, and in northern Scotland a pair bred successfully in Wester Ross in 1970 (and was suspected to have bred earlier). In the following year a Great Northern mated with a Black-throated Diver on the same loch, and they reared a youngster. Such a hybrid pair also attempted to breed in subsequent years (e.g. 1985 and 1986). Interbreeding is not uncommon when a species is attempting to extend its range; I mentioned in an earlier chapter that Mediterranean Gulls sometimes pair up with Black-headed Gulls. A more appropriate case, in the context of this chapter, is that of an adult Glaucous Gull from the Arctic seas which paired with a Herring Gull and reared hybrid young in a Shetland locality in 1975 and subsequent years.

The Black-throated Diver spread southwards after 1950 to south-west Scotland and became established on the isle of Arran and in south Ayrshire from 1956, and in Dumfries and Galloway from 1974 (Batten *et al.* 1990), while in Sweden it also increased and spread southwards during the same period (Curry-Lindahl 1959–63; Parslow 1973). It also seems to have attempted to spread to northern Germany, as a pair were thought to have bred in the then German Democratic Republic in 1968. More recently, however, evidence has become available indicating a range contraction in south-west and eastern Scotland (Batten *et al.* 1990), perhaps because of the greenhouse effect warming in the 1980s.

The southward expansion of the breeding populations of the Red-throated Diver, described in Chapter 7, has continued at least in some parts of its range; for example, in Finland the population has more than doubled since 1987, although whether this is due to climatic deterioration in the Arctic since 1950 is still far from clear. In fact, it has been increasing in numbers near the southern limits of some parts of its breeding range for more than a century.

In discussing the extension of breeding range of the Slavonian Grebe in Chapter 7, I suggested that a climatic cause should not be ruled out, and that a study of the population trends of this species and the Red-throated Diver during the current deterioration in the climate might prove illuminating. So far, however, no clear trends have emerged for either species. In northern Scotland the Slavonian Grebe has continued to increase and spread slowly towards the south. It has, however, declined recently in the Scånia province of southern Sweden (as has the Black-necked Grebe), and also in south-west Finland (Fjeldså 1973). On the other hand, it continues to spread in the centre of Sweden, and also in Norway and north-east Poland. Fjeldså considers that the more recent expansion in Scandinavia has been influenced by eutrophication of many lakes as a result of human activities. It will, however, take some time yet, I feel, before we can ascertain the factors involved with any degree of certainty, several of which involve human influences in Scotland as well as in Scandinavia.

As I mentioned in Chapter 6, the remarkable spread of the Fulmar has slowed down since 1950, and I linked this with the deterioration in the climate of the North Atlantic since that date. Nevertheless, it has continued to increase at a rate of about 7% per year, especially in those latitudes which have only just begun to be affected by the increasing cooling of the Arctic region. It has, for instance, continued to increase in southern Norway; it has spread along the northern coast of France from Brittany to Normandy, although numbers of pairs breeding remain small; and since 1972 it has been breeding on the cliffs of Heligoland, off the north-west coast of Germany, and prospecting the north-west coast of Jutland, Denmark. If the climatic deterioration in the Arctic region continues and becomes more severe, it may be anticipated, however, that the Fulmar's expansion will slow down even more or perhaps cease altogether.

Turning now to the swans, geese and ducks, we find that several species which breed in the far north have recently been retreating southwards in response to the cooling Arctic, and have been colonising new territory beyond the former southern limits of their breeding ranges.

Although there is apparently no evidence so far of a recent alteration in the breeding range of the Bewick's Swan, there is plenty in respect of the larger and more widely distributed Whooper Swan. This species has increased and spread to a considerable extent in Fenno-Scandia since about 1955, occupying breeding sites much further south in Sweden and also in Finland, where the breeding population rose from about twenty pairs around 1965 to more than 300 pairs in 1979. In both these countries the earlier retreat northwards was blamed upon human persecution, while the subsequent recovery has been attributed to the introduction and enforcement of protective measures. The latter have undoubtedly had a beneficial effect, but have perhaps tended to mask the underlying influence of the recent post-1950 climatic deterioration in the Arctic region which has probably caused many Whooper Swans to breed further south, as they did during the Little Ice Age. Whoopers are currently showing signs of attempting to recolonise northern Scotland; almost every year several pairs stay the summer (a few occasionally nest), and it seems only a matter of time before breeding is recorded regularly. In 1990, an exceptional year, as many as thirty-two pairs were present in Scotland, some of which certainly bred. In 1973 a pair nested in the Biebrza Marshes in north-east Poland, and by 1983 six pairs nested as far south as Polish Silesia. In nearby Latvia they have been breeding in small but increasing numbers since 1973. Estonia has been colonised since 1988, with a pair nesting also in 1989, while in Lithuania, following a first breeding record in 1965, a pair reared four young in 1989 near the Baltic coast.

In Chapter 7, I drew attention to the fact that the recent comeback of the Greylag Goose as a breeding bird in north and north-west Europe has very largely taken place since 1950, and I suggested that a return to a drier, less maritime climate was favouring this large goose and crowning its improved protection (plus the large number of artificial introductions) with greater success than might otherwise have occurred. Moreover, there appear to have been some perfectly natural increases and southerly extensions of breeding range since 1950, notably in Iceland, where the population was estimated to have increased from 3,500 pairs in 1960 to 18,600 pairs in 1973; in the former German Democratic Republic, where the number of breeding pairs went up from about 450 pairs in 1969 to some 2,000 pairs in 1973; and in the Netherlands (at least 800 pairs breeding by 1990), Denmark, Norway, Sweden, Finland and Estonia. It will be interesting to see if the present climatic trend continues to favour the recovery of the Greylag Goose in northern Europe.

A remarkable breeding population of Barnacle Geese has built up in the Baltic Sea area, far to the south of the normal breeding range of the species in Spitsbergen and Novaya Zemlya. Presumably they originated from Novaya Zemlya birds which pass through this region on the way to their regular wintering grounds in the Netherlands. From 1975 when the first ones nested on the southern Swedish island of Gotland, the Swedish population increased to seventy-five pairs in 1981, to about 640 pairs in 1986, and to some 1,440 pairs in 1991. The first pair to nest successfully in Estonia did so in 1981, and one or more pairs have done so annually ever since. In 1988 a pair nested successfully for the first time in West Zealand, Denmark. The male had been ringed in the Netherlands in 1986, good evidence that this pair at least was descended from the Novaya Zemlya breeding population.

Of the ducks which have been appearing and breeding further south since 1950, the Scaup (or Greater Scaup), which was forced to retreat northwards by the climatic amelioration earlier this century, has recently made considerable gains along the southern limits of its breeding range in northern Europe at least. For instance, it has recolonised the south-east coast of Sweden since 1955, the south-west coast and islands of Finland, and even northern Estonia. There have been signs, too, that it is attempting to breed again on the Faeroe Islands. It may also have renewed its earlier but short-lived attempt to colonise Scotland, but so far with indifferent success. Thus, between one and three pairs nested annually on Papa Westray in Orkney from 1954 to 1959, and up to five pairs have nested or attempted to nest annually elsewhere, including North Wales, since 1977. However, in 1990 only one pair, in northern Scotland, was suspected of attempting to breed.

Earlier in this chapter I referred to the increased frequency with which Steller's Eiders, which breed in north-east Siberia and Alaska, have been recorded in northern British waters. The increase in the numbers of these beautiful ducks wintering further south than has been recorded since the colder years of the 19th century is even more marked in the Baltic; here the wintering population has swelled annually to a remarkable extent since about 1970, prior to which only odd individuals were reported. In 1977, for instance, as many as 130 were found to be wintering off the east coast of Sweden; by 1986 the total had more than doubled. Numbers wintering off the Lithuanian coast rose from eleven in 1969–70 to c.800 by the winter of 1990–91. In July 1975, an adult male visited the west coast of Schleswig-Holstein, Germany, the first time this species has been reported there this century. Since then there have been others; they have also occurred on the coasts of Belgium, the Netherlands and Denmark, and further east off the

FIGURE 60 *King Eider*

coasts of Poland (where a peak of ninety-seven was reached in 1987), Latvia and Lithuania. The first Steller's Eiders as far west as Iceland began arriving from 1981 onwards.

The equally handsome King Eider has also begun wintering south of its usual range in greater numbers than previously this century. Off the Swedish Baltic coast, the number rose from one or two a year in the early 1960s to about ten a year by the early 1970s, and is currently running at around 100 per annum. Wintering records are becoming more frequent off the Polish coast (especially Gdansk Bay), and to a lesser extent off the coasts of Latvia, the British Isles, the Faeroes, France and the Netherlands. Two have even been recorded in Hungary since 1973. Moreover, King Eiders, mostly drakes, are summering more often than before in European waters (e.g. Scotland), usually with rafts of Common Eiders, and they sometimes pair up with them. Such hybrid pairs have bred quite regularly in Iceland and may well do so occasionally elsewhere, as they have in the Faeroe Islands since the 1980s. An undoubted pair of King Eiders, however, have already nested on at least three occasions since 1957 near Trondheim, Norway, far to the south of the normal breeding range of the species, which is mainly well within the Arctic Circle. In June 1987, a male King Eider even arrived in the Ebro Delta of north-east Spain!

The Common Scoter, which I discussed in some detail in Chapter 7, continued its increase and southward advance in Scotland until 1976 or

1977, since when it has steadily declined possibly due to the anthropogenic greenhouse effect warming; the decline in Ireland which began earlier, has been quite dramatic. In 1985, however, a pair bred as far south as northern England, and summering was reported in 1986.

The colonisation of Scottish lochs since 1970 by the Goldeneye has been quite remarkable. This easily recognised tree-nesting duck of the Boreal zone began summering regularly in the 1960s. In 1970 a pair was proved to have nested successfully in east Inverness-shire. By 1976 at least five pairs, and possibly as many as eleven, were breeding in Inverness-shire, using nest-boxes specially erected for them in lochside trees. So Goldeneyes are firmly establishing themselves in Scotland; meanwhile, summering birds are increasing on English lakes, where nest-boxes are also being provided, and they may well become established as a breeding species there too. There is no doubt that the extensive nest-box scheme has greatly helped the natural spread of this beautiful duck. In 1990, 100 pairs were confirmed as having bred in Britain, producing at least 529 young (Spencer *et al.* 1993).

Elsewhere in Europe, the Goldeneye has been increasing rapidly in Finland since 1950 (also partly due to the provision of nest-boxes, according to P. Grenquist), and has been spreading south in Norway, though declining in the west of that country. In Sweden there has apparently been no significant change in its status recently, whereas in Poland it is breeding more commonly than before. Several pairs are now nesting annually in Denmark and the number breeding in northern Germany is approaching 200 pairs, while as far south as Belgium and the Netherlands pairs started nesting for the first time from the mid-1980s. A southerly range expansion has also taken place in the western Ukraine since 1982. Therefore, a southward shift of the southern limits of the Goldeneye's breeding range appears to be taking place in north-west Europe, and although the artificial provision of nest-boxes by man has undoubtedly helped, the underlying cause is most likely the climatic deterioration since 1950. However, a rapid decline has recently occurred in the southern Bohemian (Czech Republic) population, from *c.*100 breeding pairs in 1980 to less than fifty by 1988. Heavy predation of nests by pine martens *Martes martes* seems to be the problem here, according to Czech ornithologists.

The Smew, a small merganser and another species of the northern forest lakes and pools, may also be showing signs of extending its range southwards. In the Czech Republic, a pair nested for the first time in 1984 in southern Bohemia. Another pair bred successfully for the first time in Belarus (Byelorussia) in the Minsk region in 1988, and in 1989 two pairs nested; by 1990 this population had grown to at least ten pairs. On the other hand, there apparently used to be a relict Romanian population of Smews on the Danube Delta, but in 1984 these were reported not to have bred since the early 1960s.

That magnificent white owl of the tundra, the Snowy Owl, is another of the species which has been appearing south of its normal range in Europe more frequently than usual in recent years and staying longer in the spring. According to the Swedish ornithologists, B. Nagell and I. Frycklund (1965), there was a series of large winter invasions in southern Scandinavia during the period 1960–63. Following these influxes, summering Snowy Owls were reported with increasing frequency in Estonia, Germany and the Netherlands, as well as in northern Britain. For instance they began summering regularly in Shetland from 1963, and also did so from the 1960s in the Scottish Highlands and Western Isles, where unmated females laid infertile eggs in 1982 and 1983. Eventually, a pair were discovered nesting on the Shetland island of Fetlar in 1967, and from 1967 to 1975 a total of twenty-three young were reared there; then the old male, having driven away all the young males present, left the females without a mate when he disappeared in the winter of 1975–76 (Sharrock *et al.* 1978; Batten *et al.* 1990). As a result, although as many as five females were present in 1976, in some subsequent years up to the time of writing, breeding has not occurred since, although unmated females have laid infertile eggs. It still seems possible that, if the present climatic cooling in the Arctic continues, breeding might be resumed in Shetland and perhaps elsewhere in northern Britain. However, the increasing influence of the anthropogenic greenhouse effect may prevent it, although, as we have seen, the warming due to this is uneven and has not yet significantly affected much of Europe, especially Scandinavia, which actually became cooler by about 0.6°C in the 1980s (Gribbin 1990). In this connection, it is perhaps significant that in Swedish Lapland in 1978 Snowy Owls bred in unusually high numbers; several hundred pairs, in fact. Since then, smaller influxes of breeding birds in both Swedish and Finnish Lapland have taken place in 1981, 1982, 1987 and 1988. In 1982, for instance, seventeen pairs nested in Swedish Lapland, while in Finnish Lapland thirty pairs bred in 1987 and fifteen in 1988.

During the climatic amelioration earlier this century, the numbers of Snowy Owls wintering as far south as Britain, and other European countries in similar latitudes, declined. It was, for example, counted as a more or less regular visitor to northern Britain and Ireland up to the beginning of this century. but became steadily scarcer during the warmest phase between about 1920 and 1950. Since the winter invasions of 1960–63, the frequency of such influxes to the south has continued to increase with individual wintering birds appearing as far south as south-west Germany and Hungary.

The Great Grey Owl has also been showing signs of breeding further south than usual in eastern Europe in recent years, for example in the Ukraine, and also in Sweden, where the number of breeding pairs reached record heights in the late 1980s. Pairs have also recently (in 1991) nested in Belarus (Byelorussia) and Estonia. Likewise, the Hawk Owl has been moving further south than usual since the waning of the amelioration, especially in winter. Following the winter irruption of 1983–84 into Norway, Sweden and Denmark, there was a very big irruption southwards towards the end of 1989; the total of Hawk Owls recorded in Denmark that winter was the second highest this century. Another Boreal zone owl, the Ural Owl, has increased in numbers in its breeding areas in northern Europe since 1960; in the mountain forests of central Europe it has also increased and has spread westwards, colonising those of Moravia, in the Czech Republic, since 1983.

Among the raptors, another Arctic species, the Gyrfalcon, which declined in the southern parts of its range during the climatic amelioration, is occurring further south in winter in Europe than forty to fifty years ago and is staying later than it used to, though there is little evidence so far of it extending its breeding range significantly southwards, except for a 1988 report of thirty-seven birds seen outside their north-western breeding range in Sweden.

The Osprey, on the other hand, a summer resident which breeds largely in the Boreal zone, has been increasing in Scandinavia since about 1940, in spite of direct and indirect persecution by man, and has succeeded in recolonising Scotland, apparently from Scandinavia. At least two Swedish-ringed birds have been found among the Scottish breeding pairs (Batten *et al.* 1990). Soon after 1950, reports of Ospreys spending the spring and summer in the Scottish Highlands increased, and by 1954 there was strong circumstantial evidence that a pair had succeeded in rearing two young at one location. The following year a pair nested at Loch Garten in Inverness-shire, as they have done ever since; in spite of sterling efforts by the RSPB, early attempts were unsuccessful, because of the activities of egg collectors and other selfish people, until in 1959 they managed to rear three young. Since then Ospreys nesting at this site have, under the unceasing vigilance of the RSPB, enjoyed much better fortune. Meanwhile, other pairs were beginning to nest elsewhere in the Scottish Highlands, and by 1973 sixteen or more sites were occupied. In 1992, the latest breeding season for which I have figures, more than seventy pairs nested and they succeeded in raising at least 100 young (Dorling 1992). During the past decade the species has been slowly extending its breeding range southwards in Scotland, and since at least 1988 it has been showing signs of extending it into England.

Kenneth Williamson (1975) suggested convincingly that the return of the Ospreys to Scotland was connected with one of the features of the recent climatic trend, namely the more extensive development and influence of the Scandinavian blocking anticyclone since the 1960s. At this point I feel I cannot do better than to quote, with slight modifications, from the typscript of the book on which Kenneth Williamson (1977*b*) was working at the time of his death, and which, as I have explained in the Introduction (Chapter 1), his family have most kindly placed at my disposal. After listing a number of species besides the Osprey which have begun breeding or have shown signs of doing so in Scotland or elsewhere in northern Britain, such as the Wood Sandpiper, Temminck's Stint, Redwing, Brambling, Wryneck, Red-backed Shrike, Black Redstart and Reed Warbler, Williamson goes on to remark that it is clear that these colonists, appearing so far from their customary range, are derived from Fenno-Scandian stock. This assertion has since been strongly substantiated by the discovery of Swedish-ringed birds breeding in Scotland, as we have seen.

> 'Their arrival', he writes, 'is the result of displacement by the easterly airstream flowing round the southern side of the "blocking" anticyclone, during their return migration from winter quarters in southern Europe and Africa. On reaching Scotland they will usually have flown about the same distance as would normally have taken them to their home range in Scandinavia, and their migration urge will be consequently weakened, so that they are tempted to remain "off-passage", and even settle, when they encounter suitable habitat. All individuals might not be equally affected, and some which had not quite attained this physiological condition might stay for a short time and then reorientate to the continent; and it may be significant that in the case of the first Scottish breeding records of the Bluethroat (Greenwood 1968) and Black Redstart (Lea 1974), the nest, eggs and females were found, but no male bird was ever seen.'

This explanation, incidentally, may also apply to the growing number of migrant Broad-billed Sandpipers and Terek Sandpipers appearing in northwest Europe in spring. Williamson then proceeded to discuss the recent history of the Osprey in Scotland, pointing out that the successfully-fledged broods from 1959 onwards would not have attained breeding condition until three or four years old, and therefore that it is improbable that the sudden and startling upsurge in the breeding population in the early 1970s (fourteen to sixteen known sites):

> 'was achieved solely by the progeny of the Loch Garten pair, since it is known from ringing that a high proportion of the young fall victims to *les chasseurs* when overflying France, Spain and Mauritania on their autumn journey into Africa (Mead 1973). The probability is that the consolidation is due very largely to spring migrants being deflected to Britain in the manner I have described, and this view receives strong support from the changing pattern of their spring migration, through England especially. When the records published in recent years in the county bird reports are analysed, a sharp increase is noted for 1967, maintained through to 1970, and followed by peak years in 1972 and 1973. The picture in Scotland is obscured by wandering birds, perhaps potential colonists, but if what appear to be definite passage birds (at the coast and on islands) are added to the English records, then 1969 and 1970 show up as peak years for Britain as a whole.

> 'This interpretation in no way detracts from the magnificent efforts made on behalf of the Ospreys by the RSPB, for in an age when collecting is still unfortunately rife, well-organised conservation measures are vitally important in enabling this and other species to establish a viable

breeding stock. One hopes it will only be a matter of time before similar success is achieved with the Snowy Owl, for although the protected birds on Fetlar are still the only ones to have bred, the growing incidence of summer wanderers on other Shetland islands, and in the Scottish Highlands and Hebrides, must be due in no small measure to native-born birds.'

With this assessment I completely concur. Before leaving the fascinating story of the Osprey's recolonisation of Scotland, I should mention that it was formerly a relatively common breeding bird in that country (and to a lesser extent in England, where it last nested—in Gloucestershire—in 1840), until hounded to extinction by human persecution in the early part of this century. A pair that nested on Loch-an-Eilean in Inverness-shire in 1916 is thought to have been the last to do so until 1954. Incidentally, the Osprey has also recently recolonised other parts of Europe, including Denmark, Lithuania and parts of the Ukraine where it had been absent for thirty years, and it is increasing in north-east Germany (140 pairs in 1988).

The other species listed by Kenneth Williamson as being deflected to Britain by the Scandinavian blocking anticyclone are discussed further on in this chapter, except for the Goldeneye, Long-tailed Duck, Common Scoter, Scaup, Great Northern Diver and Snowy Owl, which I have dealt with already. The Hobby seems to be another example of a raptorial bird which may have been affected in this way. It is a summer resident in Europe, breeding on the continental mainland as far north as southern Sweden, north-central Finland and similar latitudes in Russia. In Britain, however, it was until recently mainly confined to southern England and south-east Wales, but in the 1970s there were distinct signs of an attempt to colonise a part of Inverness-shire in Scotland, far to the north of its usual breeding range (Sharrock 1976).

In Chapter 8, I referred, in passing, to the recent astonishing increase and southward spread of the Hen Harrier in Britain. This has been attributed to three chief factors: firstly, the reduction in persecution during and since the Second World War, due to the employment of gamekeepers elsewhere and the smaller numbers of them still active after the war ended; secondly, a reduction in the amount of heather burning, allowing taller, thicker growth more suitable for Hen Harriers to nest in; and thirdly, the big increase in the reafforestation of moorland and other uncultivated ground which, in its early stages, has also favoured the breeding of this elegant raptor. Undoubtedly, these influences have all been real and important, but the chronology of the Hen Harrier's history in Britain inclines me to believe that recent climatic changes may also be involved.

Until the intensification of field sports and collecting in the 19th century, the Hen Harrier apparently nested quite commonly in most parts of the British Isles. The heavy persecution this and other birds of prey suffered during that century, and the early part of the present one, clearly reduced its numbers and breeding distribution. By about 1900 the Hen Harrier was virtually confined as a breeding species to Orkney, the Outer Hebrides and Ireland (Sharrock 1976; Watson 1977). It may well have been that, in addition to the adverse effects of heavy persecution, the warming of the climate during the latter part of the 19th century and the first half of the 20th was in some way unfavourable to the Hen Harrier, since the revival of its fortunes as the amelioration drew to an end is quite remarkable. In 1939 it returned to breed on the Scottish mainland, consolidating its hold there during the 1950s. Although Orkney seems the most likely source of this invasion, Richmond (1959) suggested Norway, which is quite possible as central Scotland was colonised first. Indeed it could have been deflected there in the same way as the Osprey. Then after 1960 it spread steadily southwards, crossing the border into northern England by 1968 to breed there for the first time this century. Subsequently it colonised the north Midlands and much of Wales. The British population apparently rose during the period 1973–75 to more than 500 pairs (Watson 1977), but since then it is said to have declined, in some areas at least. Yet the population was still put at about 500 pairs in 1989. Hen Harriers increased and expanded their breeding range rapidly in Ireland, too, although here the expansion was to the north; but since the mid-1970s they have declined to a marked extent there, apparently due to the maturation of conifer plantations used for breeding and the loss of feeding areas on marginal hill ground (Batten *et al.* 1990).

As I mentioned in Chapter 8, this primarily southward extension of range of the Hen Harrier took place during the very same period which saw a retreat from Britain of Montagu's Harrier; but apparently it is not due to interspecific competition, even where their breeding habitats overlap, since this harrier disappeared from its southern haunts before the Hen Harrier arrived. The sheer speed of the Hen Harrier's extension of range in the British Isles, in spite of continuing human persecution in game-preserving districts, suggests to me that the return to a drier, more continental climate since about 1950 has been more favourable to its breeding success than the maritime conditions pertaining during the climatic amelioration—at least until the summers became hotter in the 1930s and 1940s. In the neighbouring countries of Europe, the only comparable increases of which I am aware have taken place in Spain, where the northern resident population has been expanding southwards beyond Madrid since 1980, and in the Netherlands, where the breeding population grew from up to fifteen pairs in the 1950s to more than 100 pairs by 1975 (Batten *et al.* 1990); in addition a pair bred in Luxembourg in 1985 for the first time for forty years, and another pair in Denmark in 1987, the first for forty-four years. In Belgium, Hen Harriers returned to nest in 1986 after a gap of eighteen years. Elsewhere, according to Bijleveld (1974), breeding populations decreased considerably, especially in France, where the decline between 1930 and 1964 was as much as 50% (Terasse 1964). There has also been a serious decline in northern Germany in recent years. Again, hunting, loss of breeding habitat and the use of agricultural pesticides seem largely, if not entirely, to blame for the decline in these countries. Incidentally, Donald Watson, in his exhaustive and delightful book on the Hen Harrier (1977), omits any discussion of climate change as a possible factor in his consideration of the fluctuating history of this fine hawk in Britain and Ireland.

A high proportion of the waders of the Northern Hemisphere have the centres of their breeding ranges in the north, many of them in the Arctic tundra or the Boreal zone. As shown in Chapter 6, quite a number of them withdrew northwards in Europe during the pre-1950 amelioration. When that climatic phase waned, to be replaced by a cool phase over much of the continent, at least some of these species began to retreat from the increasing cold of the Arctic region, and extended the southern limits of their breeding ranges further south again. Examples of which I am aware are the Golden Plover, Dotterel, Turnstone, Broad-billed Sandpiper, Temminck's Stint, Sanderling, Purple Sandpiper, Greenshank, Wood Sandpiper, Whimbrel, Jack Snipe and Red-necked Phalarope.

In Chapter 6 I summarised the history of the Arctic-alpine Dotterel during the 1850–1950 climatic amelioration, to which, like Desmond Nethersole-Thompson (1973), Kenneth Williamson and others, I attributed its decline in that period. In support of this conclusion, I also mentioned the recovery and southward advance in Britain and apparent increase and expansion of breeding range in alpine Europe of this highly confiding bird since that amelioration waned. Recent surveys (1987–88), more accurate than ever before, have revealed that the main Scottish population now probably exceeds 860 pairs (Marchant *et al.* 1990). With justification, Sharrock (1976) suggested that it should now be searched for on suitable mountains in Ireland, where in fact a pair nested in 1975. Moreover, since about 1970 there has been some further recolonisation of the alpine regions of Europe from which it had disappeared, such as the Italian Alps and Apennines, the Czechoslovak and Polish Tatras (first since the 1940s), and also the Pyrénées, where it had never been recorded before. As with other Arctic-alpine species, it will be interesting to observe the eventual influence of the anthropogenic greenhouse effect.

The history of the Dotterel's cousin, the Golden Plover, both during the amelioration and also from 1950 to about 1980, was also described in Chapter 6. Recently, due to widespread changes in land-use, the British population seems to have decreased, especially in Wales, and breeding populations are reported to have declined in other north European countries. But in Finland Golden Plovers have extended their breeding range further south since the 1950s (Merikallio 1958), and there has also been an increase and significant southward expansion of breeding range in European Russia and, in spite of the drainage and reclamation of bogs, in the Baltic republics as well. In Latvia, for example, the breeding population grew from thirty to forty pairs in the 1950s to 300–400 pairs in the 1980s (Tomkovich 1992). Isolated pairs have also recently nested in the Netherlands and Belgium.

The Whimbrel is another example of a wader in which the trend towards withdrawal to the north has been reversed since the Arctic climate began to cool again. Confined to the northern islands of Shetland in the 1950s, it subsequently started breeding again on Orkney, Fair Isle, St. Kilda, the Isle of Lewis and the Scottish mainland (Caithness). On Shetland, where the majority breed, the population of Whimbrels trebled between the early 1950s and the 1970s (Sharrock 1976); by the mid-1980s a maximum of 471 pairs was reached (Batten *et al.* 1990). Meanwhile, its ecological competitor, the Curlew, slowed down its northward expansion in Britain and elsewhere in northern Europe, and has been declining in most countries since about 1960. Nevertheless, a pair nested in north-east Iceland in 1988 for only the second time on record.

As I mentioned in Chapter 6, the breeding population of the Greenshank in Scotland also withdrew northwards to a slight extent during the amelioration, and subsequently showed signs of returning and spreading to new areas. However, largely due to intensive afforestation of its peatland habitats recently, particularly in Caithness and Sutherland, it has since declined. The Red-necked Phalarope also withdrew northwards, and at the same time pushed the northern limits of its breeding range further towards the Arctic at the expense of its high-Arctic ecological competitor, the Grey Phalarope. Like the Greenshank, there is some evidence since 1950 that the Red-necked Phalarope is tending to breed a little further south in Europe than before, although its status as a British breeding bird has become even more parlous. Indeed, the latest reports indicate that the British breeding population as a whole is inexorably decreasing, as also is that on the Faeroes.

Some species of waders with northern distributions in Europe have actually extended the south-western limits of their breeding ranges, so that they have begun to colonise, or are attempting to colonise, the British Isles and certain other countries further to the south and west than they have ever before been recorded to breed in the annals of ornithological history. Examples, already mentioned in passing, are the Turnstone, Broad-billed Sandpiper, Temminck's Stint, Sanderling, Purple Sandpiper, Wood Sandpiper and Jack Snipe.

The first of these examples is, at first sight, a rather surprising case. As described in Chapter 6, the Turnstone retreated northwards during the height of the pre-1950 climatic amelioration from its southernmost localities around the Baltic Sea. It is in a bad way in Denmark, where it is also suffering from human disturbance of its remaining breeding haunts (Dybbro 1976). Nevertheless, there are indications that the post-1950 climatic cooling in the Scandinavian sector of the Arctic zone is causing birds from the less disturbed northern populations to move further south to breed. For some time there have been suspicions that the increasing numbers of summering Turnstones in the Faeroe Islands and northern Britain, behaving in a manner suggestive of nesting, were a prelude to an attempt at colonisation. Proof of breeding has been extremely difficult to obtain, but in 1976, at a site in Sutherland in the north of Scotland, an agitated adult Turnstone, constantly calling in alarm, led to the discovery of a downy chick a week or two old; unfortunately it could not be fully ascertained that it belonged to this species (Sharrock *et al.* 1978). Since then there have been no other reports of attempted breeding in Britain.

Pairs of Sanderlings, which normally breed no nearer Britain than the Arctic tundra of eastern Greenland and the northern coasts of Siberia, were seen spending the summer on Scottish mountains for a few years from 1973, but proof of breeding was lacking. However, there is no lack of proof in the case of the Wood Sandpiper and Temminck's Stint. Both may have been regular breeding birds in northern Britain during the Little Ice Age, but there were of course few, if any, ornithologists operating in that region until after that climatic phase had petered out. Authenticated breeding records of Wood Sandpipers at a Northumberland locality in 1853 and 1857, right at the end of the Little Ice Age, when the climate was beginning to warm up, suggest that this may have been so. During the subsequent long amelioration of the climate there was evidence of a decrease along the south-western periphery of its north European breeding range; it ceased to breed in the Netherlands after 1935, for instance, and declined in northern Germany.

It was in 1959, soon after the post-1950 natural deterioration in the climate of the Arctic region started, that a pair of Wood Sandpipers were again proved to have bred in Britain, this time in Sutherland. Since that date breeding has continued in that county, while other pairs have become established elsewhere in northern Scotland, though numbers remain low at present. The maximum total so far of known and possible breeding pairs was twelve in 1980. Numbers appear to have declined somewhat since then. Elsewhere in Europe, Wood Sandpipers have been attempting to colonise the former Czechoslovakia ever since 1935 when they were first recorded breeding; since 1981 breeding has been reported every year there. In that year breeding was confirmed for the first time ever in Iceland: at Mývatn. There are also signs of a recolonisation of north-east Poland; up to five pairs were located displaying at Bielawskie Marsh, near the Hel Peninsula, in 1977 and 1981, and a nest, the first in Poland for some 40 years, was found in 1983. In Denmark and northern Germany, however, the decline which began in the latter part of the 19th century has continued.

The closely related Green Sandpiper, whose breeding range overlaps that of the Wood Sandpiper, has also been suspected of nesting more frequently in northern Britain in recent years, but so far there is only one definite record, in Inverness-shire in 1959. Back in 1917 a pair was proved to have bred in Westmorland, and there were also sporadic instances of suspected breeding in Scotland before 1950. Like the Wood Sandpiper, Green Sandpipers have also been endeavouring to colonise the former Czechoslovakia, having bred irregularly since 1935; now they have been found nesting in growing numbers every year since 1981. They have also colonised Denmark since the mid-1950s and seem to be attempting to do the same in northern Germany, whereas in Finland they have continued to spread northwards, possibly still in response to the climatic amelioration in the first half of this century, which, as we have seen, also induced them to spread westwards and even south-westwards. In these respects the Green Sandpiper's response appears to be contrary to that of the Wood Sandpiper.

Those very few Temminck's Stints which are now nesting in the Scottish Highlands are breeding in an outpost well to the south-west of the normal breeding range, which is in the tundra from Norway to Siberia. Curiously, an isolated pair nested as far south as central Yorkshire in 1951. There were also isolated breeding records in northern Scotland before 1950, but regular breeding, or probable breeding, has only occurred since 1969, with up to six confirmed and possible pairs in 1979 and 1980. Temminck's Stints have also been breeding further south than usual in Norway recently, but have decreased considerably in Finland, especially on the west coast, apparently due to habitat changes (Cramp and Simmons 1983).

Broad-billed Sandpipers, small dumpy waders with a kink to the end of the bill, which breed chiefly in the subarctic tundra and high wet moorlands of Fenno-Scandia, have spread southwards to a quite remarkable extent since about 1950. They have established outposts as far south as the Baltic Sea coast of Schleswig-Holstein in northern Germany, and are now being seen so frequently in Britain in spring that hopes are entertained by British ornithologists that they may soon attempt to breed there too. Kenneth Williamson (1975) believed that the increasing trend towards spring occurrences in Britain of both Broad-billed Sandpipers and Terek Sandpipers is connected with the more extensive development of the Scandinavian blocking anticyclone (compare the discussion of the Osprey's recolonisation of Scotland above). As I mentioned in Chapter 6, the Terek Sandpiper, an elegant wader with an uptilted bill from the inland marshes of the Boreal zone in northern Russia and Siberia, has been spreading westwards (Figure 22) into Finland and the adjacent Baltic countries since 1930, possibly in response to the amelioration of the climate earlier this century. It is continuing to do so, having reached Latvia (in 1980) and the Gulf of Bothnia in Finland, and it has been visiting Britain, Denmark and other coastal areas of north-west Europe more frequently since 1969. The precise reasons why the Terek Sandpiper should benefit from a climatic amelioration are, however, not obvious, unless wetter conditions associated with the amelioration provided more of the flooded or marshy conifer forest habitat that it especially favours.

In June and July 1978, the Purple Sandpiper, a breeding species of the Arctic-alpine regions of northern Europe, whose range extends as far south-west as the Faeroe Islands and the mountains of southern Norway, was discovered breeding in an undisclosed locality in the Highlands of northern Scotland (Dennis 1983). This first pair produced three young. Since then breeding seems to have occurred every year, and by 1989 up to four pairs were nesting or apparently doing so in two separate localities. As the Faeroe population is reported to be declining, it is believed that the Scottish birds originated from Norway.

Finally, the Jack Snipe is also reported to be extending the boundary of its breeding range, southwards from northern Scandinavia and westwards from northern Russia. Several pairs, for example, were found nesting in the marshes of north-east Poland from the late 1970s onwards. During the 1850–1950 climatic amelioration the Jack Snipe was reported to have contracted its then breeding range north-eastwards from southern Sweden and the south side of the Baltic Sea (Harrison 1982).

Moving on now to another group of birds, the gulls and skuas, there is some evidence that the Long-tailed Skua, which breeds in the high tundra and mountains of Fenno-Scandia, Russia and elsewhere in the far north, is currently extending its range to the south. Its breeding population has been building up in Norway and Sweden recently, for instance, while in northern Britain birds have been appearing amongst colonies of Arctic Skuas during spring and summer. Moreover, Long-tailed Skuas have been occurring on passage around Britain, southern Scandinavia and elsewhere on the North Sea coasts of north-central Europe in unprecedented numbers since the end of the 1960s. Also, Arctic Skuas have been increasing in most of their breeding haunts in northern Britain since about 1940, and it looks as if they are attempting to spread south-westwards.

I mentioned above the greater frequency with which those truly Arctic gulls, the Ivory and Ross's, are appearing in British and north European waters. As O'Sullivan *et al.* (1977) pointed out, the latter are no longer almost purely winter visitors, but may occur in any month of the year. In May 1991, an Ivory Gull was even seen as far south as Bohemia in the Czech Republic. It may be that both species are extending their Arctic breeding ranges further south owing to the post-1950 climatic cooling in the Arctic zone, as may also be the case with those other Arctic breeding gulls, the Glaucous and Iceland. Both of these species are wintering further south than formerly and in greater numbers, even as far south as the Mediterranean and the Atlantic coast of Morocco! The Common Gull has continued to extend its breeding range southwards into central Europe, and during the 1980s began colonising Belgium, the former Czechoslovakia, Austria and Hungary. It is not yet clear if this is a response to the post-1950 cooling or to some other factor.

In Chapter 6 I discussed at some length the serious decreases in the southern parts of their ranges of such auks as the Puffin, Guillemot, Black Guillemot and Razorbill during the 1850–1950 climatic amelioration; I linked this with a northwards withdrawal of the fish upon which these sea birds feed, and commented on the possible cessation of the decline of the Puffin. Since I wrote that, information has come to hand that the breeding population of the Puffin has greatly increased on the Faeroe Islands in recent years, as have the numbers of Black Guillemots; also that the British population of the Guillemot doubled between 1970 and 1987 (Batten *et al.* 1990).

On the other hand, Razorbills have continued to decline in the Faeroes and in south-west England. The serious decline of the Iberian breeding Guillemot populations since the 1950s I have already described in Chapter 6, so it will suffice to say here that it has continued. At the eastern end of its European range, the Razorbill has been colonising the Estonian coast since 1988, a southerly extension of its breeding distribution. It is too early to evaluate the reasons for this.

Since the early 1980s, Little Auks have been wintering further south than before, and there has been a general shift southwards of their wintering area into the Baltic. For instance, 280,000 spent the 1987–88 winter in the Skagerrak. This strongly suggests a response to the recent cooling of the Arctic climate.

In Chapter 8 I discussed the apparently perplexing cases of the Wryneck and Red-backed Shrike which, while clearly withdrawing as breeding species from England and adjacent parts of Europe because of the more maritime and less 'continental' climate which developed there through the climatic amelioration, have paradoxically been making attempts to colonise northern Britain since the mid-1960s or possibly even earlier. These attempts have been more obvious in the Wryneck's case as several pairs have nested or attempted to nest almost annually in northern Scotland since 1969 (see Figure 34), even though there was no confirmed breeding record in England between 1973 and 1977. In the latter year two pairs were proved to have bred out of three pairs present in England. Since then, occasional pairs are said to have nested in southern England, but the Wryneck is all but extinct there as a breeding species. The Red-backed Shrike's attempt at colonisation in Scotland began with a pair which may have nested in Orkney in 1970. Subsequently, up to 6–8 pairs were present in the breeding season from 1977 to 1979 inclusive, and some of these nested successfully (Batten *et al.* 1990). In the summer of 1979 a friend of mine, Robin Prytherch, discovered a Red-backed Shrike nest with young in the top of a small ornamental conifer in a locality in the eastern Scottish Highlands. Such a nesting site is commonly chosen in its southern Scandinavian haunts and may thus indicate the source of the Scottish birds. Since 1979 only two pairs are known to have made an attempt to breed in Scotland—in 1987 (Batten *et al.* 1990) and in 1990, when a pair fledged a single youngster successfully (Spencer *et al.* 1993). At the time of writing the last pair of Red-backed Shrikes known to have nested successfully in England was in 1988.

As has been explained earlier in this chapter, the easterly airstream blowing in spring around the Scandinavian blocking anticyclone has tended to 'catch-up' summer residents, such as Red-backed Shrikes and Wrynecks, attempting to reach their normal breeding haunts in southern Norway and Sweden from their winter quarters, thus deflecting them westwards to northern Britain. In spite of the more northerly latitudes, the drier, warmer summer climate of southern Scandinavia is more favourable to these and many other species than is that of Britain. If the greenhouse warming does eventually lead to a generally more continental type of climate in England, there is a fair chance that both these species may recolonise the large areas they have vacated in the course of this century. It will be very intriguing to watch what happens in future years.

The very occasional records of breeding or possible breeding in Scotland of Black Redstarts (Orkney in 1973, Grampian in 1976) and Bluethroats (single pairs of the northern, red-spotted race bred in Inverness-shire in 1968 and 1985) also fall into the category of birds probably deflected from Scandinavia in the manner just described, as probably do the recent records of Ring Ouzels nesting in Shetland for the first time. The pair of Bluethroats which bred in 1985, for example, did so following a heavy spring passage of the species in eastern Britain associated with a period of prolonged easterly winds during May (Batten *et al.* 1990). Kenneth Williamson recorded that the numbers of Black Redstarts and Bluethroats occurring on spring migration at Fair Isle had greatly increased in the 1960s; the Bluethroat totals, for instance, swelled fourfold from 1968 to 1971, and have remained high since.

The numbers of Bramblings passing through Fair Isle, and Spurn Head in Yorkshire, have also increased since the mid-1960s. First confirmed as breeding in Britain in 1920, when a pair nested in the far north of Scotland (but the eggs were robbed), between one and ten pairs of Bramblings have bred or possibly done so in Scotland in most years since 1968 (perhaps even several years earlier). In addition, a pair apparently bred successfully in Cumbria in 1987, and others may have at least attempted to do so elsewhere in England and Wales in recent years. Certainly, the numbers of birds found holding territories is increasing. I have not, however, heard of any reports yet as to whether the region of overlap between the breeding ranges of the Brambling and its southern counterpart, the Chaffinch, has withdrawn southwards in Fenno-Scandia in response to the post-1950 climatic cooling, but according to Batten *et al.* (1990) there is no real evidence that the Brambling is spreading there. (See also pages 260–261.)

One of several exciting episodes in recent Scottish ornithological history was the apparent incipient colonisation of the Scottish Highlands in the 1970s by the Shore Lark (or Horned Lark). It will be remembered that I described this in Chapter 9. Unfortunately, ornithological hopes were soon dashed, as none have been seen since 1977. Perhaps this is not too surprising, as a marked decline in the breeding population in Finnish Lapland has occurred since the 1930s, when it was a common breeding species; by the late 1980s it was down to 'some tens of pairs, at most'. Its numbers on migration in southern Finland have also declined markedly in the same period. In this connection it is interesting that in Britain the wintering population of Shore Larks increased from about 1950 to 1975, then declined to the extent that fewer than 100 birds are now seen annually along the North Sea coast of Britain (Batten *et al.* 1990). These events may be linked to the cooling of the climate of northern Europe since about 1950.

Hot on the heels of what, at the time, was thought to be a new colonist in the Shore Lark, there came the news that Lapland Buntings had also been proved to breed in northern Scotland in the same summer of 1977. Two nests with clutches of four eggs, laid by two females apparently mated to the same male, were discovered by Ian G. Cumming (1979). In that same year no less than 14 other pairs of Lapland Buntings were thought to be nesting in six different north Scottish sites. Subsequently, two pairs were again proved to have nested in 1978, with another four pairs considered to have possibly done so, in three Scottish sites; then in 1979 no less than 11 pairs were proved to have bred, while 14 other pairs were thought to have been possibles in a total of five sites. Having reached this peak in the space of a few years, only a single pair was confirmed as breeding in 1980, with no others present. In 1981 only a single bird was seen in June; thereafter, no others were reported until 1989 when a single female was seen at one of the former Scottish sites in late July. Unless this single female was the precursor of a new attempt at colonisation of northern Scotland from, presumably, Scandinavia (where it breeds as close as southern Norway), then this recent effort seems to have failed. Unfortunately for that hope, none were seen in 1990. That the Lapland Bunting might attempt to colonise northern Britain had been anticipated as it is one of several species which has exhibited signs of extending its breeding range southwards with the return of cooler conditions to the Arctic region since about 1950. It has, for example, been found recently breeding in the Faeroe Islands.

FIGURE 61 *Snow Bunting*

Meanwhile, the Snow Bunting, which has also been showing clear signs of expanding its breeding range southwards again, has been reconsolidating its breeding population in the Faeroes and Scottish Highlands since the 1960s, having all but disappeared during the warmest phase of the climatic amelioration earlier this century. Nethersole-Thompson (1966) contended that Snow Buntings 'were more frequently recorded in summer and on more hills from 1864 to 1913 than between 1914 and 1963'. As Kenneth Williamson (1977b) put it, with regard to the latter period, in the draft papers supplied to me by his family following his untimely death:

'there has not been a definite breeding record from Ross or Sutherland since the 1920s, and no permanent breeding group remained in the Cairngorms, though occasional pairs nested. The Faeroe Islands tell a similar story: Wolley (1850) found pairs inhabiting relatively low ground, and colonies on the higher hills of the north islands; but almost a century later I retraced his steps without finding a single bird, and there can be no doubt that the retreat had become a rout in both regions by the peak of the amelioration. Coincident with the colder Arctic weather since the 1950s the Snow Bunting has improved its position among the Scottish hills.'

Sharrock (1976) considered that breeding has been fairly regular since 1945 in the Scottish Cairngorms, with at least three pairs nesting in 1947. In 1972, at least five broods were raised, and in 1975 it seemed possible that as many as 22 pairs could have nested at 13 different sites. Since then numbers and observer coverage have improved, so that by 1989 there was evidence that the Scottish breeding population numbered 31–52 pairs, spread over 11 localities (Spencer et al. 1991) in six counties. In the middle of July 1979 three adults were even seen as far south as Cumbria. Reports from the Faeroes in the mid-1970s indicated that the number of breeding pairs has grown to at least five.

Perhaps the most spectacular colonisation of Scotland by a northern species has been that of the Redwing. Apart from occasional nests discovered subsequent to the first Scottish breeding record in 1925, and more frequent ones after 1950, the first evidence of a serious colonisation came in 1967 when seven pairs were located. Next year, a BTO bird census expedition to Wester Ross discovered as many as twenty pairs in that area alone, including ten proved to have actually bred. The *Atlas of Breeding Birds in Britain and Ireland* (Sharrock 1976) suggested a possible Scottish Redwing breeding population in 1972 in the region of 300 pairs. In the decade following that major effort a decrease appeared to set in, but this may have been at least partly due to birdwatchers understandably failing to maintain every year the prodigious efforts they made during the years 1968–72, when the breeding birds of the British Isles were being intensively surveyed by an army of observers for the BTO Atlas. Nevertheless, judging by data supplied up to 1990 to the Rare Breeding Birds Panel (Spencer et al. 1993), the

259

Scottish breeding population has undoubtedly fluctuated over this period. A recovery clearly occurred from 1982 up to 1989, with a maximum of seventy-nine confirmed and possible breeding pairs recorded in 1984, following an apparent marked decline in the four years 1976–79. However, fewer birds from fewer localities were reported to the panel in 1990, but coverage was suspected to have been inadequate. This fluctuation in numbers could also be accounted for by returning Scandinavian migrants being displaced westwards to Scotland by easterly winds associated with a blocking anticyclone, as suggested by Ken Williamson (1975). Since 1976, pairs have also nested sporadically in England (Kent, Lincolnshire, Northumberland and Staffordshire). The origin of these Scottish redwings is not certain, but is thought to be Scandinavia rather than Iceland, where those which have nested in the Faeroe Islands are believed to have originated. The two races are difficult to separate in the field. Incidentally, Redwings have lately taken to nesting sporadically as far south as Austria and the former Czechoslovakia, in addition to Germany, and since 1980 have been pushing their breeding range southwards in the western Ukraine.

Fieldfares seem to be simultaneously colonising Britain from Scandinavia and the Low Countries. As I explained in detail in Chapter 9, they have been spreading westwards through northern and central Europe since about 1750; it was only a matter of time before they reached Britain, which they did in 1967. Although the rate of colonisation in Britain has been slow and the number of breeding pairs remains very low—only about a dozen pairs or less, mainly in central Scotland and north-central England—the rate may be expected to speed up as Fieldfares consolidate their recent colonisation of the Low Countries and continue their westward expansion through France. These trends will probably eventually lead to the invasion of southern England as well as the north and centre: birds are already appearing in the breeding season, and breeding may have occurred in Berkshire. Unlike the Redwing's spread, which is most probably mainly in response to the climatic deterioration since about 1950, the Fieldfare's expansion of range is more likely to be linked with the previous climatic amelioration, having been in progress for a very long time. Where I live in south-west Germany, colonising Fieldfares have adapted well to breeding in suburban parks and gardens, and they may well exploit such habitats in southern England one day.

As mentioned in Chapter 10, the Brambling has shown signs since 1960 of attempting to colonise Denmark and Scotland. In the latter country breeding by one or more pairs was suspected from at least 1968 (see page 257), but not proved until 1979 when a nest with three eggs was discovered in the Scottish Highlands. As J.T.R. Sharrock commented in Spencer *et al.* (1993), there was a time in the early 1980s when it looked as though the Brambling was becoming a regular, established breeder, but by the end of the decade hopes of this happening were dwindling. From a peak of up to ten pairs breeding or possibly doing so in ten different sites from 1982 to 1984 inclusive, numbers declined to one pair in 1988 and none since then, as far as my information goes. In some years recently, singing males or pairs have been seen in England, but they are very few. The Brambling's ecological competitor, the Chaffinch, nested for the first time in Iceland in 1986, and has been increasing in Britain, Denmark and Sweden since 1980 (Marchant *et al.* 1990).

I have also discussed in full in Chapter 10 the history during the present century of the Siskin and the Lesser Redpoll, mainly with respect to the British Isles, and I linked the rather rapid expansions of their breeding distributions since 1950 with the trend since then to a cooler, rather drier climate. Meanwhile, the high-Arctic race of the Redpoll, known variously as the Arctic, Hoary or Coues's Redpoll *Acanthis hornemanni exilipes*, may be moving south again, as parties of these finches have been appearing further south than usual in Sweden, and recently have also been encountered in winter even as far south as Romania. By 1979, the typical race of the Redpoll, the Mealy Redpoll, was reported to have extended its range southwards in Finland, and it is now breeding commonly throughout that country, while, in central Europe, the Lesser Redpoll race has continued to expand its breeding range in almost all directions.

The Twite is a close relative of the Redpoll and the Linnet, and to some extent at least replaces the latter finch on high, mountainous ground in the British Isles and in western and northern Scandinavia. In Chapter 10 I mentioned that during the climatic amelioration the Twite contracted its range somewhat towards the north, but since about 1960, by which time the amelioration was waning, it had staged a recovery in some of the southern parts of its range. For instance, in northern England it established new colonies as far south as Derbyshire and Staffordshire. More remarkably it also colonised a large part of the Welsh uplands around Snowdonia in the late 1970s (Figure 56). Lately, however, the Twite is thought to be also losing territory in some areas due to greater afforestation, overgrazing and reclamation of its upland haunts, plus more intensive agriculture. The reasons for a widespread decline in the north and west of Ireland during most of the present century are not clear, but may partly be due to some of these factors and partly due to the local effects of the changing climate over this period. There have been, incidentally, similar signs of a recovery by the Ring Ouzel in Britain since the middle of the 1960s, and it also began to breed in the Faeroe Islands and further south in Finland in the 1980s.

Finally, in Chapter 10 I mentioned that the Citril Finch, a montane forest relative of the Serin confined to south-west Europe, had been spreading northwards to a small extent in Germany since 1950. Since then it has expanded its breeding range eastwards in Austria and in the former Yugoslavia as well. Almost exactly the same has happened with the central European alpine breeding population of the Brambling since 1980. This suggests to me that climatic cooling has been responsible rather than recent warming, as in the case of the Lesser Redpoll in central Europe.

RESPONSE OF SOUTHERN SPECIES WITH A SOUTHERN OR SOUTH-EASTERN CENTRE OF DISTRIBUTION TO THE POST-1950 CLIMATIC COOLING

We have seen that during the climatic amelioration a number of species of birds (and some other animals) withdrew the northern limits of their ranges southwards, especially in north-west Europe. Among them were such birds as the Rock Sparrow, Red-backed Shrike, Roller, Wryneck and White Stork. Reptiles such as the European pond tortoise and the green lizard, and insects such as the black-veined white and mazarine blue butterflies and the field-cricket, also responded in this way. As I have already indicated, such withdrawal to the south was presumably an adverse response to the milder, more maritime-type climate which this amelioration tended to produce, with rather wetter, cloudier weather in summer. Most of these species thrive in the drier, sunnier summers which are characteristic of a continental-type climate.

It may be too early yet to determine whether or not the tendency to change to a more continental-type climate since about 1950 has had a really appreciable effect in changing the fortunes of these species, and of course the escalating greenhouse effect warming is complicating the picture. Of the birds, the White Stork continued to decline on the north-west periphery of its breeding range (e.g. the Low Countries,[1] northern Germany, Denmark and the former Czechoslovakia). On the other hand, the downward slide seems to have slowed or even halted in southern Germany and the Ukraine, while in Austria, north-east France (since 1975), Poland and Estonia increases in breeding strength have been reported. At the south-west end of its breeding range in Europe, the White Stork declined in the Iberian peninsula (by 55% between 1957 and 1981 in Spain), but has shown a recovery since then. In north-west Africa, increases have been reported in Algeria and Morocco, though the reverse has occurred in Tunisia since 1954.

The censuses available from various European countries (see Cramp and Simmons 1977) do appear to support the contention that the climatic amelioration in the first half of this century caused a shift of the White Stork breeding population in north-west Europe to the heart of the continent, where the summers have tended to remain drier and sunnier. I have already discussed the probable proximate reasons for this in Chapter 7. In addition to some of the population figures given there, I think it worth giving a few more of which I have become aware since writing that chapter. In France, where most of the breeding birds live in the north-east of the country, the numbers of pairs in Alsace alone fell from 173 in 1948 to 145 in 1960, then sharply to fifty-four in 1965, twenty-three in 1970, and to only eleven pairs in 1974 (Yeatman 1976). Since then, the French population has recovered, to reach sixty-nine pairs in 1988, 107 pairs in 1989 and 119 in 1990. Recent censuses from the Netherlands show a further decline from the fifty-eight breeding pairs for 1955 to only eight pairs in 1974 (Cramp and Simmons 1977), and in 1981 there were only four pairs.

On the other hand, the Black Stork, which also contracted its breeding range in western Europe from about 1850 (although increasing slowly since 1920 in eastern Europe), began to reverse this trend and commenced a slow expansion to the west in 1970. As suggested in Chapter 7, this may have been due to the drier summers resulting from the expanding influence in recent years of the Scandinavian blocking anticyclone, associated with the waning of the wetter conditions of the pre-1950 amelioration; and perhaps also the incipient anthropogenic greenhouse effect. In this connection it is interesting to note that Black Storks are appearing more frequently as vagrants in north-west Europe, including Britain and even Iceland: between 1957 and 1990 as many as forty-eight were reported in the British Isles, eight of them in 1989 alone; before 1958 only twenty-six had been recorded as far back as records go. In Denmark, where they had been absent since 1951, Black Storks began breeding again in 1982. Further south, in Belgium, a pair nested in 1983 for the first time since 1860, while in Luxembourg others were reported summering in 1985. Meanwhile, further east in Europe, Black Storks are continuing to increase in Germany, Poland, the former Czechoslovakia, the former Yugoslavia and Austria. In the last-named country, where breeding started in the 1930s, the breeding population grew from about five pairs in 1951 to thirty to forty pairs in 1978.

There are few, if any, signs yet of a real recovery by some other birds, such as the Hoopoe, Roller, Red-backed Shrike, Wryneck and Nightingale. The Nightingale was reported in 1978 to have returned to a number of localities in Belgium from which it had been absent for many years, and to be breeding in greater abundance than has been usual in recent years. An upward trend was also reported from the Netherlands after 1983. The Nightingale enjoyed better breeding seasons in England between 1978 and 1986, but the population has continued to fluctuate since then on an overall downward trend, and reached a particularly low level by 1991 and 1992. Numbers, however, are high in southern Europe and one wonders if the greenhouse effect warming will eventually restore the fortunes of those Nightingales which breed in southern England. I have already discussed earlier in this chapter the recent history of the Red-backed Shrike, whose relatives, the Lesser Grey and Woodchat Shrikes, have recovered slightly since 1960. Hoopoes tended to appear in larger numbers further north than usual from 1970 to 1985, for example, in Britain, south-east Sweden and Denmark. As many as four pairs of Hoopoes nested in southern England in 1977, and the same number possibly did so the following year. Subsequently, several pairs have attempted to breed, including three separate pairs in 1984, but apparently unsuccessfully. Since 1987 none have been found trying to nest, except in 1990 when two pairs were present in the breeding season, and the species has been less frequently seen in Britain on spring migration. The Hoopoe breeding population of northern Germany was also declining in 1988, while the number of breeding pairs of Rollers in the former German Democratic Republic dropped from 135 in 1961 to only three in 1988. On the other hand, in southern Germany this very handsome bird was reported to be increasing. Those two mountain thrushes of central Europe, the migratory Rock Thrush and the sedentary Blue Rock Thrush, have both exhibited some signs of extending their breeding ranges northward again since 1950.

Three species with a wide breeding range in temperate Europe, the Corn Bunting, Ortolan Bunting and Tree Sparrow, also apparently found the wetter conditions of the climatic amelioration unfavourable to them. Indeed, their histories this century in north-west Europe have much in common (see Chapter 10). Although maintaining its breeding populations here and there (e.g. the Scottish Western Isles, the English West Midlands, the East Anglian fens, Lancashire and Kent), the Corn Bunting continued to decline in most parts of Britain into the 1940s, and following a revival associated with the end of the pre-1940 agricultural recession and increased food production during and after the Second World War, which lasted until 1973, it began another increasingly steep decline which is still continuing (Marchant *et al.* 1990). Various recent studies (Harrison *et al.* 1982; O'Connor and Shrubb 1986; Thompson and Gribbin 1986) have demonstrated a close link since the 1930s between the size of Corn Bunting populations and the availability of barley. The acreage under barley was at a maximum in the mid-1960s, but has since declined as wheat production grew, thus giving a remarkably close and convincing correlation with the increase of Corn Buntings up to the 1960s and their subsequent decrease since 1973. However, as I mentioned in Chapter 10, other factors seemed to be involved, of which changes in climate is probably one, while Donald (1993*b*) also suggests the recent switch to autumn-sown rather than spring-sown cereal crops; this has possible adverse effects on the Corn Bunting's nesting habitat because of a denser sward in late May and June, and on its nesting success because the cereal is harvested earlier than before (in July). The increased use of pesticides and a reduction in the availability of winter stubble for feeding as a result of modern rotational changes are other possible factors (Macdonald 1965; Terry 1986; Donald 1993*b*). On the continental mainland, similar reductions have been reported since the 1970s over much of north-west and central Europe. I myself have noted a continuing decline of Corn Buntings in the rich farmland of the Rhine–Neckar plain since I came to live in this part of Germany in 1988.

The Ortolan Bunting, likewise, has continued the post-1960 decline outlined in Chapter 10. In the Netherlands, for example, the breeding population decreased from thirty-three known pairs in 1989 to about twenty-four in 1991. The Tree Sparrow, in contrast, staged a quite dramatic recovery from about 1960 to 1965, recolonising much previously lost territory in many parts of western and northern Britain, and especially in Ireland, where the species had apparently ceased to breed in 1959–60. After 1965 numbers fluctuated a lot, then dropped alarmingly again from 1977; according to the BTO, by 1989 they were down to 15% of their 1965 peak on farmland. They have continued downwards since. Modern farming practices are believed to be at least partly responsible, especially the increasing use of herbicides to control weeds. As I mentioned in Chapter 10, I think that these factors are now overriding the benefits of the drier climate since the late 1950s.

In the course of discussing the fortunes of certain individual species, such as the Sand Martin, the Sedge Warbler and the Whitethroat, I have made more or less passing references to the effects of the severe droughts in the Sahel region of north-central Africa over the past two decades or so. The time has come to summarise this and relate it to the current trend in the climate. A very good summary, in fact, is given in Marchant *et al.* (1990) and I cannot, I think, improve on it and therefore have no hesitation in recommending those of my readers who wish to know more to read it for themselves. What follows here is largely drawn from that account.

In 1969 it became obvious to everyone interested in birds throughout western Europe, including the British Isles, that the normally very numerous Whitethroat, a summer resident which winters in subtropical and tropical Africa, was far less common than usual that spring and summer. Following this spectacular population crash which, in Britain, made the Whitethroat three times less numerous in 1969 than it was in 1968 (BTO Common Birds Census), the population fluctuated around this low level, with particularly big drops from 1971 to 1974 and in 1984–85, until a partial recovery began in 1986.

It was soon realised that the cause of this dramatic decline in Whitethroat numbers was the failure of the rains in the western Sahel zone, the western section of a wide band of northern tropical Africa stretching from the coasts of south-west Mauretania and northern Senegal in the west to north-central Sudan and northern Eritrea. This zone varies from semi-desert along the southern edge of the Sahara to the lusher conditions of the tropical forests and savannas along its southern edge. Its rainfall averages between 1 cm and 6 cm annually, and occurs mainly in the wet season from May to October. On the amount falling depends the condition of the vegetation and, there-fore, the degree of food abundance available to migrant birds during the dry season between October and May. During this period European migrant birds are present in the greatest numbers, so lack of rain in the wet season can seriously affect their chances of survival, especially on the return migra-tion to the north in March and April when they must cross the Sahara, rely-ing on their fat reserves to see them through safely.

Between 1900 and 1967, rainfall was, with fluctuations, about average, except for the period 1950–58 when it was continuously well above aver-age. Then, in 1968, precipitation was as much as 70% below normal in some parts of the Sahel and the adjacent savanna zones (Winstanley *et al.* 1974), the lowest in any year since 1941. It was not surprising, therefore, that the Whitethroat, plus the Sedge Warbler, Redstart and Sand Martin, which all winter in the northern Sahel, suffered heavy losses and returned to their breeding haunts in such low numbers. In 1969, rainfall returned to average, but thereafter, from 1970 to 1987 inclusive, it declined again and was continuously well below average, often falling below the level of 1968; 1988, however, proved to be the wettest year since 1969. 1983 and 1984 were especially dry years and had a particularly adverse affect on Sand Martins, Swallows, Grasshopper Warblers, Sedge Warblers, Whitethroats, Chiffchaffs and Spotted Flycatchers. Altogether evidence has been produced (Mead and Hudson 1985) that fourteen species which winter in the Sahel zone were adversely affected by the especially severe drought of 1983. The situation was aggravated in the drought years by the tendency of human populations to move into the semi-desert areas and impoverished savanna with their domesticated grazing animals, thus impoverishing the habitat still more. Since the late 1980s rainfall in the western Sahel has been showing signs of returning to former average levels, with the result that the breeding populations of Whitethroat, Sedge Warbler, Redstart, Sand Martin and other affected species have been recovering in western Europe. The breeding populations of these species in eastern Europe were less seriously affected, as they winter in the eastern Sahel where the droughts tended to be less severe.

Chapters 6 to 10 were chiefly concerned with detailed accounts of a large number of species of birds (well over 200) which were enabled by the amelioration of the climate to extend their breeding ranges considerably to the north and west. With the waning of that lengthy climatic phase around 1950, it was anticipated that many of these species would be forced to withdraw southwards again. However, the effects of the post-1950 cooling of the climate in the Northern Hemisphere have been most noticeable in the northernmost latitudes; they only began to 'bite' in the region of the British Isles and neighbouring countries at a comparatively late stage. Because of this delayed reaction to the natural change in the climate, which has perhaps been encouraged by the incipient influence of the anthropogenic greenhouse effect, many of these advancing species are apparently still responding to the pre-1950 amelioration and continuing their range expansion northwards or north-westwards. This will be discussed in detail in the next chapter. Only from about 1978 onwards did some of them, such as the Fan-tailed Warbler, begin to recoil as a consequence of the severe winters of 1977–78, 1978–79, 1981–82, 1984–85 and 1986–87, especially the last two.

Meanwhile, some other species have shown signs of losing momentum now that they are really coming up against the post-1950 climatic cooling in northern Europe, while a few may be on the point of retreating. But at the time of writing there is little positive evidence of this. The Black-necked Grebe may be one species which is withdrawing from the northern limits of its European breeding range, judging by the recent steady decline in Sweden (colonised since 1919), where it has entirely disappeared from its northernmost stronghold. It has also decreased in the Netherlands (from c.240 breeding pairs in 1989 to only about 90 in 1991), although not apparently in Britain. The Bittern has declined in its few English breeding haunts since the mid-1950s; so seriously in the past few years that, in 1976, the Rare

Breeding Birds Panel decided to include detailed reports on its breeding status in its future annual reports in the journal *British Birds*. The reasons for this decline are not yet clear, but the severe winters of 1962–63, 1977–78, 1978–79, 1981–82, 1984–85 and 1986–87 have certainly had a serious effect. Other possible factors which have been suggested include reed cutting, water pollution and disturbance by coypus (Sharrock 1976). The current British breeding stock of Bitterns is thought to number fewer than thirty pairs. Breeding populations also declined in France and Spain, but, presumably because of recent mild winters since 1987, the Danish and Swedish populations reached their highest levels for many years in 1989.

As mentioned in Chapter 7, the Garganey, a summer resident in Europe, declined and contracted its breeding range south-eastwards after 1953 in Britain. This situation continued up to 1988, but since then has markedly improved, with a hundred or more possible or confirmed breeding pairs present, compared with only around fifty pairs or less (apart from ninety-four pairs in 1982 and seventy pairs in 1983). The spring and summer of both 1989 and 1990 were hot and sunny in Britain, as was the summer of 1983, and the trend since 1987 to mild winters and generally warmer and drier summers may favour the Garganey.

At the end of my account of the spectacular north-westward spread across Europe of the Collared Dove (Chapter 8), I mentioned that, since 1980, it has been declining and withdrawing from the breeding range it established in Scandinavia in the 1970s, while, on the other hand, advancing south-westwards across the Pyrénées in the 1960s to colonise the Iberian peninsula, and then moving on into north-west Africa to commence the colonisation of Morocco and Tunisia. This latest spread was to be expected, as Iberia and north-west Africa clearly provided much suitable habitat for the Collared Dove in a climate similar to that of Turkey, from which its penetration of Europe began at the beginning of this century. In Spain, it first moved west to occupy the rather humid coastal belt in the north during the 1960s and 1970s, before swinging south after 1974. Pushing down beyond Madrid from 1986, it was already breeding at Jerez de la Frontera, near Cadiz, not far from the Straits of Gibraltar, by 1989, and in Portugal by 1991. The Straits had, however, already been crossed, for pairs were discovered nesting far inland at Meknès in Morocco in 1986, and at nearby Fès in 1988. By 1990 they had arrived in Laroche and Casablanca (Morocco), and by mid-December at Bizerte in northern Tunisia.

The decline and withdrawal of the Collared Dove from its quite recently acquired breeding territory in Scandinavia suggests a response to the climatic deterioration in this sector of the Arctic zone, which is still continuing in spite of the anthropogenic greenhouse effect. In Norway, for example, it was reported in 1988 to have decreased throughout the country from being 'quite common in the Oslo area in the 1970s' to having almost disappeared. The marked decline in Finland since 1980 was specifically blamed by Hildén (1989) on the severity of the 1984–85 and 1986–87 winters there.

Two woodpeckers which have been retreating from the northern limits of their breeding range in Europe, presumably in response to the post-1950 climatic cooling, are the Middle Spotted Woodpecker and the White-backed Woodpecker, the first of which has the more westerly distribution. The White-backed Woodpecker, which was once common over much of Sweden, the southern half of Finland and south-east Norway, decreased to a serious extent in these countries during the 1980s, although in Norway, where it is approaching extinction in the east, modern forestry practices are blamed. The species is also apparently in danger of extinction in Sweden where only fourteen nesting pairs were known in 1988 and eleven in 1989, while in Finland the population in the latter year was estimated at only thirty to fifty pairs and as few as fifteen pairs by 1993.

In Chapter 8 I linked the disappearance of the Middle Spotted Woodpecker from many parts of Europe before 1950 with the deforestation of broad-leaved woodlands, while suggesting that a subsequent slight recovery and northerly extension of range indicated that climatic change may also have been involved. The speed with which it has declined and retreated south-eastwards in Sweden during the 1970s reinforces my belief in the climatic factor (see Crested Lark below). The last pair to breed successfully did so in 1980, and as a resident the Middle Spotted Woodpecker finally became extinct in Sweden in 1982.

The Crested Lark also contracted its breeding range in south-west Sweden and south-east Norway, and has now vanished from the latter country which it colonised after 1900. In Sweden it decreased from the mid-1960s to the point of extinction in its last stronghold around Malmö in the extreme south by 1989. Its decline there, and that of the Middle Spotted Woodpecker, may be connected with a cooling in Scandinavia which is out of step with neighbouring parts of the Northern Hemisphere. A marked decline has also taken place in the Netherlands, where the breeding population of the Crested Lark dropped from 3,000–5,000 pairs in 1979 to 1,000–2,000 pairs in 1985–86, and to less than 400 in 1991. Meanwhile in Switzerland the breeding population died out in August 1990. Crested Larks have also disappeared as a breeding species from the northern edge of their former range in European Russia, and have decreased in France, Germany, Denmark, Poland and much of central Europe as well as the Netherlands (Simms 1992).

Similarly, the Woodlark has declined in southern Scandinavia and also in Denmark, where only a few pairs were left in the 1980s. Both species almost certainly retreated in the face of the post-1950 climatic deterioration in the north. In Britain, following an increase which more than doubled the known breeding population of the Woodlark between 1975 and 1981 to 400–430 pairs, the population dropped again to 210–230 pairs in 1983 after the severe winter of 1981–82. It was still about the same in 1986 when the BTO censused the breeding population. So the Woodlark population seemed to become more or less stable between the breeding seasons of 1982 and 1987 in spite of the severe winters of 1984–85 and 1986–87, which had some adverse effects, and numbers levelled out at around 200 pairs. However, following the very mild winters of 1988–89 and 1989–90, the British breeding population rose by 1991 to a total of about 350 pairs (Gibbons *et al.* 1993). This improved situation may be correlated (see Chapter 9) with the distinctly warmer springs and fine, hot summers of the rest of the 1980s and 1990–91. Thus it might well have been the cold springs of past years and their negative influence on the Woodlark's particular insect prey that were chiefly responsible for its decreases, the Woodlark being an early nester.

Although there is little real evidence of climate change having any obvious marked effect on the range and numbers of Meadow Pipits, either last century or this century in Europe, the BTO's Common Birds Census has shown a steep decline between 1981 and 1986 in the British breeding population. This seems to have been linked to the series of hard winters and cold, wet springs in those years in Britain and elsewhere in western Europe, including the south-west where many British Meadow Pipits spend the winter. With the milder winters and warmer springs since 1987 the British population has been increasing again, and a recent study (Crick *et al.* 1993) suggests that they are now nesting earlier and laying bigger clutches.

Since the mid-1970s, at which point I broke off my discussion of the histories in Europe of the Fan-tailed Warbler and Cetti's Warbler in Chapter 9, north-west Europe has experienced the hard winters of 1977–78, 1978–79, 1981–82, 1984–85 and 1986–87. Following a large influx in Belgium in 1975, the numbers of Fan-tailed Warblers there declined annually to such an extent that none were reported in 1978. In France and the former Yugoslavia the population crashed in the 1984–85 and 1986–87 winters, with the result that the Fan-tailed Warbler became almost extinct in both countries. During May 1973, Eric Simms and I encountered this species in high numbers everywhere we went in the Camargue and adjacent parts of southern France where suitable habitat existed; in September 1987 I saw hardly any. These hard winters, especially those of 1984–85 and 1986–87, were disastrous for it there, and also for Cetti's Warbler, although, as we have seen, both species have recovered remarkably quickly from such setbacks before. For details of the fortunes of Cetti's Warbler in north-west Europe since 1975, see Chapter 12.

After I completed the first draft of my account of the Dartford Warbler in Chapter 9, the severe heath fires of the extremely hot summer of 1976 and the effects of the two severe winters of 1977–78 and 1978–79 heavily reduced its then high breeding population in England. As a result of the first of those winters, for example, the reduction was in the order of 60%, and that was a less severe winter than the succeeding one. Nevertheless, the population recovered remarkably quickly, and by 1980 it was already larger than it had been in 1978. Thereafter, in spite of temporary setbacks (most pronounced in Surrey and less so in south-west England), it grew, as a result of the mild winters of 1987–88 and 1988–89, to a maximum total of between 500 and 640 pairs by 1989. However, exceptionally thorough surveys were carried out by ornithologists in both 1988 and 1989, so earlier counts and estimates may have been comparatively on the low side (Spencer *et al.* 1991). The latest figures available to me put the breeding population in 1990 at a maximum of 928 pairs (Spencer *et al.* 1993), and Colin Bibby in Gibbons *et al.* (1993) considered the population to have probably risen by 1991 to 950 pairs, the highest totals ever reported in Britain. Breeding pairs were reported from Cornwall, Devon, the Isle of Wight, Wiltshire, Surrey and Sussex in addition to the strongholds in Dorset and Hampshire. It remains to be seen how soon the present cooling phase in the climate of northern Europe will eventually be counteracted, and perhaps replaced, by the increasing global warming due to the anthropogenic greenhouse effect. The serious loss and fragmentation of the heathland habitat favoured by this perky little warbler during the present century has made it much more diffi-cult for it to recover from frequent bad winters (still possible in the climatic upheaval caused by the onset of the greenhouse warming) and serious heath fires than was the case in the Little Ice Age, when vast and continuous areas of suitable habitat existed virtually unspoiled in southern England.

A severe setback was also suffered by the Firecrest in its recent attempt to colonise Britain in force. Following a very successful breeding season in 1975 with a possible maximum population of 121 pairs, there was a serious decrease in both the number of breeding pairs and sites in 1976 when not more than 27 pairs were thought to have bred. The population remained much the same in 1977 with a possible maximum of thirty-one pairs, but then dropped to only eleven pairs in seven sites in 1978, following a hard winter. The succeeding winter was also hard, but surprisingly the breeding population recovered to a possible seventy-three pairs in 1979 in twenty-five sites, then to seventy-eight in 1980 in thirty sites. By 1983, a peak of 175 pairs in seventy-five sites was attained, but was immediately followed by a steep decline to only twenty-nine pairs in 1986. However, the very next year (following the severe winter of 1986–87), a rapid recovery set in so that by 1989 a total of 131 confirmed or possible pairs in fifty-two sites were recorded. Unfortunately, as mentioned in Chapter 9, following a poor spring, the 1990 total dropped to ninety-seven pairs in forty-eight localities (Spencer *et al.* 1993).

I have already discussed in detail in Chapters 4 and 10 the Cirl Bunting's decline and contraction of range since 1910, and linked it with the climatic changes of the present century. Suffice it to say here that it continues to decline and withdraw southwards on the north-western edge of its range in England, northern France and southern Germany, and may go on doing so until the recent cooling trend in northern Europe is counteracted by the current greenhouse warming becoming more powerful. By 1989, the English breeding population was almost entirely restricted to south Devon, the area which the species probably originally colonised in the 18th century. However, since that year the Devon breeding population has recovered to a remarkable extent, trebling from 114 to more than 350 pairs in 1993! Much credit for this is due to the RSPB for persuading local farmers to delay the autumn sowing of cereals, thereby increasing the number of fields lying fallow in winter with a rich supply of cereal grains and weeds, important for Cirl Bunting winter survival; but the mainly mild winters since 1987 must also have helped the recovery.

Finally, it looks as if the protracted northward spread of the Serin through Europe (see Chapter 10) has slowed to a halt, and that the long-held hopes of British ornithologists that southern England was next on its list for successful colonisation will be unfulfilled, in spite of continuing efforts. I will be surprised if it succeeds in becoming established, although the growing greenhouse effect may change the situation.

[1] A pair, however, nested successfully in Belgium in 1979 for the first time since 1973.

Chapter 12

SPECIES EXPANDING NORTH, NORTH-WEST OR WESTWARDS

In response to greenhouse effect warming

In the previous chapter I discussed those species of birds whose changes in breeding range during the 1850–1950 climatic amelioration have been reversed (or appear to have been so) by the post-1950 cooling of the climate in the Arctic region, which has particularly affected northern Europe and north-east Canada, especially the northernmost latitudes. As we have seen, this has caused a number of Arctic and subarctic birds (and other animals) to retreat southwards and appear, and even breed, in parts of Europe where we have not been accustomed to seeing them previously this century. Meanwhile, the great majority of species which were prompted by the pre-1950 climatic amelioration to expand their breeding ranges northwards, north-westwards or westwards have continued to do so in spite of the subsequent climatic cooling.

Thus we have been witnessing a seemingly paradoxical situation in which at the same time as 'northern' species have been retreating before the Arctic cooling, 'southern' species have still been responding to the residual benefits of the comparatively recently concluded climatic amelioration, and probably now also to the strengthening greenhouse effect. So here we have an explanation for the present somewhat extraordinary spectacle of, for instance, birds characteristic of the very warm Mediterranean region, such as the Mediterranean Gull, Collared Dove, Serin, Fan-tailed and Cetti's Warblers, colonising or attempting to colonise countries such as Britain and Belgium from the south, while simultaneously others, such as the Redwing, Lapland Bunting, Snowy Owl, Whooper Swan, Purple and Wood Sandpipers, characteristic of cold northern regions such as Lapland, have been attempting to colonise from the north.

If by any chance the climatic deterioration in northern Europe should continue, and its influence become more pronounced and widespread, one would anticipate a gradual halt to the northward advance of these 'southern' species, followed by a retreat—at least in many cases. But it now seems more likely that the anthropogenic greenhouse effect will counteract or supersede the natural cooling process, and obviously a lot will depend on how soon this occurs. Meanwhile, the advance to the north, north-west or west of most of those species discussed in Chapters 6 to 10, plus several additional ones such as the Black-shouldered Kite, goes on. The following notes give the most recent information I have available on them. (The reader will find that in some cases this overlaps with data in earlier chapters.) See also Appendices 1, 2.1 and 2.2.

Great Crested Grebe

In Norway, which was first colonised in 1904, the breeding population increased from about fifty pairs in 1970 to about 200 pairs by 1978, and it is still expanding northwards there and elsewhere in Fenno-Scandia. It is also still increasing over most of its European range, including Britain and Ireland where it is spreading more and more on to canals and rivers (Marchant *et al.* 1990).

Red-necked Grebe

Since the 1970s birds in full breeding plumage have been summering on suitable waters in England and Scotland. In 1987 a pair was seen in eastern England with a 'very young juvenile'; subsequently, in 1988, 1989 and 1990, two pairs (in England and Scotland) attempted to breed, but were unsuccessful (Spencer *et al.* 1989–91, 1993). This apparent attempt at colonisation of England and Scotland represents a westward extension of range, presumably from Denmark or northern Germany. These two countries are probably also the source of the Red-necked Grebes which have been summering regularly in the Netherlands in recent years, and occasionally breeding.

Black-necked Grebe

Continuing to increase steadily in Britain, but declining in Sweden.

Fulmar

The slowing rate of expansion has been referred to in Chapters 6 and 11. Nevertheless, in a mere two decades the estimated British population has grown from around 305,600 pairs in 1970 to possibly as many as 530,000 pairs by 1989.

Gannet

Still increasing at a rate of 2.5–3% per annum, and founding new colonies. Between 1969 and 1985 the British population, which represents 61% of the world population and 72% of the European, grew from about 107,000 pairs to 161,000, a rate of increase of 2.6% per annum (Wanless 1987; Batten *et al.* 1990).

Cormorant

The breeding population along the south coast of the Baltic from Denmark to Poland has been greatly increasing since 1980, with the result that the Cormorant's range has been expanding northwards into Sweden and the Baltic republics, south-eastwards into Belarus (Byelorussia) and the Ukraine, and southwards into the former Czechoslovakia and Austria. A westward colonisation of Spain and a northward one of north-west Italy have also begun, both presumably from Corsica and Sardinia. A big population increase has also occurred at the same time in the Netherlands. The reasons for this population explosion are not yet clear, but may be partly related to a relaxation in human persecution and perhaps partly to an improvement in the food supply since the 1920s.

Pygmy Cormorant
Since the 1980s this very small cormorant has been extending its wintering range northwards into Lower Austria, and its breeding range into Hungary (since 1988) and also further north in Moldova, where breeding was first proved in 1982; up to 400 pairs were present there by 1990.

Bittern
Although serious declines have occurred, and are continuing, in the breeding populations in England, France and Spain in recent years, those of Denmark and Sweden have 'boomed' since 1987, apparently due to the recent run of mild winters.

Night Heron
Still increasing and extending its range in southern Germany and Lower Austria. Temporarily colonised Switzerland 1967–71. Small colony became established in southern Poland in 1964. In France it is increasing again, following a decline between 1968 and 1974; by 1981 the breeding population was 60% up on 1974. The Spanish breeding population increased from 1,300 in 1988 to 1,600 pairs in 1989.

Squacco Heron and Little Egret
The continued range expansion of these two species since earlier this century has probably been largely due to improved protection, not climatic change. Nevertheless, one wonders if climatic warming due to the growing influence of the greenhouse effect since about 1975 is becoming involved, as Europe's breeding populations are rapidly increasing and showing definite signs of spreading northwards. For example, pairs of Little Egrets nested for the first time in the Netherlands in 1979 and in northern France in 1978. The Little Egret population in France as a whole in fact increased between 1974 and 1981 by 23%. The Spanish breeding population of the Squacco Heron rose from fewer than 250 pairs in 1988 to about 400 in 1989.

Cattle Egret
Still spreading northwards in Portugal and Spain, with around 54,000 pairs reported breeding in 1989 in the latter country. In France, the Camargue breeding population increased from two pairs in 1968 to 125 in 1975; pairs are now breeding as far north as Loire-Atlantique (1981), Brenne and north

of Lyon (1992). During the late 1980s, breeding began in Piedmont, north-west Italy.

Great White Egret
Following a steady increase of the breeding population in Austria, summering birds have appeared in the Netherlands since 1976, and a pair nested in 1978; by 1992 about five pairs were breeding. In southern Sweden in the spring and summer of 1989 as many as twelve were reported, a record number. Otherwise this species is declining in other parts of its European range, where fluctuations are probably not connected with climatic changes.

Grey Heron
The northward spread in Norway and elsewhere in Fenno-Scandia apparently continued up to about 1970, when the cooling of the Arctic zone inhibited it. Breeding populations in Italy and Spain are rapidly increasing.

Purple Heron
Was still increasing in the Netherlands and Germany in the early 1970s, but then declined in the Netherlands from c.900 pairs in 1977 to only c.190 by 1991. Marked decreases have also been reported from France, Switzerland, Hungary and Romania. During the 1970s, the Purple Heron became a more frequent visitor to Britain, and summering birds were recorded in 1975, 1976, 1977 and 1979, but not since. Recent decline may not be due to climate change.

Black Stork
The westward range expansion mentioned in Chapter 7 has continued, and for convenience is described in Chapter 11, together with the White Stork.

Mute Swan
Still increasing and spreading in countries around the Baltic Sea, such as Finland and Estonia, but northward expansion in Sweden frustrated by heavy losses during recent severe winters.

Shelduck
The big increase of this species in north-west Europe (including the British Isles) during the present century, and which is still continuing, is thought to have been due to greatly reduced persecution and increasingly effective conservation measures. However, the recent colonisation of the Finnish coast and spread northwards up the Gulf of Bothnia coasts suggests that climatic amelioration may also have played some part.

Gadwall
Still extending breeding range northwards (e.g. Estonia) and westwards (e.g. Belgium and France) to at least a limited extent.

Shoveler
Extension of northern limits in Estonia, Finland and Sweden, and perhaps elsewhere, continued at least up to about 1970.

Marbled Teal
This southern duck, characteristic of the drier parts of the Warm Temperate zone, has been increasing and expanding its breeding range in Morocco in recent years. In 1991 a pair bred in central Spain, the furthest north yet.

Red-crested Pochard
Still colonising new areas and expanding breeding populations in some parts of Europe, such as Austria, the former Czechoslovakia, Hungary, Poland and Switzerland.

Pochard
Still increasing and spreading in many parts of western Europe, such as France, Belgium, the Netherlands, Austria (first bred 1957) and Switzerland (first bred 1952). Breeding population of the Ukraine has expanded up to 8,000 pairs during the 1980s. Since 1954 a few pairs have nested regularly in Iceland, and since 1976 in Norway at Akershus, where by 1982 the breeding population on the lake there had built up to about fifteen pairs.

Tufted Duck
Spread apparently still continuing (e.g. Britain, the Low Countries, Estonia and southern Sweden). Breeding in Faeroes since 1966, France regularly since 1964 and Austria since 1960. Also spreading south in the Ukraine since 1977, where around 340 pairs were breeding annually by the later 1980s.

White-headed Duck
From the end of the last century, this stiff-tail duck of the Warm Temperate zone gradually contracted its range in the Mediterranean region, withdrawing from most of its former haunts in south-east Europe and some of those in the south-west. However, since the late 1980s it has been increasing again in its haunts in southern Spain and expanding its range, so that by 1990 breeding pairs had penetrated northwards into central Spain. Although protective measures may be playing a part, it is difficult to avoid the conclusion that the White-headed Duck's northward expansion is in some way connected with the warming due to the greenhouse effect. Recently, vagrants have been appearing in winter as far north as the Netherlands and Schleswig-Holstein in northern Germany.

Black-shouldered or Black-winged Kite
Increasing in its Iberian strongholds and expanding its range; a pair were present in south-west France in 1983 and 1984, and breeding was proved

there for the first time in 1990. It is also being seen more frequently else-where in central Europe, even as far north in Germany as Lower Saxony.

Black Kite
Spread continues, although it almost disappeared from Finland during the 1960s (Cramp and Simmons 1980); it has been nesting or attempting to nest in Flanders (Belgium) since 1976, and since 1980 has become regular with at least two successful breeding pairs in 1982. Breeding pairs in Luxembourg increased from 5–7 in 1983 to seventeen pairs by 1993. It is appearing much more frequently in Britain as a vagrant since 1970; between 1958 and 1992 inclusive 170 were satisfactorily recorded, compared with only five before then. Doubtless it will soon attempt to breed in Britain, if it has not already done so.

Red Kite
Signs of another recovery and possible spread in north-west Europe. For example, pairs have nested in Belgium since 1973 (growing to about twenty pairs by 1989) and in the Netherlands since 1976. Meanwhile, the recoloni-sation of Denmark has continued to prosper, with some twenty breeding pairs by 1986. In Sweden there was a breeding population of about 200 pairs in 1989; by 1990 about 1,000 individuals were estimated to be present at the end of the breeding season. The numbers wintering in 1982–83 were three times as high as in the early 1960s. The Swiss, Slovak and Polish popu-lations are also continuing to increase. These changes may be connected with increased protection rather than with climatic change. In Germany, following a decrease in the 1960s and subsequent recovery, the population was again reported to be declining, due it is thought to new methods of refuse disposal (Evans and Pienkowski 1991). The isolated Welsh popula-tion continues to increase and experienced its best ever season in 1990 when, of eighty-four pairs, sixty-five bred successfully.

Montagu's Harrier
Although continuing to decline in many parts of Europe, it has increased in Sweden since 1974, with fifty-five pairs reported nesting in 1981. The Danish breeding population also increased from twenty or thirty pairs in the 1970s to about fifty pairs in 1981. Increasing in the Netherlands as well. A pair nested in Finland in 1987, the fourteenth breeding record. A recovery has occurred in England since 1976, with up to fifteen pairs now breeding, or attempting to do so, annually. Most English pairs are now nesting in cereal crops.

Red-footed Falcon
Also still spreading west in Europe, and occurring much more frequently in Britain as a vagrant—386 recorded since 1958–92, of which 107 in 1992. Numbers of passage birds have also greatly increased in France, the

FIGURE 62 *Red-footed Falcon*

Netherlands, Denmark, Sweden, Poland, Germany and elsewhere in north-west Europe, especially in spring. A few have even reached Iceland.

Hobby
Has recolonised south-east Norway since 1956 after apparently becoming extinct at the end of the 19th century. This species has increased its breeding population in England and Wales fivefold since 1960 and, according to the Rare Breeding Birds Panel (Spencer *et al.* 1991), more than doubled it in the decade 1979–89. It has also extended its range considerably northwards and north-eastwards in recent years, reaching Scotland by 1990. The warming, apparently due to the greenhouse effect, since 1975 seems to be helping this small falcon, which feeds to a large extent on active flying insects, such as dragonflies, in fine weather.

Saker Falcon
Apparently extending its breeding range northwards in Russia and westwards in central Europe in recent years as far as eastern Austria, although this may now have halted.

Spotted Crake

As mentioned in Chapter 7, northward spread first noticed in Sweden after 1950, but in 1979 and again in 1991 decreases were shown in national censuses. In Britain, however, Spotted Crakes continue to increase and spread on a small scale, although with irregular fluctuations; the number of singing males has grown from fewer than ten in about four localities in the early 1980s to between ten and twenty-one spread over six to fourteen localities in the late 1980s (Spencer *et al.* 1991).

Oystercatcher

Still increasing since *c.*1980 over much of north-west Europe.

Black-winged Stilt

Still continuing to breed sporadically well north of its main range in southern Europe, even north of Bremen in Germany as well as in the Netherlands in the 1980s. As mentioned in Chapter 6, this species bred in several countries north of the normal range as a result of the 1965 irruption. In Britain, since the first breeding record in 1945, single pairs have nested in Cambridgeshire in 1983 (unsuccessfully) and Norfolk in 1987 (successfully). In 1988 a pair appeared at a Buckinghamshire site from 7 to 18 June, displaying and copulating, but did not breed. (See Figure 21.)

Avocet

Still increasing and spreading in northern Europe, including Britain, where the breeding population currently varies between 300 and 500 pairs.

Little Ringed Plover

Since 1950 has continued to increase and spread northwards in northern Europe, including Scotland and Fenno-Scandia, helped by the provision of new habitat in the form of gravel- and sand-pits. (See Figure 19.)

Greater Sand Plover

Since 1950 has been extending its breeding range westwards from its former western limits around the Caspian Sea and Armenia, breeding in Syria since 1952, in Jordan since 1963, and Asiatic Turkey since 1967. In July 1992, an adult in breeding plumage spent a week near the south-west shore of Neusiedlersee, just inside Hungary's frontier with Austria. This plover is also appearing more frequently elsewhere in Europe as a summer vagrant.

Spur-winged Plover

Has consolidated its colonisation of Greece, especially around the Evros and Nestos deltas where many pairs are now breeding.

Sociable Plover

Since the late 1980s has begun to appear more frequently as a vagrant in north-west Europe, particularly in Fenno-Scandia. One female attempted unsuccessfully to hatch two eggs in Pori, south-west Finland, in May–June 1990, far to the north-west of this species normal breeding range.

White-tailed Plover
In the 1960s this species began to extend its breeding range westwards in the Middle East through the southern Caucasus and Syria into Asiatic Turkey, where it nested for the first time in 1971.

Ruff
Since about 1960 has been increasing and expanding its breeding range northwards in southern Scandinavia and westwards into eastern England, probably due to population shifts in the north-central mainland of Europe caused by the drainage of former wet grasslands and the subsequent change to arable farming. So climatic change is probably not involved in this case.

Black-tailed Godwit
Still increasing along north-west limits of its mainland European range. From a peak of eighty-seven pairs in 1976, the British breeding population has, however, currently declined to about fifty or sixty pairs.

Curlew
Having nested for the first time in north-east Iceland in 1988, two pairs bred for the first time in the Faeroe Islands in 1989.

Marsh Sandpiper
A pair of this central Asiatic species bred in Finland for the first time in 1978, and then again in 1983 and subsequent years; in eastern Poland for the first time in 1988, and in southern Sweden in 1992; also in Latvia in 1974 and 1975, and increasingly since 1982. Appearing with growing frequency on migration in Finland, and also elsewhere in Europe, notably in Sweden, Britain, Denmark and the Netherlands. Therefore, its expansion of range north-westwards and westwards is continuing, perhaps more in response to the greenhouse effect than the pre-1950 climatic amelioration.

Terek Sandpiper
Continues to colonise Finland and, since 1980, Latvia in the course of its westward range extension, and has now reached the coast of the Gulf of Bothnia. Has been visiting Britain and Denmark more frequently since 1969. (See Figure 22.)

Pomarine Skua
North-westward extension of range: breeding on Spitsbergen since 1984.

Little Gull

Still increasing in some areas, but fluctuating; new colony found in northern Poland in 1978, and breeding attempts have occurred in southern Norway since 1976. Following the increasing frequency of summering birds in England since the late 1950s, four pairs have laid eggs since 1975, but none succeeded in rearing young.

Mediterranean Gull

Has greatly increased and spread in the former USSR and since 1988 is breeding as far north as Belarus (Byelorussia). Pairs have been nesting in Austria since 1977, and a colony of about twenty-five pairs was found in Italy in 1978, the first evidence of breeding there; they subsequently increased to ninety pairs by 1981 and then to c.1,000, but in 1990 were reported to have decreased to 867 pairs. In the former Czechoslovakia, Poland, southern Sweden, Hungary, Germany, France, Belgium and the Netherlands (125 pairs in 1990, half of them in Zealand), breeding has become regular or almost so since the 1950s and 1960s. Also breeding on the Ebro Delta, Spain, since 1987. One pair bred in southern England in 1977, only the third instance of a pure pair doing so. Subsequently, the numbers of pure pairs breeding there annually have grown slowly, but have fluctuated, reaching some six to sixteen pairs in up to ten sites by the early 1990s. (See Figure 17.)

Black-headed Gull

Still increasing over much of its European range.

Slender-billed Gull

The isolated colonies in south-west Iberia and the south of France increased during the 1980s, and in May 1985 two adults appeared on the Neusiedlersee in Austria. Another adult in breeding plumage visited the Hortobágy fishponds in Hungary from 3 to 7 May 1992. An adult was also present on Furilden Island, Gotland, in Sweden, from 16 to 22 May the same year. Two more in breeding plumage appeared in Austria in May 1993.

Common Gull

Extending breeding range southwards into inland waters in Belgium and central Europe since the early 1980s, as well as in the British Isles. Still increasing in parts of north-west Europe.

Lesser Black-backed Gull

Still increasing in parts of north-west Europe (e.g. Britain and the Faeroes), but not everywhere in its breeding range.

Herring Gull and Great Black-backed Gull

Status as Lesser Black-backed Gull above.

Caspian Tern
Westward extension of breeding range to east Spain? A pair bred in the Ebro Delta in 1988. However, in Sweden the breeding population declined from about 900 pairs in 1971 to fewer than 450 pairs in 1991.

Lesser Crested Tern
A pair of this species, which normally breeds no nearer to Europe than the east coast of Egypt and the Arabian Gulf, nested in 1990 in the Comaccio Valley on the Adriatic coast of north-east Italy. From 1984, at least up to 1990, a lone female Lesser Crested Tern paired with a male Sandwich Tern in a colony of the latter species on the Farne Islands, off the Northumberland coast of north-east England. They apparently successfully reared a hybrid chick in 1989 but apart from possible success in 1986, not in other years despite attempts at nesting (Spencer *et al.* 1991, 1993).

Whiskered Tern
Continuing evidence of a north-westerly spread from the Ukraine through Belarus (Byelorussia) to Poland since mid-1980s. By 1992 some 130 pairs were breeding in Poland and five pairs bred for the first time in Latvia.

Black Tern
Although isolated pairs have bred or attempted to breed in south-east and eastern England since 1966, with up to seven pairs doing so in 1969 (one pair successfully), and single pairs in Ireland in 1967 and 1975, no more have done so in the British Isles up to the time of writing since 1978. Remarkably, single pairs attempted to nest in Iceland in 1983 and 1984, far to the north-west of their normal breeding range.

White-winged Black Tern
Marked increase and expansion of breeding range westwards since 1970. For instance, there have been big increases in breeding colonies in Latvia and Poland since 1979, with about 1,000 pairs breeding in the latter country by 1992. Also appearing frequently in Britain: more than 600 reported since 1957.

Woodpigeon
Apparently continuing to spread northwards in northern Europe (e.g. Fenno-Scandia) and consolidating its breeding populations in newly colonised areas, such as the Faeroes (1969). Pairs have even nested in Iceland on odd occasions since 1964.

Collared Dove
Continued to spread and consolidate its colonisation of new territory (see Figure 32), such as Iceland, up to *c*.1980, but since then has declined seriously throughout its range in Norway, Sweden and Finland, especially in the late

1980s, due it is thought to losses in the hard winters of 1984–85 and 1986–87. Now, however, colonising Spain, Morocco and Tunisia (see Chapter 11).

Turtle Dove
This summer resident has been increasing and expanding its range north-wards in Latvia and Estonia since about 1960, and also in Denmark to the west. Birds are also being seen more frequently in southern Sweden, although breeding has yet to be proved (Marchant *et al.* 1990). Turtle Doves also increased slowly in Britain between 1963 and 1979, before falling into a decline which lasted at least up to 1992, when what may be the start of a new increase was detected by the BTO's Common Birds Census. With the increasing warmth due to the greenhouse effect, it may be hoped that the incipient new upturn in this dove's fortunes in Britain may continue.

Laughing (Palm) Dove
A widespread species in the Warm Temperate and Tropical zones that has been spreading northwards from its strongholds in north Africa, and began the colonisation of Morocco in 1982. It has also appeared on several occasions recently in Italy and Greece, and at least once in France; it may soon attempt the colonisation of Mediterranean Europe, perhaps in response to the greenhouse warming.

Great Spotted Cuckoo
Wanderers are now being reported more frequently as far north as Norway, Sweden, Germany, the Netherlands and the British Isles. Moreover, the first instance of known breeding in the former Yugoslavia occurred in 1982, while in 1992 a juvenile was watched being fed by Carrion Crows as far north in France as La Faute sur Mer, Vendée.

Eagle Owl
North-westward extension of breeding range: a few pairs have been breeding in Belgium and Luxembourg since 1982, following many years' absence and a reintroduction of the species in adjacent parts of Germany in that year; also a pair bred in south Jutland, Denmark, in 1985, the first Danish breeding record for more than a century.

Tengmalm's Owl
Since 1950 has expanded its breeding range westwards in central Europe (e.g. western Germany, and the Bourgogne plateau and forests of Lorraine in France) in the wake of the westward spread of the Black Woodpecker, for whose nest-holes this owl displays a particular partiality (Yeatman 1976). Also, during the 1980s, it was found in both the French and Spanish Pyrénées. The northward spread in southern Germany is apparently continuing. Also spreading north in the Balkans, having colonised south-west Romania since 1966.

Pallid Swift
A northerly extension of breeding range has occurred in north-west (former) Yugoslavia since 1986, with at least one colony now established in Istria. The first record in Switzerland was of a colony at Locarno in 1987.

Alpine Swift
Northward spread in southern Germany apparently continuing. Also spreading north in the Balkans, having colonised south-west Romania since 1966.

White-rumped Swift
Since it began colonising southern Spain near Cadiz in the early 1960s, this species has extended its range northwards into Extremadura in west-central Spain, 400 km north of Cadiz. It is following in the wake of the Red-rumped Swallow, whose old nests it uses.

Little Swift
Now showing signs of attempting to colonise the extreme south of Spain from Morocco.

Bee-eater
Now breeding regularly as far north in France as Brittany, Normandy and near Paris. In 1982 a pair bred near Calais. In Austria, widespread breeding in eastern areas since the 1970s, plus isolated nesting elsewhere. In 1964, 1965 and 1983 breeding was reported in the Netherlands, while in Denmark, following breeding records in 1961, 1962, 1966 and 1973, three pairs nested in south Jutland in 1985. More recently, in late spring 1992, as many as fifty-five Bee-eaters were reported in Denmark. In Sweden a small colony established near Jönköping in 1976 was reoccupied in 1978. Two pairs bred on the coast of Poland in 1986 and 1987. A single bird even appeared as far north as Iceland in June 1989. Since 1981 more Bee-eaters, often in parties of four or more, have been seen in Britain in the summer months than in the previous two decades, and some of them may have nested undetected (Batten *et al.* 1990). This is certainly a species that can be expected to respond to the greenhouse warming.

Hoopoe
The Hoopoe was in decline in north-central Europe for much of this century, but since the end of the maritime-influenced climatic amelioration around 1950 and the gradual return of generally drier, more 'continental' summers, it has shown some signs of returning to those areas. Sporadic breeding records have been more frequent in, for example, England, northern France and Denmark since 1970.

Grey-headed Woodpecker
Continuing to spread slowly westwards in Belgium and France.

Green Woodpecker
Still spreading northwards to some extent in Scotland, Denmark and Sweden in the 1980s, but this may now have ceased.

Black Woodpecker
Since 1984 has extended breeding range in France as far west as the Atlantic coast of Brittany.

Great Spotted Woodpecker
Following a stabilisation of the British population at the end of the 1850–1950 climatic amelioration, a marked increase occurred again in the 1970s and 1980s and was convincingly correlated with the outbreak of Dutch elm disease (see Chapter 8). Since then the population has stabilised again. However, a pronounced decrease has been reported from Sweden, following a population increase from 1970 to a peak in 1980 and 1981 (Hustings 1988).

Syrian Woodpecker
North-westerly and northward advance of this woodpecker revived in the 1970s. In Poland it has been breeding as far north as the Baltic coast since 1986. Since it first appeared in the former Czechoslovakia in 1949, it has spread north and north-westwards, reaching eastern Bohemia by 1974. It has also been spreading westwards in Austria, reaching Linz by 1982. In 1991 the Ukrainian breeding population was said to number about 1,000 pairs. The first breeding record in Belarus (Byelorussia) was in 1980 in the south-east, and it has since occurred in the south-west. (See Figure 38.)

Middle Spotted Woodpecker
Reported to be increasing in south-west Germany since 1960; declined rapidly from 1970s to extinction as a breeding species in the south of Sweden by 1982, where a century ago it was locally quite common (see Chapters 8 and 11).

Dupont's Lark
Previously considered to be absent from Spain as a breeding species, it was found to be widespread there during the 1980s, and as far north as Burgos near the Bay of Biscay in 1984. In 1988 the Spanish breeding population of this African lark of arid scrub and grass steppes was estimated at 7,000–8,000 pairs! It is not clear whether it is a recent northward range extension or a case of it being overlooked.

Shore Lark
Has spread northwards in Romania from its breeding range in the southern Balkans since the 1950s. Continued to spread northwards in Scandinavia during the 1950s and 1960s, although disappearing as a breeding species in Finnish Lapland.

$\mathcal{F.D.}$

FIGURE 63 *Citrine Wagtail*

Crag Martin

Since 1980 this southern European species has been gradually extending its breeding range northwards in southern Europe, notably in Bulgaria and Switzerland since 1980, Bavaria (1981), Austria (1982), Romania (1986) and the former Yugoslavia (1988). Birds also being reported from northern Europe, including Britain and Finland. This recent spread may well be influenced by the increasing anthropogenic greenhouse effect since about 1980.

Swallow

Continued to spread northwards in Scandinavia during 1950s and 1960s, but subsequently decreases were reported from Estonia, Latvia, Lithuania, northern Poland, northern Germany, Denmark, the Netherlands, Britain and Sweden. This may be connected with the post-1950 climatic deterioration in the Fenno-Scandian sector, but Marchant *et al.* (1990) pointed out that these populations may be wintering in, or migrating through, large parts of Africa affected by periodically below-average rainfall over the past three decades. In Denmark, Møller (1989) found a positive correlation between Swallow mortality and the amount of rainfall in South Africa. He also produced evidence that drought in Africa impairs the breeding condition of birds returning in spring and leads to a fall in the number of eggs they lay.

Red-rumped Swallow

Bred in Romania for the first time in 1976; since then it has colonised the south-west of that country. Since the 1950s it has occurred more frequently in countries north of its current breeding range, including western France, the British Isles, the Low Countries, Switzerland, Germany, Poland, Norway, Sweden, Finland and, in 1988, even Iceland. In Spain it continues to spread towards the north coast, while in France it is spreading steadily northwards from the coastal départements.

Yellow Wagtails

The ashy-headed race *cinereocapilla* consolidated and increased its breeding strength and range in Austria in the 1970s, while the black-headed race *feldegg*, which bred for the first time in Hungary in 1970, was reported breeding in Romania in 1972 for the first time. The dark-headed races in Denmark, and in northern Finland and Sweden, have recently been reported to have increased (Marchant *et al.* 1990).

Citrine Wagtail

Since 1950 the Citrine Wagtail has extended its breeding range very considerably to the west and south-west in the former USSR, penetrating as far west as Sumy and Kharkov in the Ukraine by 1976 (Wilson 1979) and still spreading west and south in 1992. From 1976 Citrine Wagtails paired with Yellow Wagtails have been found nesting in Estonia, Finland, Sweden and even in England (Essex, 1976). In 1991 the first pure pair bred successfully in Finland. Following the appearance of a pair at Žuvintas in 1985, a pair

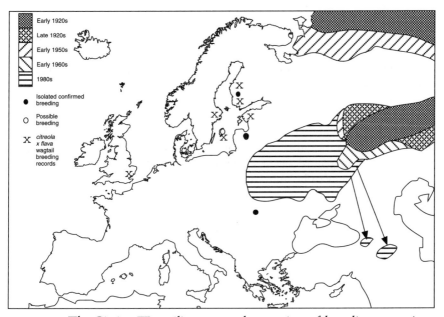

FIGURE 64 *The Citrine Wagtail's westward extension of breeding range into Europe (adapted from Wilson 1979; Harrison 1982; Cramp and Simmons 1988; Barthel 1990; Simms 1992).*

nested there in 1986, the first breeding record for Lithuania and the most westerly pure breeding record so far; in 1992 six pairs bred in Lithuania. This species is also being recorded more frequently in many European countries, including the Baltic republics, Poland, the former Czechoslovakia, Hungary, the former Yugoslavia, Greece, Cyprus, Italy, Austria, Germany, Denmark, the Netherlands, the British Isles, France and even Spain and Iceland. (See Figure 64.)

Grey Wagtail

Increasing and spreading again in Denmark by 1975, it had by then nested in Estonia and Finland for the first time, and was even nesting in Swedish Lapland, 1,000 km north of the previous limit. It has also colonised most of Norway even, since 1973, as far north as Finnmark. In 1991 it nested for the first time in Latvia. This represents a remarkable northward extension of breeding range, following recovery from the frequent severe winters of the late 1970s and early 1980s.

Thrush Nightingale

The westward spread of this species reached the Dutch coast in 1978, when singing males were reported from the Friesian islands of Texel and Vlieland. In central Europe a pair nested as far west as southern Bohemia, Czech Republic, for the first time in 1989.

Bluethroat

Breeding populations in central and southern Europe have been increasing and expanding in some areas (e.g. the former Czechoslovakia, Italy and Spain). Both the white-spotted and red-spotted races have increased rapidly in the former Czechoslovakia since 1978.

Red-flanked Bluetail

Still slowly advancing west, with breeding in Estonia since 1980 and signs that Norway may be on the point of being colonised. In 1992 the second Finnish breeding record occurred in Kuusamo, although actual breeding population thought to be quite substantial.

Black Redstart

Expansion of breeding range has continued, especially in Denmark, where it has become common even in small towns; at least 150 pairs were breeding in southern Jutland alone by 1978. It has also colonised south-west Finland since about 1972 and south-east Norway since 1983.

Blackbird

Has consolidated its breeding strength in recently colonised areas, such as the Faeroe Islands, 1950–78, and is continuing to spread northwards in Fenno-Scandia.

288

Fieldfare

The westward advance across central Europe has continued since 1950, with some three to a dozen pairs breeding annually in Britain in the years 1977 onwards. The Fieldfare has also extended its breeding range southwards in recent years into Austria, Romania and the former Yugoslavia. By 1989 at least 700 pairs were said to be nesting in the south-east Dutch province of Limburg alone.

Cetti's Warbler

Subsequent to information given in Chapter 9, the breeding population established in Britain since 1972 continued to grow and expand—in 1977 some 154 pairs bred or were suspected of doing so in forty-one different sites. In 1982, in spite of a temporary set-back due to the hard winters of 1978–79 and 1981–82, the estimated breeding population was put at about 200 pairs. Two summers later it peaked at 316 pairs, but following the hard winters of 1984–85 and 1986–87 the breeding population decreased. However, the mild winters since then have allowed a quick recovery and by 1990, following a very mild winter, it reached the highest peak so far recorded of 345 pairs (Spencer *et al.* 1993). The hard winters had been most severe in south-east England and they caused a shift of the population to south-west England, and the temporary extermination of the once flourishing Kent population. Expansion of breeding range also continued on the European mainland, at least up to and including 1984. However, the severe winters of 1984–85 and 1986–87 not only caused a population crash in France, but also affected the Low Countries and other parts of north-west Europe. (See Figure 44.)

Fan-tailed Warbler

As mentioned in Chapter 11, a decline occurred in Belgium after 1975, no birds being observed in 1978. The colonisation of the former Yugoslavia from Italy continued, with breeding reported from Dalmatia in 1978, but after the severe winters of 1978–79, 1984–85 and 1986–87 the population crashed both there and in France. Very recently, following subsequent mild winters, a recovery has commenced and a pair nested in the Netherlands in 1990 for the first time since 1984.

Lanceolated Warbler

Since the beginning of the 1980s this *Locustella* warbler has also shown signs of extending its breeding range westwards: singing males have appeared most years in south-east Finland.

Grasshopper Warbler

Continued to spread north and north-westwards in southern Scandinavia up to the 1980s at least, but with marked fluctuations.

River Warbler

Still extending its range westwards and north-westwards. Around 200 singing males have been recorded in Sweden in most years since 1988, when a second breeding record was confirmed. In the Netherlands, the number of males singing each spring continued to increase in the early 1990s.

Savi's Warbler

Has continued to expand its range north and west. Bred for first time in Latvia in 1972 and Estonia in 1977; increasing in Sweden. Singing males appeared in southern Finland during the 1980s, a pair bred in 1984, and in 1986 about seventy-five singing males were reported. Singing males have also appeared in southern Norway. Following the discovery of a singing male at Wicken Fen, Cambridgeshire, in the summer of 1954, the first to summer in England for almost a century, Savi's Warbler was subsequently discovered breeding in 1960 in a huge reed-bed in east Kent (Stodmarsh), and has done so ever since, varying in numbers from two to twelve pairs or singing males. Since the mid-1960s singing males have been found in East Anglian reed-beds, and more recently in other suitable sites elsewhere in England, such as Devon, and as far north as Lancashire and Yorkshire. In the late 1970s up to thirty singing males were reported annually from up to fifteen sites, but their numbers have declined since 1980 to an average of fifteen in twelve sites.

Moustached Warbler

There are signs that this south European warbler may be attempting to extend its range northwards in central Europe. A pair may have bred in Bavaria, south-east Germany in 1981 and 1982, and a pair certainly did so in 1984, successfully raising young. In May 1989 a singing male was found in southern Poland, near the Slovak border.

Paddyfield Warbler

There has been some evidence since about 1975 that this species is extending its breeding range along the northern edge of the Black Sea and in west-central Asia, and beginning to spread north-westwards. Vagrants have been appearing more frequently since then in north-central Europe and an instance of breeding was reported from Finland in 1991.

Blyth's Reed Warbler

Following its colonisation of south-east Finland in the mid-1930s, this warbler showed little sign of advancing further west until the end of the 1970s. By the beginning of the 1980s, however, the breeding population in southern Finland had built up, and it began the colonisation of southern Sweden and south-east Norway. Then, by the spring of 1988, the Finnish population was reported to exceed 5,000 singing males (pairs). Meanwhile, in 1981 the first singing males appeared in Poland and by 1983 were thought to be nesting. The colonisation of Latvia also began in the 1980s with numbers still increasing in 1988. Singing males have also been appearing in Denmark in recent years.

Marsh Warbler

Has started to colonise southern Norway since 1978, and in June 1982 a singing male was reported on one of the Faeroe Islands. In England it seemed to be nearing extinction as a breeding species by the late 1980s when it almost disappeared from its former stronghold in Worcestershire. However, at about the same time, south-east England appeared to be in the process of being colonised from the adjacent parts of the European mainland, where the Marsh Warbler is flourishing (Spencer *et al.* 1991).

Reed Warbler

Has continued to consolidate and expand its breeding range in southern Norway since its arrival in 1947, and also in Sweden, especially since 1975 (Hustings 1988). However, decreases have been reported from Denmark and Finland (Marchant *et al.* 1990).

Great Reed Warbler

Big increase in southern Sweden in recent years, with, for example, 208 singing males in 1980 and 305 in 1988. However, there has been a marked decrease in eastern Germany and the former Czechoslovakia since the late 1970s, apparently due to habitat destruction. Occasional singing males continue to appear in spring in England, among the latest being individuals in Berkshire, Lincolnshire and Northumberland in 1990, Lancashire in 1991, and in Devon, Somerset (Avon) and Suffolk in 1992.

Olivaceous Warbler

Still advancing north-westwards in the 1980s.

Booted Warbler

Since 1981 this species has shown signs of extending its breeding range north-westwards, with singing males in southern Finland in most years and in Estonia in 1987.

Olive-tree Warbler

The northward spread of this species in the Balkans has now extended to the Dobrogea region of Romania.

Icterine Warbler

The evidence for a recovery of this species in northern Europe has subsequently been borne out by a northward expansion of the Norwegian population in the 1980s. A pair laid eggs in 1970 in Yorkshire, but the species was not suspected of breeding again in Britain until 1989 when a male was found singing from 5 to 10 June in the Scottish Highlands (Spencer *et al.* 1991). On 15 May 1992, a male sang in a Bristol suburban garden (Dadds 1993).

Melodious Warbler

The northward and north-eastward expansion of range by the Melodious Warbler continued during the 1980s and since into north-east France, southern Belgium, the Netherlands, Luxembourg, northern Switzerland, southwest Germany and northern (former) Yugoslavia. Since May 1986, singing males have even appeared in eastern Poland, close to the border with Belarus (Byelorussia), while in eastern England a male was discovered singing in suitable breeding habitat on 28 June 1989 in Suffolk, but was not seen subsequently (Spencer *et al.* 1991).

Barred Warbler

Following the decline of the 1960s and 1970s, the breeding population in Denmark has recovered since 1983, probably due to the warm, dry summer of that year and some subsequent ones. In the summer of 1990, a pair bred successfully in Telemark, southern Norway, the sixth breeding record for that country.

Lesser Whitethroat

Continuing to increase in Britain in the 1980s and 1990s, colonising new areas further north in Scotland, such as Speyside and near Aberdeen. Also increasing and spreading into new areas in the extreme south-west of England.

Whitethroat

I discussed in Chapter 11 the effect on European breeding populations of this and other species, such as the Sedge Warbler, of the severe droughts in the western Sahel region of northern tropical Africa which lasted from 1968 to 1987, with little respite. After the population crash of the Whitethroat in 1969, caused by the particularly severe Sahel drought of 1968, the western European populations remained low, but they have shown signs of recovering since the improvement in the Sahel rains from the mid-1980s, especially those of 1991. By 1992 the Whitethroat recovery was becoming quite strong, judging by the BTO Common Birds Census in Britain.

Blackcap

The marked long-term increase of this familiar species in north-central Europe, including the British Isles, since the mid-1950s and 1960s is continuing, but so far I am unaware of any evidence that it is extending its breeding range northwards, although this may be happening. The factors involved in this widespread increase are at present uncertain, but if climatic change is an underlying cause, it would appear that the Blackcap has benefited more from the generally drier, continental-type conditions which have set in since

the waning of the milder and wetter pre-1950 long-term climatic amelioration. The increasing warmth since about 1975 due to the growing influence
of the anthropogenic greenhouse effect may also be acting in the species'
favour, perhaps through the increase in activity of flying insects in the generally warmer summers.

Greenish Warbler
Since 1970 this species has become abundant throughout Finland, and since
1978 has been attempting the colonisation of the former Czechoslovakia,
northern Germany, Norway, Sweden and Denmark (about fifty singing males
in 1988 and breeding suspected). In Poland it is also increasing and spreading.

Wood Warbler
Apparently continuing to extend its range northwards in Fenno-Scandia
(Marchant *et al.* 1990).

Firecrest
Appearing more frequently in the breeding season in south-west Finland
since the late 1980s and continuing to spread north-westwards in France.
Breeding population increasing again in southern Britain. (See Figure 47.)

Red-breasted Flycatcher
In 1978 a male occupied a territory near Brussels, Belgium, during the
breeding season, as did another male at Bargen, Switzerland in 1981. Since
then there has been an expansion of its breeding range in Austria, and in
southern Norway pairs bred successfully in 1982 and 1989.

Collared Flycatcher
As mentioned in Chapter 9, has continued to extend its breeding range very
slowly northwards since 1950.

Semi-collared Flycatcher
Spreading westwards and northwards in Greece and Bulgaria (Matvejev
1985).

Bearded Tit
The first record of this species in Estonia occurred when two males appeared
in April 1978 in the south-west, near the Latvian border. See Chapter 10 for
details of its range expansion since 1950. Following the severe 1978–79
winter, marked decreases were reported from Latvia, Denmark and southern Sweden, but a good recovery followed within a few years, and by 1989
the Swedish breeding population was reported to be at its highest ever. By
1984 the colonisation of southern Finland began, with the first breeding
record reported in 1987; by 1991 Bearded Tits were breeding in several
places. Since the mid-1970s there has been an expansion of breeding range

in Italy. The species is also now breeding along the old course of the Rhine in south-west Germany.

Willow Tit
This species, which had been curiously absent from Denmark, began colonising Jutland in 1977.

Blue Tit and Great Tit
Northward extension of range in northern Britain and Fenno-Scandia continued at least up to the early 1970s.

Azure Tit
Belarus (Byelorussia): signs of a westward extension of breeding range towards the Polish frontier with a pair nesting near Brest.

Nuthatch
Apparently still spreading northwards in southern Scotland. From 1975 until 1991 there was a marked increase in England and Wales, following which a decrease occurred.

Short-toed Treecreeper
The spread of this species into Denmark has continued since 1950, and it even showed signs of attempting to colonise England from 1969. Individuals appeared in southern Sweden for the first time in 1983.

Penduline Tit
The range expansion of this species to the north and west in Europe has continued unabated: it bred for the first time in Denmark in 1964, Sweden in 1964, Finland in 1985 and Belgium in 1989. Meanwhile, breeding pairs are increasing and spreading in Spain, France, western Germany, Denmark, the Low Countries, Sweden, the Baltic republics and the former Yugoslavia. Following the colonisation of the Netherlands in 1981, fifty-five pairs were reported to be breeding in 1989 and about 130 pairs in 1990. Since Penduline Tits first appeared in Britain in the 1980s, about fifty have so far been reported. In the spring of 1990 an apparently lone male in Kent went so far as to build one complete nest and three-quarters of a second nest (Spencer *et al.* 1993). The first breeding record is now confidently expected. (See Figure 65.)

Golden Oriole
Further to the account given in Chapter 10, the 1976 breeding population in southern Britain has been maintained since 1977 and numbers about forty pairs at the time of writing. Elsewhere in northern Europe the Golden Oriole steadily continued to increase and extend the northern limits of its breeding range up to about 1980, but apparently not since. In some parts of its northern breeding range it has, in fact, been reported as declining recently.

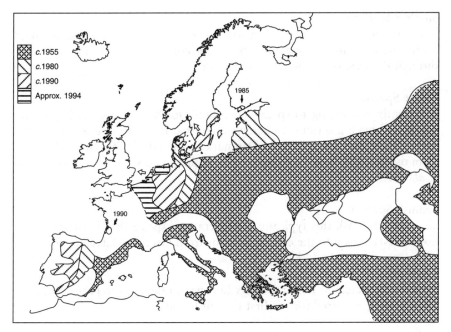

FIGURE 65 *The Penduline Tit's western extension of breeding range in Europe (adapted and updated from Voous 1960; Harrison 1982; Eesti Atlas 1982; Perrins and Simmons 1993).*

Lesser Grey Shrike
This central and south-east European species has spread northwards into the Baltic republics since 1950 and may have bred in Estonia in 1979 and 1980. It has been colonising north-east Spain (Catalonia) from southern France, breeding being proved for the first time in 1962.

Magpie
Continuing marked increase over much of its European breeding range.

Jackdaw
Continuing to spread northwards in some parts of northern Europe.

Rook
Apparently continued its northward range expansion in Fenno-Scandia, at least as late as 1978. In the late 1980s, as in Britain, there is evidence of a recovery of the north European populations following the general decline in the 1960s and early 1970s.

Carrion Crow
Northward and westward advance apparently continuing at the expense of the Hooded Crow.

Spotless Starling
Recently this west Mediterranean species began rapidly extending its range northwards in northern Spain and began the colonisation of southern France through the Pyrénées-Orientales and Aude.

House Sparrow
Apparently continuing to spread north and north-west after 1950, on the whole, although a serious decline was reported in the Faeroe Islands in the early 1980s, and lesser ones in Denmark in the late 1970s and in Sweden from the late 1980s. Also thought to be declining now in Britain (Marchant *et al.* 1990).

Spanish Sparrow
Since about 1950, this typically Mediterranean species has been spreading north-westwards into the central Balkans from Greece (Matvejev 1985).

Chaffinch
Apparently increased in much of its European range in the past decade. In 1986 nesting occurred for the first time in Iceland.

Serin
Rapid expansion has slowed down since about 1970, and it is only making slow progress in extending the present north-western limits of its breeding range (for example, in southern England, as described in detail in Chapter 10).

Goldfinch
Northward spread in Fenno-Scandia continued up to the 1970s, and is still continuing in Scotland.

Trumpeter Finch
The marked northward spread of this species from north-west Africa into southern Spain continues; breeding now confirmed in the Cartagena Mountains.

Scarlet (Common) Rosefinch
Since the 1970s the Scarlet Rosefinch has continued its colonisation of such countries as Austria, Switzerland, the former Yugoslavia, Germany, Denmark, Norway, Sweden, Hungary, the former Czechoslovakia (up to 500 pairs) and the Netherlands (since 1987; about fifty breeding pairs in 1992), while advance-guards have even reached Bulgaria (breeding since 1993), Greece, the Channel coast of northern France (nested in Pas de Calais in 1993), Spain, Belgium, the Faeroes and Britain. The first pair nested in Britain in 1982 (in the Scottish Highlands), and a few pairs are now breeding annually in Scotland. In the spring of 1992 there was an unprecedented influx into Britain, and some stayed to nest, including the first confirmed

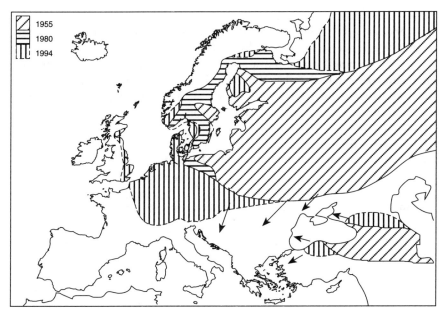

FIGURE 66 *The Scarlet Rosefinch's spread westward through Europe (updated from Stjernberg 1985).*

pairs to do so in England. In the western Ukraine it is spreading southwards. (See Figure 66.)

Bullfinch

Range expansion and increases continued in the British Isles up to the mid-1970s, and perhaps elsewhere in northern Europe, but not in Scandinavia since the early 1980s.

House Bunting

Since the 1960s has been spreading northwards in Morocco, and, if the current global warming grows, may begin the colonisation of Spain across the Straits of Gibraltar.

Rustic Bunting

Since 1978 has also shown signs of colonising Estonia, with males regularly singing, and pairs having bred in 1979 and 1986. A pair also nested in Latvia in 1985. The colonisation of Sweden has continued apace with a rapid extension of its range to the south as well as the west.

Yellow-breasted Bunting

Westward advance continues, with signs that Estonia is about to be colonised. (See Figure 53.)

Pallas's Reed Bunting

In recent years this Asian species of reed bunting has dramatically extended its breeding range westwards to breed for the first time in Europe in 1981 in the Bol'shezemel'skaya tundra area of Russia. It appears to be a bird of drier and cooler habitats than the Reed Bunting (Harrison 1982). Its westward spread may therefore be connected with cooler conditions in the tundra of northern Europe since about 1950.

Black-headed Bunting

Since the late 1960s, as mentioned in Chapter 10, has been extending its breeding range northwards through Bulgaria and into Romania.

To summarise the scale of the impact of the changes in climate on the 435 European breeding species of birds since the beginning of the 20th century, at least 309 species (71%) have undergone alterations in their geographical distributions to a greater or lesser extent (see Appendix 1), probably at least partly due to climate change. Of these, 196 species (45.1%) characteristic of the Temperate zone have at some time or other during this period spread northwards or north-westwards (Appendix 2.1). Thirty-six of these, plus another 28 species not included in that list are spreading westwards as well, making a total of 64 (14.7%) (Appendix 2.2). These additional 28 appear to be moving westwards only. Thus, altogether, 224 species (51.5%) of the 435 European breeding species have moved northwards, north-westwards and/or westwards in the course of this century. Since 1950 or later, at least 160 species (36.8%), perhaps as many as 192 (44.1%), have bucked the post-1950 natural climatic cooling trend by continuing or starting to spread in these directions, presumably encouraged by the 'unnatural' warming produced by the anthropogenic greenhouse effect, notably since about 1975 (Appendices 2.1 and 2.2). A further 32 species (7.4%) retreated south or south-eastwards from 1900 or so (Appendix 2.3). Between 1900 and 1950, 50 species (11.5%) characteristic of the Arctic/subarctic zone retreated northwards or probably did so (Appendix 2.4), but since 1950, 55 species (12.6%) from this zone have advanced southwards again or appear to have done so (Appendix 2.8).

The preciseness of the above figures is subject to the fact that a small number of species included have shifted their ranges first one way during the period since 1900, then the other, or may have advanced in, for example, the western part of their range and retreated in the east, or *vice versa*.

INSECT SPECIES CONTINUING TO SPREAD NORTHWARDS OR NORTH-WESTWARDS

It is of comparative interest that many moths and other insects have also continued to expand their ranges northwards and north-westwards in north-west Europe in the same way: for example, the large thorn, varied coronet, toadflax brocade, Blair's shoulder-knot, Dewick's plusia (silver spangle) and Vine's rustic moths, the map butterfly, the hover-fly *Volcella zonaria* and the ladybird *Harmonia quadripunctata*. In fact, since 1850 more than forty different species of the larger Lepidoptera (butterflies and moths) from the European mainland have appeared naturally (i.e. unassisted by man) in the British Isles for the first time. Of these, nearly thirty have become either permanently or temporarily established, while at least a few of the others seem to be attempting to do so. Many other animals are spreading northwards too, though specific information in many instances seems to be lacking or has so far escaped my notice. It will certainly be tremendously interesting to observe and record what happens in future years, which brings me to some consideration of future climatic trends.

THE PRESENT AND LIKELY INFLUENCES OF THE ANTHROPOGENIC GREENHOUSE EFFECT WARMING ON EUROPEAN BIRDS

Here and there in previous chapters I have already made frequent references to the probable influences of the anthropogenic greenhouse effect on various species of birds; now I will endeavour to summarise them and draw general conclusions.

There can be little doubt that within the last decade at least, the global warming due to the greenhouse effect has begun to exert an increasingly marked influence on birds and other wildlife in Europe, especially in the central and southern parts of the continent. Many of the species which advanced their breeding ranges northwards and north-westwards during the 1850–1950 natural climatic amelioration have resumed their inexorable advance, following a recession or halt (relatively brief in some cases) during the natural deterioration from about 1950 to 1980 to a generally colder, drier continental climate, particularly in the Arctic. They have recently been joined by some additional species, notably including some basically African and south-western Asiatic birds, such as the Black-winged Kite, Spur-winged Plover, Lesser Crested Tern, Laughing Dove, White-rumped Swift, Little Swift and Trumpeter Finch; and insects, such as the African monarch butterfly *Danaus chrysippus*. Moreover, other birds from that part of the world, such as the Pink-backed Pelican, Western Reef Heron, Crab Plover, Greater Sand Plover, Egyptian Plover and White-crowned Black Wheatear, have been appearing more frequently in Europe in recent years, although still very rare vagrants. Of significance in respect of these developments, Gribbin (1990) stated that the present warming is strongest over parts of southern Asia, northern Africa and south-western Europe, among other regions.

Many European insects, including some butterfly and many moth species, have also continued the north or north-westward expansion of their breeding ranges begun during the 1850–1950 amelioration, or have resumed it since the end of the cool phase from 1950–80. Some, such as the social wasps *Dolichovespula media* and *D. saxonica*, and the sphecid wasp *Nitela borealis*, have only very recently crossed the sea and begun the colonisation of south-east England; 1980 in the case of the two social wasps, and 1987. Meanwhile, some species of Orthoptera, notable indicators of a warming environment, have rapidly been extending their ranges significantly northwards in southern Britain since 1975 (Burton 1991). Before that year, one of these, the long-winged conehead bush-cricket *Conocephalus discolor*, was entirely confined to scattered, mainly coastal localities from Sussex west to Dorset. Intensive searches in the late 1970s, and since, resulted in the discovery of many more sites for the species, some of them well inland. In the 1980s it became clear that not only was a population explosion taking place, but that *discolor* was extending its range northwards and westwards. During the hot summer of 1990 it even reached the Isles of Scilly and the southern fringe of London, and by 1992 was found as far north as Oxfordshire. During very hot summers, and in response to high population overcrowding, some long-winged coneheads with extra long wings appear, which fly rapidly and are capable of extended flights, thus enabling them to colonise new territory more rapidly.

Plants, too, are moving northwards. In Britain, botanists at Sheffield University's Unit of Comparative Plant Ecology, headed by Professor Philip Grime, have been studying the spread since 1980 of, for example, the Canadian fleabane *Conyza canadensis*, the prickly lettuce *Lattuca serriola*, and the formerly extremely rare lizard orchid *Himantoglossum hircinum*.

If the greenhouse warming eventually leads to drier as well as hotter conditions in the future, we may possibly expect the return as breeding species to Britain and other parts of north-west Europe, from which they virtually vanished in the course of this century, of such birds as the Red-backed Shrike and Wryneck; and perhaps eventual colonisation by other lovers of a very warm, dry climate such as the Bee-eater and Roller. So far, however, there is very little evidence of that happening.

The fortunes of Arctic and subarctic species will also change as the greenhouse warming increasingly takes hold. Some, such as the Scaup, Long-tailed Duck, Temminck's Stint, Wood Sandpiper, Gyrfalcon, Snowy Owl, Redwing and Brambling are already showing signs of retreating northwards again since 1980. On the other hand, some other species which one would have expected to do likewise have either shown no signs yet of doing so, or are even, like the Slavonian Grebe and Whooper Swan, still consolidating their southern breeding outposts. Furthermore, those truly Arctic gulls, the Ivory and Ross's, and Arctic ducks, the King Eider and Steller's Eider, plus the Great Northern and White-billed Divers, continue to winter remarkably far south. Presumably, if the anthropogenic greenhouse effect is not checked through international efforts, they too will eventually retreat northwards.

As we have seen, Scandinavia is currrently climatically 'out-of-step' with, for example, Siberia, and is continuing to experience relatively cool conditions in spite of the warming in the far east and south. In fact, mean temperatures over Scandinavia in recent years have dropped by around 0.6°C, compared with a rise of 0.31°C over the Northern Hemisphere as a whole during the same period, and Scandinavia is thus almost a full degree cooler than the adjacent latitudes (Gribbin 1990). It is therefore perhaps not surprising that the Crested Lark, which colonised southern Sweden and south-east Norway from 1900, began decreasing after 1950 and finally vanished as a breeding species from both countries by 1990. Another example is the Middle Spotted Woodpecker, which declined at the north-western limits of its breeding range in Europe before and during the 1950s, becoming extinct in Denmark in 1960 and in Sweden in 1982.

It is, of course, still early days as far as assessing the influence of the greenhouse effect on European birds is concerned, but if it continues unchecked there is little doubt that their responses to it will become more obvious and perhaps even more dramatic than those I have described in this book. The situation is, of course, complicated by the present unstable state of the weather in the North Atlantic sector; this is often characteristic of marked climatic change. For example, as described in Chapter 4, the transition from the Medieval Warm Period to the Little Ice Age was accompanied by very unsettled, sometimes turbulent weather, such as the tempestuous storms of the 13th century. In recent years, those of us living in north-west Europe have endured some similar weather extremes: very severe gales, storms and hurricanes; unusually heavy spells of rainfall with serious flooding alternating with droughts; very warm summers and occasional very cool, wet ones; and severe winters alternating with extremely mild ones. North Americans have suffered similar, even worse extremes. Extreme weather has, of course, occurred from time to time throughout the 20th century, but what we are now experiencing is a remarkable variety of weather extremes condensed into a relatively short span of years, not spread out over many. The weather machine is currently becoming tempestuous, but presumably, in due course, a more stable climatic pattern will become established.

Chapter 13

WHAT OF THE FUTURE?

What are we heading for? An overheating earth because of the greenhouse effect or, in spite of it, a new ice age? It is generally accepted by most earth scientists that we are actually still living in an ice age—in an interglacial of the so-called Last Glaciation. But for the anthropogenic greenhouse effect, the indications from about 1950 were that we were entering a natural cooling phase with an eventual readvance of the polar ice sheets, possibly even a major one. Recent research has indicated that such advances of the ice sheets could be much more rapid than had been believed hitherto.

As I mentioned in the Introduction (Chapter 1), I am not a climatologist, but primarily a zoologist and naturalist interested in the effects of climatic change on animals and plants. Therefore, I largely depend upon the conclusions and prognostications of others far better qualified than myself in interpreting the climate's trends. In fact, since I first contemplated writing this book in 1971, numerous books and many articles and scientific papers have been published which discuss, to varying degrees, current climatic trends and prospects. Among the books may be mentioned those by H.H. Lamb (1972–77), Nigel Calder (1974*b*), John Gribbin (1975*a*, 1976*a,b*, 1989 [with Mick Kelly], 1990), Brian S. John (1977), Ronald Pearson (1978), Georg Breuer (1980), Hermann Flohn (1980), Michael J. Ford (1982) and Bert Bolin *et al.* (1986). There seems little point in repeating in detail here much of what they and others have written on the subject. Instead, I have thought it more sensible in this final section to confine my remarks mainly to summarising the known and likely effects on the geographical distribution of European birds (and to a limited extent other animals) should on the one hand the present cooling of the climate in some parts of the Arctic region of the Northern Hemisphere intensify and become protracted, or should on the other hand the greenhouse effect gradually exert an opposing warming influence. Europe, in fact, faces a looming paradoxical situation in which two opposing forces are currently counteracting each other; which will eventually triumph, or whether they will cancel each other out, remains to be seen. A lot depends upon man's determination to reduce the emission of greenhouse gases, and his success or otherwise in doing so. So far as he is represented by politicians, he has shown a minimum of determination; to make any real effort is all so inconvenient to most politicians, economists and businessmen, who show such little real concern for the natural world. Later in this concluding chapter I will also consider the role that wildlife plays in indicating and helping us to interpret climatic changes.

Although most climatologists, and many other kindred scientists, accept that we are actually living in an interglacial period (the Flandrian)—a warm interlude in what some of us at least were taught at school to regard as the 'last' Ice Age—there is a good deal of uncertainty and argument among them as to whether or not the post-1950 climatic cooling foreshadows the approach of a new glaciation (or would have foreshadowed it, were it not for the waxing greenhouse effect). But perhaps sooner or later, now most probably later, the ice sheets will eventually again expand outwards and southwards from the Arctic to cover huge areas of the land surface of Europe, Asia and North America, unless, of course, the greenhouse effect really can halt them. Faced with the two options, warming due to the greenhouse effect would not really be preferable owing to various accompanying disadvantages: it will bring severe droughts (and as a consequence famines), and has already done so in subtropical regions; and it will cause a significant rise in sea level and therefore flooding in low-lying areas, to say nothing of causing excessive, killing heat in local areas (e.g. in Greece).

Some interglacial periods have lasted many thousands of years, as indicated in an early chapter; for example, the last major one, the Ipswichian (Eemian), occupied some 50,000–60,000 years. At their height the climate was very much warmer than at any time since the final retreat of the ice which began our interglacial, only some 10,000–13,000 years ago. It could be that we are still living in an early phase of such a major interglacial; but it is also possible that we are living in one of those relatively minor warm periods (interstadia) which occurred at times during the Last Glaciation; perhaps similar to the earliest one which may have lasted for as much as 20,000 years, and whose climatic optimum seems to have been warmer than that attained some 5,000–7,000 years ago during the present so-called Postglacial Period—now renamed the Flandrian Interglacial. It *is* possible, therefore, as some scientists believe, that since 10,000 years or so have passed since the last glacial phase ended, we could have been approaching relatively soon a new major advance of the ice sheets, but for the anthropogenic greenhouse effect.

There are indeed those who think that there is definite evidence—although this is dismissed as 'misleading' by Brian S. John (1977) and others—that the climatic oscillations since the post-glacial (Flandrian) climatic optimum (Holocene Warm Period), which have produced alternating cold and warm phases of greater or lesser extent, are all contained within a broad downward sweep or spiral of climatic deterioration leading inexorably to the next glaciation. Moreover, they believe (or believed) that once the cooling really gets going on a global scale it will accelerate and a new glaciation could arrive remarkably quickly (in 1992 convincing evidence was produced that this has happened in the past)—perhaps, as Nigel Calder (1974*a,b*) has suggested, within a few hundred years. An increasing tendency for accumulating snow to fail to melt leads to the formation of permanent snowfields, which in turn become icefields, and finally ice-sheets. Permanent snow- and ice-covered areas, especially on high ground, create their own cold. For instance, the reflection of solar radiation from a snow surface is so great that about 80% is reflected back. Therefore, if permanent snow- and icefields become established, then a steadily escalating expansion of the area of snow and ice resisting the effects of the sun, warm winds and summer rains will result, causing the steady development of an ice-cap (Manley 1972). To quote Professor Manley's own words:

> 'Radiation on clear winter nights leads to the cooling of the adjacent surface air, and the slow building up of the resistant "cushions" of cold air that we recognise as a characteristic of the present-day Siberian winter. These cold anti-cyclones are noted for their propensity to fend off the surface incursions of warmer air.'

Until quite recently however, some scientists, while conceding that we do not yet know enough to rule out the possibility of a relatively imminent glaciation, saw little reason to believe from the available data that this was likely. They considered that the post-1950 climatic cooling, like those others which have occurred since the Last Glaciation, was primarily a local fluctuation which would do nothing to upset the overall stability of the earth's climatic pattern. On the whole, they said, climatic oscillations in one part of the world, or even within one of its hemispheres, are balanced by opposite changes in another part. As Brian S. John (1977) put it: 'Cooling of quite a different order is probably required to introduce a global glaciation.'

Another argument that has been made in favour of the view that a new glaciation was 'not just around the corner' is based on the believed existence of at least some more or less regular climatic oscillations. Major ice ages, for example, seem to have occurred every 150 million years or so, while little ice ages lasting between 500 and 900 years, such as the comparatively recent one described in Chapter 4, appear to take place at intervals of about 2,500 years. As the last Little Ice Age ended only about a century and a half ago, Brian S. John suggested that the next one should not be due until around AD 4000.

THE GREENHOUSE EFFECT

At this point we must consider more fully than hitherto the impact (albeit unintentional) of man's activities on the earth's climate, and whether the climate changes they bring about are of any significance in comparison with natural climatic fluctuations. There are many earth scientists who are convinced that they *are* capable of significant, even major, effects. In the early 1970s already Willi Dansgaard (1975) of Denmark warned that recent natural climatic changes could be drastically altered by man-made atmospheric pollution. But it took another decade before the fact of growing global climatic warming, due to the polluting activities of man adding significantly to the natural, balanced greenhouse effect to produce the now well-known supplementary anthropogenic greenhouse effect, became widely accepted. Politicians and national governments were among the last to be convinced, and some of them are apparently still unconvinced.

One of the chief sources of this pollution is the greatly increasing quantities of carbon dioxide (CO_2) released into the atmosphere by the burning of fossil fuels, such as coal and oil, by human industries. Motor transport is another important source of this gas. The recorded increase in the carbon dioxide level in the atmosphere between 1880 and 1940 was 11% (incidentally neatly fitting within the peak period of warmth of the 1850–1950 climatic amelioration, but not considered to be the cause of it as shown by the Camp Century ice core isotope studies!). Since then it has escalated and it has been calculated that between 1958 and 1989 it rose by a little over another 11%, an average atmospheric concentration increase of 35 parts per million by volume to just over 350 ppm, the highest level for 160,000 years and greater than at any time in the present interglacial or the previous one (the Eemian). From the ice core studies it has been shown that the natural carbon dioxide concentration in the air before man exerted an influence was 280 ppm; so the concentration in 1989 showed a rise of almost exactly 25% (Gribbin 1990). The rate of increase in the CO_2 concentration since 1950 is sufficient to raise the world's average temperature by about 3°C by AD 2000. Between 1880 and 1950, for comparison, the average global temperature rose by about 0.5°C while that of the Northern Hemisphere alone rose by 0.8°C. Over the twenty years from 1967 to 1986 the total warming of the land surface of the Northern Hemisphere was 0.31°C (Gribbin 1990). The process by which this greenhouse effect warming takes place is as follows: the water vapour and carbon dioxide concentrations (together with other greenhouse gases like methane) allow most of the sun's heat to penetrate to the earth's surface, but tend to trap and absorb into the atmosphere some of the earth's outgoing heat, especially infra-red radiation, which would otherwise escape into space. Many climatologists have warned that if this man-induced escalation of what is otherwise a natural, balanced process were to continue unchecked, returning increasing amounts of heat to the earth's surface, it will eventually result in, among other adverse consequences, serious melting of the polar ice-caps, more especially that of the Arctic, with the consequent flooding of many of the world's coastal towns and cities, including Britain's. Gribbin (1990) points out that if the Arctic ice-cap were to melt completely it would not re-form 'until the carbon dioxide concentration has fallen far below the critical value at which the ice-cap disappeared when the concentration was increasing'. Gribbin, incidentally, gives an excellent account of the natural and anthropogenic greenhouse effects in his 1990 book, and the consequences of failing to check and control the latter.

However, the different kinds of pollution produced by our industrialised society can produce conflicting effects, which to some extent tend to cancel each other out. For example, man is putting an increasing quantity of dust and other filth from factories and the like into the atmosphere which operates in quite the opposite way to the greenhouse effect, by allowing the earth's heat to escape into space, but reducing the amount of heat radiation reaching the earth's surface from the sun by absorption and by reflecting some of these incoming rays back into space. Thus it tends to have a cooling feedback effect upon the earth's climate additional to that resulting from natural dust clouds produced by volcanic activity (of which there has been an increasing amount in recent years), or that due to the passage of our Solar System through dust lanes which, as I have discussed in Chapter 2, are thought by some to be a likely cause of ice ages. Nevertheless, the cooling produced in this way is currently being more than negated by the strength of the greenhouse effect.

Other by-products of man's activities which are considered to have serious effects on climate include the creation of clouds by jet aircraft which may reflect some of the sun's radiation away from the earth, in addition to the CO_2 and other pollutants they emit; also the appalling scale of the destruction of the forests, especially the tropical rain forests, which formerly clothed so much of the earth's surface, and which play such an important role in recycling oxygen and in conserving water. There is also particular concern about the depletion of the ozone layer in the stratosphere by the release into the atmosphere of chlorofluorocarbon gases (CFCs) from such sources as the millions of aerosprays, such as hair-sprays, in use in modern society, although these are currently being superseded by less damaging substitutes. The ozone layer surrounds the earth at an altitude of between fifteen and fifty kilometres, and protects us and all other living creatures from the harmful effects of large doses of ultraviolet radiation which would otherwise reach us from the sun. CFCs also contribute to a very considerable extent to the anthropogenic greenhouse effect warming. Between 1955 and 1985 alone they contributed a warming 30% as large as that caused by the amount of carbon dioxide released over the same period (Gribbin 1990).

Climatic changes and their causes are certainly worthy of the attention and interest of everyone, not least politicians and governments. Fortunately, partly due to recent highly publicised trends and disasters, such as the awful years of drought and famine in the Sahel Desert of West Africa, and those of Ethiopia with their horrific tide of human misery, the attention of the world authorities has indeed been focused upon the subject, and more public money is available than ever before for research into the climate. In view of, for instance, man's rapidly expanding populations, and the need to feed more and more mouths, it is vitally important to understand and if possible predict climatic trends. Moreover, we must discover and if necessary control the impact of man on the whole of his environment, which, of course, includes the climate, and the flora and fauna.

DO BIRDS AND OTHER WILDLIFE PLAY ANY ROLE IN PREDICTING CLIMATIC CHANGE?

The answer, I believe, is not much, if any, but they do provide clear evidence and confirmation that climatic changes are occurring. I believe the evidence I have amassed in this book makes it obvious that many birds and other animals are very sensitive to, and respond to, even relatively slight climatic changes. However, even though, as we have seen, some species are quick to respond to climatic shifts which give them, for example, an advantage over related species with whom they may be in competition, and thus enable them to colonise new territory, they do not seem to predict or give us any really significant advance warning of impending changes of climate earlier than evidence available to us from other more reliable sources. But when a wide assemblage of birds, insects and other animals alter their normal distributions, apparently more or less in unison, they certainly do indicate that such changes are in progress, and provide confirmation of the reality of the changes noted by climatologists from other data.

For instance, the recent retreat southwards of a number of Arctic and subarctic birds to breed or winter far to the south of their accepted normal breeding and wintering ranges (e.g. Scotland) was indicative of the reality and extent of the climatic deterioration in the Arctic between 1950 and about 1980. Yet, on the other hand, animal species are often slow to react to the cessation of hitherto favourable climatic changes, as shown by the continued northward advance of many southern species in response to the post-1850 natural climatic amelioration, in spite of the fact that it began waning about 1950; they only tend to recoil when conditions become really adverse for them. However, this marked continuation of the northward spread of many animal species, especially since 1980, is now a clear demonstration, in my view, of the reality of the increasing warming due to the anthropogenic greenhouse effect. Of course other factors such as habitat loss may be involved, but they sometimes mask the underlying influence of climatic change and are not therefore easy to separate.

What changes in the distribution and movements (e.g. migration routes) of birds, insects and other animals can perhaps tell us more accurately, although not as precisely as modern scientific methods using computer models, is the likely magnitude of impending climatic oscillations; for example, were it not for the warming due to the anthropogenic greenhouse effect, the fact that some northern birds were appearing and breeding as far south in Europe as they were in the period 1950–80 seemed to indicate the return of a natural cooling of the severity and length of the Little Ice Age. Now, the

307

magnitude of the current northward expansion of many species is demonstrating the growing pace and strength of the current greenhouse warming.

Of course, if the anthropogenic greenhouse effect continues to increase in pace and strength as predicted by many scientists in this field, by the 2030s the earth could warm up by an additional 4°C (in any case it is almost certain to do so within the next hundred years), sufficient to melt all the ice over the North Pole and the Arctic Ocean (Flohn 1980). Flohn calculated that, as this happens, the subtropical climate zone of the Northern Hemisphere will shift at least 800 km north of its present position in winter, although only about 200 km north of its present summer position. Obviously, the speed and extent of these changes will prove disastrous for many plant and animal species, especially those of the high Arctic. For instance, many Arctic birds, such as the Snowy Owl, Rough-legged Buzzard, Goshawk and Gyrfalcon, have long been adapted to preying upon cyclic-fluctuating populations of Arctic lemmings and other small mammals, and the snowshoe or varying hare. None of these, predators or prey, may be able to adapt quickly enough to rapidly warming conditions in an increasingly ice-free Arctic and, moreover, cope with growing competition from related species as these advance northwards. This will, of course, be true of all Arctic and subarctic species, not just those involved in cyclic relationships.

Such a marked and rapid warming (fifty times faster than at the start of the present interglacial) may be too much for the northward spread of forest trees, which will probably be unable to respond so quickly. It has been estimated that they will need to advance at a rate of at least 600 km per century; the fastest rate recorded during the warming following the Last Glaciation was 200 km per century by spruce. An average rate is around 20 km a century (Gribbin 1990). So the speed with which woodland and woodland-edge species of birds and insects, although in the main highly mobile and quick to exploit and colonise vacant habitats as they become available and conditions are right, are able to spread northwards into the Arctic zone in this predicted scenario next century will depend upon the speed with which the trees can advance. Already, in the 1990s, we are apparently rapidly approaching a climatic phase comparable to that of the Medieval Warm Period, and by next century will be experiencing one like that of the so-called Climatic Optimum, now preferably called the Holocene Warm Period or Hypsithermal, around 6,000 years ago. But it's all happening much faster than it did then! Birds and other wildlife, let alone man, will be hard put to adapt quickly enough.

Important from the point of view of man's economic interests is the necessity of a proper understanding of the part both short-term and long-term climatic fluctuations play in altering the numbers and distributions of plants and animals, especially those species of economic importance. If we are to manage such natural resources effectively we must try, as Robert M. May (1979), then of Princeton University and now of Oxford University, has urged, 'to achieve skill in forecasting such climatic changes; and to convince resources managers to include these factors in their deliberations'. It would then be possible to make allowances for such events as the movement north during the climatic amelioration earlier this century of the whole ecosystem which seriously affected the economies of Greenland and Iceland. Thus the seal-hunting economy of western Greenland was replaced by a cod-fishing economy, while the fishermen in Icelandic waters had to search for their quarry further to the north than before. When the Arctic climate began deteriorating again from about 1950, herrings reappeared off Iceland's northern coasts, while cod retreated southwards to the detriment of the west Greenlanders, who started to turn to seals once more.

But whether or not the long-term reactions of plants and animals can foretell what the climate has in store for us, and warn of possibly serious consequences for the livelihoods of human communities, it is a fascinating and satisfying end in itself to study those reactions as precisely as possible. Kenneth Williamson (1975) wrote:

'It is not the individual summers and winters that count in the fortunes of mammals, birds and butterflies, since normal powers of recovery can combat the vicissitudes of short-term changes, but the cumulative effects of long runs of years which are either "maritime" or "continental" in type, dominated either by balmy westerlies or cold north easterlies—in short, years accumulated as periods of so-called "amelioration" or "deterioration".'

Yet too many naturalists fail to look beyond the effects of short-term climatic changes. It is this failure that I have attempted in some measure to rectify in this book, while at the same time trying to show just how interesting some of the cumulative effects of long-term climatic changes have been, and still are. I hope it will succeed in encouraging and assisting many more people, especially in Britain (the Finns and Scandinavians have long shown greater interest in the subject), to take up the study of the reactions of wildlife to these alternating periods of ameliorating and deteriorating climate, and especially to the growing threat of the greenhouse effect. Imperfect and speculative as this book is in parts, I hope nevertheless that it will stimulate others not only to investigate the current and future responses of animals due or partly due to climate changes, but also to throw new light, from other aspects and perhaps also from previously overlooked data, on what may have happened in the distant or not so recent past.

Appendix 1

CHANGES IN EUROPEAN BREEDING DISTRIBUTIONS

Considered due to climatic changes since 1850

Symbols: ↑ = *north* ↗ = *north-east* → = *east* ↘ = *south-east*
↓ = *south* ↙ = *south-west* ← = *west* ↖ = *north-west*

	Advancing		Retreating	
SPECIES	1850–1950	1950–1993	1850–1950	1950–1993
Red-throated Diver	↙↓	↓	↑ ?	
Black-throated Diver		↙↓	↑ ?	
Great Northern Diver		↓↘	↑ ?	
White-billed Diver		↙↓	↑ ?	
Little Grebe	↑	↑ ?		
Great Crested Grebe	↘↑	↘↑		
Red-necked Grebe		←		
Slavonian Grebe	↙ From 1908	↙↓	↑ Until *c.*1860	
Black-necked Grebe	←↘↑	↘↑ ?		
Fulmar	↙↓	↙↓		
Cory's Shearwater				
Manx Shearwater				
Storm Petrel				
Leach's Storm Petrel				
Gannet	↘↑↗↓	↘↑↗		
Cormorant		←↑↗↓↘		
Shag	↑↗	↑↗ ?		
Pygmy Cormorant		↘↑↗		
White Pelican				
Dalmatian Pelican				
Bittern	↘↑	?		?
Little Bittern				
Night Heron	↘↑	↘↑		

SPECIES	Advancing		Retreating	
	1850–1950	*1950–1993*	*1850–1950*	*1950–1993*
Squacco Heron	Began increasing early 1900s	↑	Decreased in 19th century	
Cattle Egret	↑	↖↑		
Little Egret	Decrease before *c*.1900; increase since	↖↑	Decreased in 19th century	
Great White Egret		↖ Since 1976		
Grey Heron	↖↑	↑		
Purple Heron	↖↑	↖ Till 1979		From 1979 ↓
Black Stork		← From *c*.1970	W Europe only ↘→	Till *c*.1970 ↘→
White Stork			↘	↘
Glossy Ibis				
Spoonbill				
Greater Flamingo				
Mute Swan	↑↗	↑↗		
Bewick's Swan				
Whooper Swan		↓	↑	
Bean Goose				
Pink-footed Goose				
White-fronted Goose				
Lesser White-fronted Goose				
Greylag Goose		↓	↑	
Canada Goose	↑	↑		
Barnacle Goose		↓ From 1975		
Brent Goose				
Ruddy Shelduck				
Shelduck	Population increase from 1900	↑ Population increase & range expansion		
Mandarin Duck	Increasing in GB	Increasing in GB		
Wigeon	↓			
Gadwall	↖↑	↖↑		
Teal				
Mallard	↑	↑		
Pintail	↙↓			?
Garganey	↖			↘ Up to 1988
Shoveler	↖	↖ Up to 1970		
Marbled Teal		↑		
Red-crested Pochard	←↖↑	←↖↑		
Pochard	←↖↑	←↖↑		
Ferruginous Duck	←			→
Tufted Duck	↖	↖		

SPECIES	Advancing 1850–1950	Advancing 1950–1993	Retreating 1850–1950	Retreating 1950–1993
Scaup		↓	↑	
Eider	↓	↙↓		
King Eider		↓	↑	
Steller's Eider		↓	↑	
Harlequin Duck			↖	
Long-tailed Duck		↘	↖	
Common Scoter	↙	↙ Up to 1970		↗ Since 1970
Velvet Scoter				
Barrow's Goldeneye				
Goldeneye		↓	↑	
Smew		↓↘	↑ ?	
Red-breasted Merganser	↓	↓		
Goosander	↓	↙↓		
Ruddy Duck		↑ Introduced in 1960		
White-headed Duck		↑ From c.1985	↙↓ From c.1900	↙↓ Until c.1980
Honey Buzzard	↖↑	↖		
Black-shouldered or Black-winged Kite		↑↗		
Black Kite	↖	↖		
Red Kite	↑ After c.1900	↖↑	Before c.1900 ↓	
White-tailed Eagle				
Lammergeier				
Egyptian Vulture				
Griffon Vulture				
Black Vulture				
Short-toed Eagle	From c.1920 ↑	?	Until c.1920 ↓	
Marsh Harrier		↖ Improved protection + ? climatic warming since c.1975		
Hen Harrier	↓ From 1939	↙↓	↑ Until 1939	
Pallid Harrier	← From c.1940	← ?		
Montagu's Harrier	↖↑ From c.1900	↑ Renewed since 1974		↓ 1953–1974
Goshawk				
Sparrowhawk				
Levant Sparrowhawk		↖		
Buzzard		Increasing		
Long-legged Buzzard				
Rough-legged Buzzard		↓ ?	↑ ?	
Lesser Spotted Eagle				
Spotted Eagle				

SPECIES	Advancing		Retreating	
	1850–1950	1950–1993	1850–1950	1950–1993
Steppe Eagle				
Imperial Eagle	?	↘		
Golden Eagle				
Booted Eagle				
Bonelli's Eagle				
Osprey		↙	↗	
Lesser Kestrel	From 1947 ↗	↗		
Kestrel				
Red-footed Falcon	←From c.1930	←↘		
Merlin				
Hobby		↘↑	↓↘	
Eleonora's Falcon				
Lanner Falcon				
Saker Falcon		←↘↑		
Gyrfalcon		↓	↑	
Peregrine				
Hazel Hen				
Willow/Red Grouse		?	↑	?
Ptarmigan		↓ ?	↑	
Black Grouse				
Capercaillie				
Chukar Partridge				
Rock Partridge				
Red-legged Partridge				
Barbary Partridge				
Grey Partridge				
Quail		↘	↘ Until 1942	
Pheasant				
Andalusian Hemipode				
Water Rail	↑	↑ ?		
Spotted Crake		↑	↓↘	
Little Crake				
Baillon's Crake				
Corncrake				
Moorhen	↑	↑ ?		
Purple Gallinule				
Coot	↑	↑ ?		
Crested Coot				

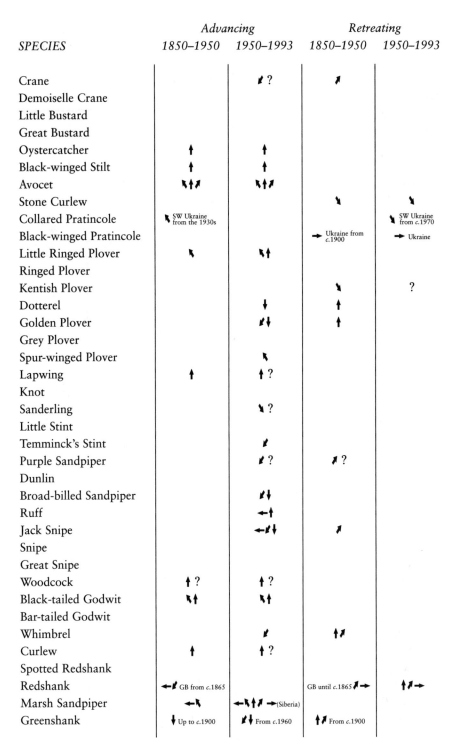

| SPECIES | Advancing | | Retreating | |
	1850–1950	1950–1993	1850–1950	1950–1993
Crane		↙ ?	↗	
Demoiselle Crane				
Little Bustard				
Great Bustard				
Oystercatcher	↑	↑		
Black-winged Stilt	↑	↑		
Avocet	↖↑↗	↖↑↗		
Stone Curlew			↘	↘
Collared Pratincole	↘ SW Ukraine from the 1930s			↘ SW Ukraine from c.1970
Black-winged Pratincole			→ Ukraine from c.1900	→ Ukraine
Little Ringed Plover	↘	↖↑		
Ringed Plover				
Kentish Plover			↘	?
Dotterel		↓	↑	
Golden Plover		↙↓	↑	
Grey Plover				
Spur-winged Plover		↘		
Lapwing	↑	↑ ?		
Knot				
Sanderling		↘ ?		
Little Stint				
Temminck's Stint		↙		
Purple Sandpiper		↙ ?	↗ ?	
Dunlin				
Broad-billed Sandpiper		↙↓		
Ruff		←↑		
Jack Snipe		←↙↓	↗	
Snipe				
Great Snipe				
Woodcock	↑ ?	↑ ?		
Black-tailed Godwit	↖↑	↖↑		
Bar-tailed Godwit				
Whimbrel		↙	↑↗	
Curlew	↑	↑ ?		
Spotted Redshank				
Redshank	←↙ GB from c.1865		GB until c.1865 ↗→	↑↗→
Marsh Sandpiper	←↘	←↖↑↗ →(Siberia)		
Greenshank	↓ Up to c.1900	↙↓ From c.1960	↑↗ From c.1900	

	Advancing		Retreating	
SPECIES	1850–1950	1950–1993	1850–1950	1950–1993
Green Sandpiper	←↙ ↑(Finland)	←↙↓ ↑(Finland)		
Wood Sandpiper		↙↓	↗	
Terek Sandpiper	←	←		
Common Sandpiper				
Turnstone		↙	↑	
Red-necked Phalarope	↑ From c.1900 to c.1925	↓ From c.1950 to c.1975	↑ Until c.1900 From 1930 to c.1950	↑ From c.1975
Grey Phalarope			↑	?
Pomarine Skua		↖ Since 1984		
Arctic Skua	↙ From c.1940	↙		
Long-tailed Skua		↙↓	↑ ?	
Great Skua				
Great Black-headed Gull				
Mediterranean Gull	↖	↖↑		
Little Gull	←	←↖		
Sabine's Gull				
Black-headed Gull	↖↑	↖↑		
Slender-billed Gull		↗ ?		
Audouin's Gull				
Common Gull	↖↑↓↘	↖↑↘		
Lesser Black-backed Gull	←↖	←↖		
Herring Gull	↑	↖↑		
Glaucous Gull		↓ ?	↑	
Great Black-backed Gull	↑	↖↑		
Kittiwake	↓	↓		
Ivory Gull		↓	?	
Gull-billed Tern	↖	↖		
Caspian Tern		← ?		
Sandwich Tern	Contraction before 1900; increase and expansion from 1920	Increase continuing		
Roseate Tern	↖ From c.1900	↖ Until c.1965		↘ From c.1965
Common Tern	↑	↑ ?		
Arctic Tern			↑ From 1900	↑
Little Tern	↖↑	↑		
Whiskered Tern	↖	↖		
Black Tern	↖	↖ ?		
White-winged Black Tern	←	←		
Guillemot		↓	↑	
Brünnich's Guillemot				
Razorbill		↓ ?	↑	

SPECIES	Advancing		Retreating	
	1850–1950	1950–1993	1850–1950	1950–1993
Black Guillemot		↘ ?	↖	
Little Auk		↓	↑	
Puffin		↓	↑	
Black-bellied Sandgrouse				
Pin-tailed Sandgrouse				
Pallas's Sandgrouse	← Irruptions more frequent & larger	← Irruptions infrequent & smaller		
Rock Dove				
Stock Dove	←↘↑			↓ ?
Woodpigeon	↘↑	↑		
Collared Dove	↘↑	↗↘↑ Up to c.1980		↓ From c.1980
Turtle Dove	↘↑	↑ E Europe		↘ W Europe
Laughing (Palm) Dove		↑		
Great Spotted Cuckoo	↗ From 1943	↑		
Cuckoo	↑ ?			↘ ?
Oriental Cuckoo				
Barn Owl	↑			↓
Scops Owl	↑	↑		
Eagle Owl		↘		
Snowy Owl		↓	↑	
Hawk Owl		↓ ?	↑ ?	
Pygmy Owl				
Little Owl	↘↑			↑↘
Tawny Owl	↑	↑ ?		
Ural Owl		←↘↓		
Great Grey Owl		↗↓		
Long-eared Owl	↑	↑ ?		
Short-eared Owl	↑	↑ ?		
Tengmalm's Owl		←↘		
Nightjar		↘ Since c.1981	↘	Until c.1981 ↘
Red-necked Nightjar				
Swift	← Ireland from 1932			Ireland & Scotland →
Pallid Swift		↑		
Alpine Swift	↑	↑		
White-rumped Swift		↑		
Little Swift		↑		
Kingfisher		↑ ?		
Bee-eater	↘↑ From 1930s	↘↑		
Roller	Increased in E Europe	?	W Europe ↘	↘ ?

SPECIES	Advancing		Retreating	
	1850–1950	1950–1993	1850–1950	1950–1993
Hoopoe		↖↑ ?	↓↘	
Wryneck			↘	↘
Grey-headed Woodpecker	←	←		
Green Woodpecker	↑ From c.1925	↑	↓ Before c.1900	
Black Woodpecker	←	←		
Great Spotted Woodpecker	↑	Until c.1980 ↑		↓ ? Since c.1980
Syrian Woodpecker	↖↑	↖↑		
Middle Spotted Woodpecker		↖↑ Central Europe	↓	↓ In Sweden
White-backed Woodpecker				↘
Lesser Spotted Woodpecker		↑		
Three-toed Woodpecker				?
Dupont's Lark		↑ ?		
Calandra Lark				
Short-toed Lark				
Lesser Short-toed Lark				
Crested Lark	←↖↑			↓
Thekla Lark				
Woodlark	↖↑ c.1920–1950	↖↑ 1983+	↓↘ c.1850–1920	↓↘ 1952–1982
Skylark				
Shore Lark	←	←↑		
Sand Martin				
Crag Martin		↖↑		
Swallow	↖↑	↖↑		
Red-rumped Swallow		↖↑		
House Martin				
Tawny Pipit		↖↑	↓↘	
Tree Pipit	↑			
Meadow Pipit				
Red-throated Pipit		↓	↑	
Rock/Water Pipit				
Yellow/Blue-headed Wagtail	Mainland Europe ↑	Mainland Europe ↑ ?	British Isles since 1930s →	
Citrine Wagtail	←	←		
Grey Wagtail	↑↗	↑↗		

	Advancing		Retreating	
SPECIES	1850–1950	1950–1993	1850–1950	1950–1993
Pied/White Wagtail	← Ireland since 1900	↑ Shetland & Orkney		
Waxwing				
Dipper				
Wren				
Dunnock	↑	↑ ?		
Alpine Accentor				
Rufous Bushchat				
Robin				
Thrush Nightingale	←↖ From c.1930	←↖	Up to c.1850 →	
Nightingale	←↖ From c.1930 to c.1950		↘→ Before 1930	↘ Since c.1950
Bluethroat		←↙ Both races	?	
Red-flanked Bluetail	← From c.1930	←		
Black Redstart	↖↑	↖↑		
Redstart				
Whinchat				
Stonechat				
Isabelline Wheatear				
Wheatear	↑	↑ Until 1976		
Pied Wheatear				
Black-eared Wheatear		↖ Balkans only		
Black Wheatear			↓	
Rock Thrush		↑ ?	↓	
Blue Rock Thrush		↑ ?	↓	
Ring Ouzel		Expanding from c.1960–1980	↑	Contracting since c.1980
Blackbird	←↖↑↗	←↖↑↗		
Fieldfare	←↙↖	←↙↖↑		
Song Thrush	Fluctuating	Fluctuating	Fluctuating	Fluctuating
Redwing	↙↓ From 1920s	↙↓		
Mistle Thrush	↖↑	↖↑ Until c.1980		↓↘ ? Since c.1980
Cetti's Warbler	↖↑	↖↑↗		
Fan-tailed Warbler	↑	↑		
Lanceolated Warbler		←		
Grasshopper Warbler	↑	←↖↑ Up to c.1980		
River Warbler		←↖		
Savi's Warbler	↖↑	↖↑		
Moustached Warbler		↑ ?		
Aquatic Warbler				
Sedge Warbler	↖↑	↑		

SPECIES	Advancing 1850–1950	Advancing 1950–1993	Retreating 1850–1950	Retreating 1950–1993
Paddyfield Warbler		↘		
Blyth's Reed Warbler	←	←		
Marsh Warbler	↑	↑		
Reed Warbler	↑	←↘↑		
Great Reed Warbler	↘↑	↘↑		
Olivaceous Warbler	↘	↘		
Booted Warbler		↘		
Olive-tree Warbler	↘	↘		
Icterine Warbler		↑	From *c.*1930 ↘	
Melodious Warbler	↑↗ From *c.*1930	↑↗		
Marmora's Warbler				
Dartford Warbler		↑↗	↙↓	
Spectacled Warbler				
Subalpine Warbler				
Sardinian Warbler				
Rüppell's Warbler				
Orphean Warbler			↓	↓ ?
Barred Warbler	↘	↘		
Lesser Whitethroat	←↘↑	←↘↑		
Whitethroat	↘↑	↘↑ Up to 1969		From 1969 ↘
Garden Warbler				
Blackcap	←↘	↘↑		
Greenish Warbler	←↘	←↘		
Arctic Warbler	←	← ?		
Bonelli's Warbler	↑	↑ ?		
Wood Warbler	←↘↑	←↘↑		
Chiffchaff	←↘↑	←↘↑		
Willow Warbler	←↘↑	?		
Goldcrest	↑	↑		
Firecrest	↘↑	↘↑		
Spotted Flycatcher	↑ Since 1940	↑ Until 1964		Since 1964 ?
Red-breasted Flycatcher	←↘↑	←↘↑		
Semi-collared Flycatcher		←↘↑ In the Balkans		
Collared Flycatcher	↑	↑		
Pied Flycatcher	←↘↑	←↘↑		
Bearded Tit		←↘↑↗↓	↘→	
Long-tailed Tit				
Marsh Tit	↑	↑ ?		

SPECIES	Advancing		Retreating	
	1850–1950	*1950–1993*	*1850–1950*	*1950–1993*
Sombre Tit				
Willow Tit	↑ Scotland			↓ ? Scotland
Siberian Tit			↑	↑ ?
Crested Tit	↑ From 1929	↑ ?		↓ ?
Coal Tit	↑	↑		
Blue Tit	↑	↑		
Azure Tit	← Until c.1900	← Since c.1980	c.1900–c.1980 →	→ Until c.1980
Great Tit	↑	↑		
Corsican Nuthatch				
Nuthatch	↑ From 1927	↑	↓ Until c.1920	
Rock Nuthatch				
Wallcreeper				
Treecreeper (Common)	↘ ?	?		?
Short-toed Treecreeper	↑ Since c.1920	↑		
Penduline Tit	←↘↑ From c.1920	←↘↑	↓ 1870–1890	
Golden Oriole	↑	↘↑		
Red-backed Shrike			↘	↘
Lesser Grey Shrike		↙↘↑	↘	
Great Grey Shrike				
Woodchat Shrike			↘	↘ ?
Masked Shrike				
Jay	←↘↑	←↘↑		
Siberian Jay			↑	↑
Azure-winged Magpie				
Magpie	↘↑	↑		
Nutcracker	←↘↑	↘↑ ?		
Alpine Chough				
Chough	Increasing in Ireland since c.1975		↙↓	↙↓ ?
Jackdaw	↘↑↘	↘↑		
Rook	↑	↑		
Carrion Crow	↘↑	↘↑		
Hooded Crow			↘↑↗	↘↑↗
Raven				
Starling	↘↑			↓↘
Spotless Starling		↑↗		
Rose-coloured Starling				
House Sparrow	↘↑	↘↑ Until c.1980		From c.1980 ↓↘
Spanish Sparrow		↘ In the Balkans		

321

	Advancing		Retreating	
SPECIES	1850–1950	1950–1993	1850–1950	1950–1993
Tree Sparrow		←↖↑ From *c*.1959 to *c*.1975	→ From *c*.1890–1950 →	Since *c*.1976 →
Rock Sparrow			↓	?
Snowfinch				
Common Waxbill		↗→		
Chaffinch	↑	↑		
Brambling		↙	↑	
Serin	↖↑	↖↑↗		
Citril Finch		↑ 1950s & 1960s		
Greenfinch	↑	↖↑		
Goldfinch	↑	↑		
Siskin	↙↓	↙↓		
Linnet	↖↑			↓↘
Twite		↓	↑	Declining in Ireland from *c*.1970
Redpoll	1900–1910 ↓	↓↖→	↑ Before 1900? Then from 1910	Declining in Ireland from *c*.1970
Arctic Redpoll		↓ ?	↑	
Two-barred or White-winged Crossbill				
Crossbill				
Scottish Crossbill				
Parrot Crossbill				
Trumpeter Finch		↑		
Scarlet or Common Rosefinch	←↖↑	←↖↑		
Pine Grosbeak				
Bullfinch	←↖↑	←↖↑ Up to *c*.1980		↓↘ Since *c*.1980
Hawfinch	↖↑	?		?
Lapland Bunting		↙	↑	
Snow Bunting		↓	↑	
Yellowhammer				↘ Since *c*.1970
Cirl Bunting	↖↑↗ Until 1910		From 1910 ↙↓	↙↓
Rock Bunting			↓	
Ortolan Bunting	↖↑			↓↘
Cretzschmar's Bunting				
Rustic Bunting	←	←↙		
Little Bunting	←	←		
Yellow-breasted Bunting	←	←		
Reed Bunting	↑	↑		
Pallas's Reed Bunting		←1981		

| | Advancing | | Retreating | |
SPECIES	*1850–1950*	*1950–1993*	*1850–1950*	*1950–1993*
Black-headed Bunting		↑		
Corn Bunting			Since *c*.1900 ➜	➜

Appendix 2.1

TEMPERATE BREEDING SPECIES ADVANCING NORTHWARDS

In response to climatic warming

SPECIES	1900–1950	Since 1950	Remarks
Little Grebe	X	X	
Great Crested Grebe	X	X	
Black-necked Grebe	X	X?	
Gannet	X	X	
Cormorant		X	
Shag	X	X?	
Pygmy Cormorant		X	Since 1980
Bittern	X	X	
Night Heron	X	X	
Squacco Heron	X	X	
Cattle Egret	X	X	
Little Egret	X	X	
Great White Egret		X	Since 1976
Grey Heron	X	X	
Purple Heron	X	X	Until 1979
Mute Swan	X	X	
Canada Goose	X	X	
Shelduck	X?	X	
Gadwall	X	X	
Garganey	X		
Shoveler	X	X	Until 1970
Marbled Teal		X	
Red-crested Pochard	X	X	
Pochard	X	X	
Tufted Duck	X	X	
Ruddy Duck		X	Introduced in Britain 1960

SPECIES	1900–1950	Since 1950	Remarks
White-headed Duck		X	From c.1900
Honey Buzzard	X	X	
Black-shouldered Kite		X	
Black Kite	X	X	
Red Kite	X	X	
Short-toed Eagle	X		
Marsh Harrier		X	
Montagu's Harrier	X	X	
Levant Sparrowhawk		X	
Imperial Eagle	?	X	
Lesser Kestrel	X	X	From 1947
Red-footed Falcon		X	
Hobby		X	Since 1975; greenhouse effect?
Saker Falcon		X	
Quail		X	
Water Rail	X	X?	
Spotted Crake		X	
Moorhen	X	X?	
Coot	X	X	
Oystercatcher	X	X	
Black-winged Stilt	X	X	
Avocet	X	X	
Collared Pratincole	X		From the 1930s in the Ukraine
Little Ringed Plover	X	X	
Spur-winged Plover		X	
Sociable Plover		X?	
Lapwing	X	X	Until c.1970; declining since
Ruff		X	
Woodcock	X?	X?	
Black-tailed Godwit	X	X	
Curlew	X	X?	
Marsh Sandpiper	X	X	
Wood Sandpiper	X	X	In Finland
Red-necked Phalarope	X		From c.1900 to c.1925
Pomarine Skua		X	Since 1984
Mediterranean Gull	X	X	
Little Gull		X	

SPECIES	1900–1950	Since 1950	Remarks
Black-headed Gull	X	X	
Slender-billed Gull	X	X?	
Common Gull	X	X	
Lesser Black-backed Gull	X	X	
Herring Gull	X	X	
Great Black-backed Gull	X	X	
Gull-billed Tern	X	X	
Sandwich Tern	X?	X?	
Roseate Tern	X	X	Until *c.*1965
Common Tern	X	X?	
Little Tern	X	X	
Whiskered Tern	X	X	
Black Tern	X	X?	
Stock Dove	X		
Woodpigeon	X	X	Until *c.*1980
Turtle Dove	X	X	Eastern Europe since *c.*1960
Laughing (Palm) Dove		X	
Great Spotted Cuckoo	X	X	From 1943
Cuckoo	X		
Barn Owl	X		
Eagle Owl		X	
Little Owl	X		
Tawny Owl	X	X?	
Long-eared Owl	X	X?	
Short-eared Owl	X	X?	
Tengmalm's Owl		X	
Nightjar		X	From *c.*1981
Pallid Swift		X	
Alpine Swift	X	X	
White-rumped Swift		X	
Little Swift		X	
Kingfisher		X?	
Bee-eater	X	X	
Hoopoe		X?	
Green Woodpecker	X	X	
Great Spotted Woodpecker	X	?	
Syrian Woodpecker	X	X	
Middle Spotted Woodpecker		X?	
Lesser Spotted Woodpecker		X	

SPECIES	1900–1950	Since 1950	Remarks
Dupont's Lark		X?	
Crested Lark	X		
Woodlark	X	X	From 1920–1950; then from 1983
Shore Lark		X	
Crag Martin		X	
Swallow	X	X	
Red-rumped Swallow		X	
Tawny Pipit		X	
Tree Pipit	X		
Yellow Wagtail races	X	X?	
Grey Wagtail	X	X	
Pied/White Wagtail		X	
Dunnock	X	X?	
Thrush Nightingale	X	X	
Black Redstart	X	X	
Black-eared Wheatear		X	Balkans only
Rock Thrush		X?	
Blue Rock Thrush		X?	
Blackbird	X	X	
Fieldfare	X	X	
Mistle Thrush	X	X	
Cetti's Warbler	X	X	
Fan-tailed Warbler	X	X	
Grasshopper Warbler	X	X	Up to c.1980
River Warbler		X	
Savi's Warbler	X	X	
Moustached Warbler		X?	
Sedge Warbler	X	X	
Paddyfield Warbler		X	
Marsh Warbler	X	X	
Reed Warbler	X	X	
Great Reed Warbler	X	X	
Olivaceous Warbler	X	X	
Booted Warbler		X	
Olive-tree Warbler	X	X	
Icterine Warbler		X	
Melodious Warbler	X	X	
Dartford Warbler		X	

SPECIES	1900–1950	Since 1950	Remarks
Barred Warbler	X	X	
Lesser Whitethroat	X	X	
Whitethroat	X	X	Until 1969
Blackcap	X	X	
Greenish Warbler	X	X	
Bonelli's Warbler	X	X?	
Wood Warbler	X	X	
Chiffchaff	X	X	
Willow Warbler	X	?	
Goldcrest	X	X	
Firecrest	X	X	
Spotted Flycatcher	X	X	From 1940–1964
Red-breasted Flycatcher	X	X	
Semi-collared Flycatcher		X	Balkans only
Collared Flycatcher	X	X	
Pied Flycatcher	X	X	
Bearded Tit		X	
Marsh Tit	X	X?	
Willow Tit	X		In Scotland
Crested Tit	X	X?	From 1929
Coal Tit	X	X	
Blue Tit	X	X	
Great Tit	X	X	
Nuthatch	X	X	From 1927
Treecreeper (Common)	X?	?	
Short-toed Treecreeper	X	X	Since *c.*1920
Penduline Tit	X	X	From *c.*1920
Golden Oriole	X	X	
Lesser Grey Shrike		X	
Jay	X	X	
Magpie	X	X	
Nutcracker	X	X?	
Jackdaw	X	X	
Rook	X	X	
Carrion Crow	X	X	
Starling	X		Decline since *c.*1970
Spotless Starling		X	
House Sparrow	X	X	Until *c.*1980
Spanish Sparrow		X	Balkans only

SPECIES	1900–1950	Since 1950	Remarks
Tree Sparrow		X	From 1959–c.1975
Common Waxbill		X	
Chaffinch	X	X	
Serin	X	X	
Citril Finch		X	1950s & 1960s
Greenfinch	X	X	
Goldfinch	X	X	
Linnet	X		
Redpoll	X	X	*Flammea* and *rostrata* races
Trumpeter Finch		X	
Scarlet Rosefinch	X	X	
Bullfinch	X	X	Until c.1980
Hawfinch	X	?	
Cirl Bunting	X		Until 1910
Reed Bunting	X	X	
Black-headed Bunting		X	

Appendix 2.2

TEMPERATE BREEDING SPECIES ADVANCING WESTWARDS

In response to climatic warming

SPECIES	1900–1950	Since 1950	Remarks
Red-necked Grebe		X	From Denmark & north Germany to Britain
Black-necked Grebe	X		
Cormorant		X	
Black Stork		X	From c.1970
Pochard	X	X	
Ferruginous Duck	X		
Pallid Harrier	X	X	From c.1940
Red-footed Falcon	X	X	From c.1930
Saker Falcon		X	
Greater Sand Plover		X	
White-tailed Plover		X	
Ruff		X	
Jack Snipe		X	
Redshank	X		
Marsh Sandpiper	X	X	
Green Sandpiper	X	X	
Terek Sandpiper	X	X	
Little Gull	X	X	
Lesser Black-backed Gull	X	X	
Caspian Tern		X?	
White-winged Black Tern	X	X	
Pallas's Sandgrouse	X		
Stock Dove	X		
Ural Owl		X	
Tengmalm's Owl		X	

SPECIES	1900–1950	Since 1950	Remarks
Swift	X		Ireland from 1932
Grey-headed Woodpecker	X	X	
Black Woodpecker	X	X	
Crested Lark	X		
Shore Lark	X		
Citrine Wagtail	X	X	
Pied Wagtail	X		Ireland since 1900
Thrush Nightingale	X	X	
Bluethroat		X	White-spotted race
Red-flanked Bluetail	X	X	
Blackbird	X	X	
Fieldfare	X	X	
Lanceolated Warbler		X	
Grasshopper Warbler		X	Until c.1980
River Warbler		X	
Blyth's Reed Warbler	X	X	
Reed Warbler		X	
Lesser Whitethroat	X		West Britain
Blackcap	X		
Greenish Warbler	X	X	
Arctic Warbler	X	X?	
Wood Warbler	X	X	
Chiffchaff	X	X	
Willow Warbler	X		
Red-breasted Flycatcher	X	X	
Semi-collared Flycatcher		X	
Pied Flycatcher	X	X	
Bearded Tit		X	
Azure Tit	X	X	
Penduline Tit	X	X	
Jay	X	X	
Nutcracker	X		
Tree Sparrow		X	From c.1959–c.1975
Scarlet Rosefinch	X	X	
Bullfinch	X	X	British Isles until c.1930
Rustic Bunting	X	X	
Little Bunting	X	X	
Yellow-breasted Bunting	X	X	
Pallas's Reed Bunting		X	

Appendix 2.3

TEMPERATE BREEDING SPECIES RETREATING SOUTHWARDS

In response to climatic warming

SPECIES	1900–1950	Since 1950	Remarks
Black Stork	X		
White Stork	X	X	
White-headed Duck	X	X	Until *c*.1980
Short-toed Eagle	X		Until *c*.1920
Hobby	X		Due to wetter summers
Quail	X		Until 1942
Spotted Crake	X		
Stone Curlew	X	X	
Collared Pratincole		X	Ukraine from *c*.1970
Kentish Plover	X	?	
Nightjar	X	X	Until *c*.1981
Roller	X	X?	
Hoopoe	X		
Wryneck	X	X	
Middle Spotted Woodpecker	X	X	
Woodlark	X	X	From *c*.1850–1920 & from 1952–1982
Tawny Pipit	X		
Nightingale	X	X	From *c*.1900–1930
Black Wheatear	X		
Rock Thrush	X		
Blue Rock Thrush	X		
Icterine Warbler	X		From *c*.1930–1950
Dartford Warbler	X		
Orphean Warbler	X	X?	
Bearded Tit	X		

SPECIES	1900–1950	Since 1950	Remarks
Red-backed Shrike	X	X	
Lesser Grey Shrike	X		
Woodchat Shrike	X	X?	
Chough	X	X?	
Rock Sparrow	X		
Cirl Bunting	X	X	From 1910
Rock Bunting	X		

Appendix 2.4

TEMPERATE BREEDING SPECIES RETREATING EASTWARDS

In response to climatic warming

SPECIES	1900–1950	Since 1950	Remarks
Black Stork	X	X	West Europe until c.1970
Black-winged Pratincole	X	X	Ukraine from c.1900
Yellow Wagtail	X		British Isles only
Bearded Tit	X		
Azure Tit	X	X	Until c.1980
Tree Sparrow	X	X	From c.1890–1950 & again from c.1976
Corn Bunting	X	X	

Appendix 2.5

TEMPERATE BREEDING SPECIES RETREATING SOUTHWARDS

In response to climatic cooling

SPECIES	1900–1950	Since 1950	Remarks
Purple Heron		X	From 1979
White Stork		X?	
Garganey		X	Until 1988
Montagu's Harrier		X	
Roseate Tern		X	From c.1965
Stock Dove		X?	
Collared Dove		X	From c.1980
Turtle Dove		X	Western Europe only since 1960
Cuckoo		X?	
Barn Owl		X	
Little Owl		X	
Great Spotted Woodpecker		X?	
Middle Spotted Woodpecker	X	X	
White-backed Woodpecker		X	
Crested Lark		X	
Nightingale	X	X	Since c.1950
Mistle Thrush		X	Since c.1980
Willow Tit		X?	In Scotland
Starling		X	
House Sparrow		X	Since c.1980
Linnet		X	
Bullfinch		X	Since c.1980
Yellowhammer		X	Since c.1970
Ortolan Bunting		X	

Appendix 2.6

NORTHERN BREEDING SPECIES RETREATING NORTHWARDS

In response to climatic warming

SPECIES	1900–1950	Since 1950	Remarks
Red-throated Diver	X?		
Black-throated Diver	X?		
Great Northern Diver	X?		
White-billed Diver	X?		
Slavonian Grebe	X		In North America only
Whooper Swan	X		
Greylag Goose	X		
Scaup	X		
King Eider	X		
Harlequin Duck	X		
Long-tailed Duck	X		
Common Scoter		X	Since 1970
Goldeneye	X		
Smew	X?		
Hen Harrier	X		Until 1939
Rough-legged Buzzard	X?		
Osprey	X		
Gyrfalcon	X		
Willow/Red Grouse	X	?	
Ptarmigan	X		
Crane	X		
Dotterel	X		
Golden Plover	X		
Purple Sandpiper	X?		
Jack Snipe	X	X	
Whimbrel	X		

SPECIES	1900–1950	Since 1950	Remarks
Greenshank	X		
Wood Sandpiper	X		
Turnstone	X		
Red-necked Phalarope	X	X	From 1930 to c.1950; and again since c.1975
Grey Phalarope	X	?	
Long-tailed Skua	X		
Glaucous Gull	X		
Ivory Gull	X?		
Arctic Tern	X	X	
Guillemot	X		
Razorbill	X		
Black Guillemot	X		
Little Auk	X		
Puffin	X		
Snowy Owl	X		
Red-throated Pipit	X		
Ring Ouzel	X		Contracting range again since c.1980
Siberian Tit	X	X?	
Siberian Jay	X	X	
Hooded Crow	X	X	
Brambling	X		
Twite	X		
Redpoll	X		
Arctic Redpoll	X		
Lapland Bunting	X		
Snow Bunting	X		

337

Appendix 2.7

NORTHERN BREEDING SPECIES ADVANCING SOUTHWARDS

In response to climatic warming

SPECIES	1900–1950	Since 1950	Remarks
Fulmar	X	X	Slowing down since 1950
Wigeon	X		
Pintail	X		
Eider	X	X	
Common Scoter	X	X	Up to 1970
Red-breasted Merganser	X	X	
Goosander	X	X	
Common Gull	X	X	
Kittiwake	X	X	

Appendix 2.8

NORTHERN BREEDING SPECIES ADVANCING SOUTHWARDS

In response to climatic cooling

SPECIES	1900–1950	Since 1950	Remarks
Red-throated Diver	X	X	
Black-throated Diver		X	
Great Northern Diver		X	
White-billed Diver		X	
Slavonian Grebe	X	X	From 1908
Whooper Swan		X	
Greylag Goose		X	
Barnacle Goose		X	From 1975
Scaup		X	
King Eider		X	
Steller's Eider		X	
Long-tailed Duck		X	
Goldeneye		X	
Smew		X	
Hen Harrier	X	X	From 1939
Rough-legged Buzzard		X?	
Osprey		X	
Gyrfalcon		X	
Ptarmigan		X?	
Crane		X?	
Dotterel		X	
Sanderling		X	
Temminck's Stint		X	
Purple Sandpiper		X	
Jack Snipe		X	
Broad-billed Sandpiper		X	

SPECIES	1900–1950	Since 1950	Remarks
Whimbrel		X	
Greenshank		X	
Wood Sandpiper		X	
Turnstone		X	
Red-necked Phalarope		X	From 1930 to c.1950
Long-tailed Skua		X	
Glaucous Gull		X?	
Ross's Gull		X	
Ivory Gull		X	
Guillemot		X	
Razorbill		X?	
Black Guillemot		X?	
Little Auk		X	
Snowy Owl		X	
Ural Owl		X	
Great Grey Owl		X	
Red-throated Pipit		X	
Bluethroat		X	Red-spotted race
Ring Ouzel		X	c.1960–1980 in Britain
Fieldfare	X	X	
Redwing	X	X	From the 1920s
Brambling		X	
Siskin	X	X	
Twite		X	
Redpoll		X	
Arctic Redpoll		X?	
Lapland Bunting		X	
Snow Bunting		X	
Rustic Bunting		X	

Appendix 3

![divider]

CLIMATIC EFFECTS ON THE EVOLUTION OF MIGRATORY BIRDS

By Kenneth Williamson

![divider]

Climate has exercised a powerful effect on the evolution of bird species in all regions of the world. The alternations of pluvials and droughts in the African continent are too remote in time for the complex intricacies of their contribution to the Ethiopian avifauna to be unravelled, although Hall and Moreau (1970), and many others, have attempted a shrewd evaluation in the case of certain groups of species.

Easier to contemplate, because of its simplicity, is the evolutionary adaptation of the life-styles of the Wandering and Royal Albatrosses and the Giant Petrel which, employing a slope-soaring technique between the troughs and crests of the great ocean rollers, encircle the globe between their breeding seasons in the one-way traffic system of the Roaring Forties. By comparison the appearance of a British-ringed Manx Shearwater on the southern coast of Australia in November 1961, and the occasional recovery in that continent of English-bred Common and Arctic Terns, are ephemeral events. Nevertheless, they are striking testimony of the power of the west wind belt to affect the normal life of individual birds.

The west winds of the Northern Hemisphere have also played a part, though perhaps a less significant one, in evolutionary processes there. The Bonxie or Great Skua has its closest relatives in the Antarctic and subantarctic climate zones, where several forms exist; the one inhabiting Tristan da Cunha and Gough Island is so similar in structure and plumage to our own as to suggest that the Bonxie's arrival in the north (though it certainly took place before the Viking era) must be a recent event in evolutionary time. Its present distribution appears to be restricted by the North Atlantic storm-track to Iceland, the Faeroe and Shetland Islands, and northern Scotland, with recent fringe extensions to south-east Greenland and Bear Island.

Leach's Storm Petrel is a characteristic seabird of the American offshore islands, and the few colonies on remote island groups in the north-east Atlantic may owe their origin to the transportation of birds by depressions derived from the hurricanes which sometimes engulf vast numbers in their winter quarters south-east from the Caribbean. Trapped in the calm air of the storm's eye, but probably unable to feed because of the violence of the sea, they are borne along by the hurricane, becoming progressively weaker, and are not infrequently deposited many miles inland if the storm crosses the coast. Residual hurricanes occasionally turn and swing eastwards across the Atlantic as huge depressions, eventually spilling hundreds of petrels over the British Isles, as happened on a big scale in 1891 and 1952.

In the study of natural history there are well-known ecogeographical (sometimes called 'climatic' or 'ecological') rules relating certain character-istics of warm-blooded animals to the conditions of their environment. One of the best known in its application to birds is Bergmann's Rule, which states that in a polytypic species (that is, one with a number of geographical forms) the body size of a subspecies tends to increase with the decreasing mean temperature of its habitat. Thus body size, generally indicated by an average lengthening of the wing, increases with higher latitude; and the wings of populations which live at high altitudes also tend to be longer than those of the same species residing in lowland regions. This is really a restatement of the general principle, as Huxley (1942) pointed out, that in warm-blooded animals body surface relative to bulk tends to diminish with decreasing mean temperature of the environment; it is an adjustment compensating the organism, since the smaller the body the more heat it radiates to the colder air. This correlation has been criticised by Scholander (1955) on the grounds that there are more efficient adaptations for conserving body heat; this may well be true, but it does not necessarily invalidate the hypothesis as a partial explanation of Bergmann's Rule. There are, however, some striking excep-tions, and the reasons for them are not always clear. There are many aspects of the problem, and the selection pressures which have combined to produce (or, conversely, suppress) this 'latitude effect' are many. Mayr (1956) wisely insisted that Bergmann's Rule must be regarded as a 'purely empirical find-ing which can be proven or disproven no matter to what physiological theory one might ascribe this size trend', and he warned strongly against the search for an all-or-none solution to this complicated biological problem.

Some consideration has been given to the nature of this latitude effect in migratory as distinct from resident species, though much of it has concerned the bird in its winter quarters, rather than the actual journeys which take it there and bring it home again in spring. Rensch (1939) showed that a size correlation could often be found with the minimum winter temperature, selection being exerted by the most rigorous life conditions. This suggests that in migratory species the body size of a northern population is likely to be a function of selection due to factors in the off-season environment, rather than any attribute of high latitude in the breeding area.

Hemmingsen (1951) also proposed that if we are to test the validity of Bergmann's Rule in migratory species, it is to the winter rather than the summer range that we must look for an answer. He further pointed out that the timing of migration might be involved, since some subspecies—and also members of closely-allied species—return earlier in the season, under colder weather conditions, than others. He gave as examples the Curlew and

Whimbrel, and the Great and Common Knots. In these couples an association is apparent in the first named, between greater body size and early spring movement from relatively northern and colder winter quarters; and in the second named, between smaller body size and a later migration in milder weather from relatively warmer and more southerly wintering grounds.

Salomonsen (1955) also showed that there are cases in which a marked correlation exists between Bergmann's Rule and the rigour of conditions in the winter environment. This is especially so with those races of a species which practise allohiemy, or which (in other words) are segregated in different regions outside the breeding season. Once such allohiemy has been established, differences in the selective factors operating in the two regions will tend to encourage subspecific differentiation. The Ringed Plover affords an excellent example, for here we find a reversal of the normal latitude effect in that the smallest race *tundrae* nests furthest north in the total range; however, its migration leap-frogs the range of the typical race of middle Europe, and its smaller size is really a response to the warmth of a tropical wintering area. Salomonsen also saw the Redshank as a good illustration, Scandinavian birds wintering in a Mediterranean climate being slightly smaller than the 'stationary' Dutch and British birds; but there is a difficulty, which I shall refer to later, in the case of the Icelandic race.

All these contributions, Salomonsen's especially, are of great importance to our understanding of evolution and subspeciation in birds; but there are certain populations, notably among the landbirds of Greenland and Iceland, whose divergence in size from their continental counterparts cannot be satisfactorily accounted for on his or Hemmingsen's hypotheses. I believe that the most important contributory factor to be taken into account in their case is the selection pressure exerted during the very brief span of their biennial migrations. For, as Mayr has emphasised, the working of natural selection is particularly efficient during catastrophes and other periods of acute environmental stress, so that even the short-term influence of a long migratory flight may be expected to play its part. In comparison with their continental cousins, which can follow overland routes and make do with short sea crossings, these Greenland and Iceland birds are called upon twice yearly to make a mandatory trans-oceanic flight, with all its attendant hazards of wind drift away from the most direct route. The rigours of these long flights are such that survival favours the most robust individuals, possessing the greatest resources of energy (in the form of stored glycogen and fats); so that in the course of numerous generations the stock has developed greater body dimensions, primarily apparent in weight, but also secondarily in an allometric increase in the length of wing and tail.

It is worth looking more closely at the evidence derived from studying the migration of these Greenland and Iceland birds, whose brief summer is passed under conditions no worse (and often markedly better) than those obtaining at similar latitudes on the continent of Europe. The fullest discussion will be given to the most striking example, the Wheatear, which ranges from middle and northern Europe through Iceland and Greenland and into Arctic Canada. A number of other species will also be examined to see to what extent they fall into line with this hypothesis.

Wheatears

The Greenland race of the Wheatear, according to Salomonsen (1950–51), 'is much more common in the interior fjord country with a drier and warmer climate, and a rich insect-life, than in the coastal areas, where it is rather scarce. It has its greatest population density in the south (Julianehåb District) and becomes less numerous towards the north'. The Julianehåb District is actually subarctic in climate, and at about 61°N is at much the same latitude as the Shetland Islands where the Wheatears belong to the typical race *oenanthe*. Wynne-Edwards (1952) found the species abundant at the head of Clyde Inlet on Baffin Island at abut 70°N, and it nests sparingly in other parts of eastern Canada westwards to within about 200 miles (322 km) of the Mackenzie River. All these birds travel to West Africa where they spend the winter with birds of the typical race, and they must cross the North Atlantic to Europe on the first stage of their autumn migration.

In Iceland and the Faeroe Islands Wheatears are intermediate in weight, and in length of wing and tail, between British and European *oenanthe* and the Greenland race *leucorrhoa*: thus they have been named (after a famous Danish ornithologist) *schioleri*. Table I gives wing measurements of museum specimens collected on the various breeding grounds:

TABLE I *Wing measurements of different populations of Wheatears*
(after Salomonsen 1950–51)

Origin	N	Males Range (mm)	Average	N	Females Range (mm)	Average
Greenland	36	102–110	105.00	37	100–108	103.37
Iceland	49	99–107	102.55	23	96–103	98.81
Faeroe Is.	22	97–103	99.61	13	95–101	97.54
Scandinavia	56	92–99	96.38	36	89–97	93.22

The majority of the Greenland birds arrive on their breeding grounds in middle and late May, but many do not reach the northern limits till early June (Salomonsen 1950–51). This agrees well with the passage dates of big Wheatears through Fair Isle, where movement proceeds during early and mid-May. Ticehurst (1909) gives the last week of April as the customary

FIGURE 67 *Greenland (left) and Common Wheatears held in the hand. (Photograph: Eric Hosking)*

arrival time in southern England. This means that the biggest birds are subject to warmer weather during their spring journey than the smallest birds, which reach Britain towards the end of March and continue to pass throughout April. Thus the Wheatear provides an exception to Hemmingsen's Rule that in related forms it is the smaller one which moves later and under the more congenial conditions. In autumn there is a difference of up to a fortnight in the passage through Fair Isle, local and Shetland birds departing in bulk from mid-August, the intermediate kind passing during the last days of the month and the big *leucorrhoa* appearing in September.

It is interesting to consider the migration in relation to the greater body weight of the Greenland race (Figure 67). Spring passage through Fair Isle is steady and protracted, with birds much less numerous than in autumn, when of course a large part of the movement comprises young of the year. A proportion in spring undoubtedly leaves from Ireland, the Outer Hebrides and other parts of western and northern Scotland, and perhaps also from the Norwegian coast. At both seasons specimens of *leucorrhoa* show some variation in weight but there is wider variation in autumn, and in general the heaviest pass by in spring and the lightest appear (often in fairly homogeneous groups) in the fall. It is possible that the light spring birds have sustained a wind drift to Fair Isle when attempting to cross the North Atlantic from some Irish or west Scottish starting point, while some appear to arrive with displaced continental migrants on easterly winds.

For the most part, however, the spring birds are pursuing a northward course through Britain under the optimum conditions of anticyclonic weather. They need make only negligible sea crossings from Scotland *via* Orkney to reach Fair Isle, and may be presumed to have expended very little of the reserves of energy built up by an adequate food supply whilst travelling at a leisurely pace northwards. As would be expected, the best weights are recorded in spring, with individuals sometimes exceeding 40 gm. A detailed analysis made by Alec Butterfield of the weights of Wheatears captured in the traps at Fair Isle in spring reveals a striking difference between *leucorrhoa* (on passage) and local *oenanthe* (Table II).

TABLE II *Comparison of spring weights of Greenland Wheatears (leucorrhoa) on passage at Fair Isle and local birds (oenanthe)*

	Males		Females	
	leucorrhoa	*oenanthe*	*leucorrhoa*	*oenanthe*
Number of specimens	30	45	24	35
Mean weight (gm)	30.98	24.11	30.35	23.80
Standard deviation	5.62	2.38	4.41	2.08
Ratio of mean weight to standard deviation	5.51	10.13	6.86	11.44

No weights are available for Greenland birds before they set out in autumn, but we should expect them to be at least as heavy as spring birds trapped under the best weather conditions. There is one tell-tale case of a Greenland bird coming on board a weathership at Station India, 300 miles (483 km) south of Iceland and 800 miles (1,287 km) east of Cap Farvell; it was caught by a member of the meteorological staff who is also a good ornithologist, Ivor McLean, and he kept the bird in his cabin and fed it on mealworms. Exhausted on arrival, it weighed 21.24 gm; but when he released it in the Clyde six days later it scaled 36.45 gm—a 72% increase. Otherwise, our discussion must be limited to the samples of *leucorrhoa* trapped on passage at Fair Isle. Some birds, passing in anticyclonic weather, have presumably island-hopped along the Iceland–Faeroes–Shetland route, and these may reach 35 gm; but the majority are much lighter, and weights between 20 and 25 gm are not unusual. The average of thirteen birds trapped 1–4 September 1953, for example, was as low as 22.69 gm, with a range of 18.98 to 27.18 gm.

I have shown that the first stage of the autumn migration of the Greenland Wheatear may take it on a non-stop flight (since it cannot rest or feed at sea) across more than 1,000 miles (1,609 km) of ocean, wind-borne around the western and southern sides of big Atlantic depressions (Williamson 1958a, 1961a). These birds leave land under the benign influence of the Greenland anticyclone—a shallow high supported by the vast ice-cap—but get involved with an entirely different weather system out at sea. There is a fairly constant succession of such depressions in the autumn, moving north-east towards Denmark Strait and Iceland, frequently with gale or storm force winds on their periphery. These systems move fast, surrounded by a backing airstream at first north-westerly, later westerly, and ultimately perhaps south-westerly in the approach to Britain. The Wheatear's route is therefore a circuitous downwind one, and with its own flight speed added to wind speed it is conceivable that it may travel for much of the distance at 60–70 mph (97–113 km/ph).

There is abundant evidence in the meteorological correlation of Wheatear records at sea, mainly at the weather ships, and in the circumstance of their arrival at Fair Isle, of this 'cyclonic approach' migration, and a number of cases were documented in the above-mentioned papers. A particularly interesting one which bears repetition was reported to me by James Fisher. When *HMS Vidal* was returning from the 'annexation' of Rockall on 20–21 September 1955, and was about half-way between southern Ireland and Cornwall, two dozen birds came on board ship soon after midnight. He recognised them as *leucorrhoa* and suggested they were cyclonic migrants from Greenland. Before 0800 hours GMT the previous day three *leucorrhoa*, which must have been involved with the same depression, were trapped at Fair Isle weighing between 20.4 and 23.2 gm. It is apparent from the synoptic weather chart that all these birds were finishing a long haul round an eastward-moving low, with winds so strong (forces 8–9) that a direct flight from Greenland to Shetland would have been quite impossible. They may well have travelled between 1,500 and 2,000 miles (2,414 and 3,219 km) to reach *HMS Vidal*.

Cyclonic approach is characteristic of the autumn migration to the British Isles not only of the Wheatear but also of other Greenland and Iceland birds ranging in size from Redpolls to geese, and there is no problem in associating the major arrivals with the movement of appropriate depressions. Survival favours the fittest, those capable of carrying the biggest fuel reserves in their stored glycogen and fats. The Greenland Wheatear is a classic example of what Verne Grant, the American ecologist, has called 'catastrophic selection', the smaller and weaker ones being weeded out, the more robust ones surviving to carry on the race. The time-scale of the evolution of this well-defined and quite remarkable subspecies may be reckoned in centuries rather than the millenia usually attributed to subspecies formation.

Redwings

The Icelandic race of the Redwing, named *coburni*, breeds at approximately the same latitudes as the Scandinavian or typical race, *iliacus*. It is a heavier bird, and has a longer wing, tail and tarsus on average (Table III). In general it is darker than the continental bird, having a stronger suffusion of buff and olive-brown on the underside. The breast spots tend to be more confluent, creating a 'clouded' rather than a distinctly spotted pattern, and the sides of the breast and flanks are heavily washed with brownish olive. The mantle coloration, like the chestnut of the underwing, is individually variable in both races. Most Icelandic birds have a buffish tone in the eye-stripe, throat, breast and especially the dark-centred under tail-coverts. The legs are horn colour and only rarely flesh or pinkish-flesh as in *iliacus*. As with nearly all geographical races the extreme types are rather easy to identify: at Fair Isle, where numbers of one or the other race are likely to arrive together on the same day, one can feel confident of the source by the impression gained from samples trapped; but isolated individuals do sometimes present difficulty, though this can usually be resolved if measurements as well as plumage are taken into account.

TABLE III *Measurements of Redwings*

iliacus

	N	Range (mm)	Average	Variance
Wing	537	109–126	116.66	8.8139
Tail	323	69–89	79.89	12.3310
Bill	437	16–25	20.89	1.4655
Tarsus	437	26–34	30.26	1.5513

coburni

	N	Range (mm)	Average	Variance
Wing	246	113–133	121.33	7.7630
Tail	216	78–93	85.22	9.6899
Bill	230	17–24	21.36	1.3248
Tarsus	321	30–36	32.41	1.2257

A substantial difference in weight can be demonstrated from the laboratory data for birds trapped at Fair Isle. In autumn 1956, for instance, over 300 were weighed and the average difference was 5.5 gm in favour of *coburni*. The mean weight of a sample taken on any one day varies according to the kind of weather and the length of the journey, as with the Wheatear, and in 1956 we were dealing with Scandinavian birds which for the most part had made a short sea crossing from Norway in ridge or col conditions, and with Icelandic birds which, as cyclonic migrants, may well have travelled three times as far. In other autumns (with fewer trapped, and *coburni* arriving in more congenial weather) the disparity has been as much as 10 gm. There are fewer weights recorded for the spring, and these suggest a norm for Scandinavian birds of about 70 gm compared with about 89 gm for Icelandic.

Continental Redwings spend the winter over most of western and southern Europe, including the British Isles. Theirs does not appear to be a strictly south-west oriented migration, however, since many reach Italy and countries much further east. This is true even of birds ringed in a previous year whilst wintering in Britain. Indeed, there seems to be something of an 'irruption' quality about their movements in some years. Recoveries of birds ringed as young in Iceland have been made in the off-season in Ireland and the Outer Hebrides. One *coburni*, marked at Fair Isle, reached Antwerp in Belgium in November, but such continental 'overspill' seems to be slight.

Selection pressure due to spatial separation in winter cannot be great. The dispersal of Scandinavian birds over the breadth of Europe, with many alternating between Atlantic and Mediterranean types of climate, must keep the adaptive variation under the influence of winter to a minimum, tending to diminish any gap in size and weight. A study of the meteorological background to the autumn passage of Icelandic Redwings at Fair Isle reveals much in common with that of the Greenland Wheatear; for whilst an island-hopping movement *via* the Faeroe and Shetland Islands occurs in col and anticyclonic weather, there are frequently much longer cyclonic journeys. Indeed, because the Icelandic Redwing is a late migrant, the majority of the movements fall into this category, since the weather in the north-east Atlantic deteriorates rapidly during October as the Azores High declines and the low pressure systems gain ascendancy.

During our years at Fair Isle, much the most fascinating for Icelandic Redwing migration was 1956. An unprecedented invasion took place in three major waves, interspersed with smaller influxes. After a minor one on the late afternoon of 11 October, Redwings were present in fair strength at dawn on the 12th, and twenty-five *coburni* were trapped at an average weight of 71.3 gm. There were no big flocks, but parties numbering up to thirty or so were dispersed widely over the hill and crofting ground, and there were certainly not fewer than 500 birds. A low moved across Iceland on the 11th, leaving a strong west wind and cloudy sky; the Atlantic winds were westerly at force 6 behind the southward-moving cold front.

The second big wave, of well over 1,000 birds, appeared after mid-day on the 18th, and during a brief but busy spell in the afternoon we caught fifty-nine (average weight 71.1 gm). The earlier trapping rounds had been unproductive, with nothing to indicate an arrival of birds until five Redwings were captured together at 1230 hours GMT, just as the full force of a cold front with strong wind and rain squalls put a temporary halt to field work. When the skies cleared in mid-afternoon it was at once obvious that a great arrival had taken place; flocks streamed down off the hill and followed the long drystone wall towards the village, many getting embroiled with the traps as they did so. Meanwhile, several skeins of Pink-footed Geese went over, and a few Whooper Swans and Merlins, and there was a Gyrfalcon. As in the earlier case the Atlantic weather was cyclonic, a deepening low passing between the Faeroes and Iceland; these conditions persisted overnight as the low moved towards the Norwegian coast, with a backing airstream to westwards of Fair Isle. Redwings were abundant next morning when fifteen were caught (average weight 66.8 gm), a further eleven coming to hand later in the day (average 69.9 gm).

The third wave began late on 24 October. Redwings were few early in the day but there was a strong goose passage, including a big skein of Barnacle Geese; but from early afternoon Redwings were at least as numerous as they had been on the 12th, and by nightfall twenty-two *coburni* (average 71.9 gm) had been trapped. Redwings were still common next morning and more had probably come in, since a few birds were heavy, 80–89 gm, and others were very light, 52–60 gm. The late arrivals of the previous day could not possibly have put on 10 gm or more—indeed, they are more likely to have lost weight whilst roosting—and it is likely that the heavy ones had had a shorter and more direct journey round a depression centred over south-east Iceland (Figure 68).

An interesting feature of these movements is that interpretation as a simple downwind drift from Iceland is unsatisfactory for all except the arrivals of the 25th, if we are constrained to regard this as their starting point. An alternative view is that the Redwings, leaving Iceland in anticyclonic weather at an earlier stage, had got across the Denmark Strait to Greenland and then coasted south. If they had flown downwind from southern Greenland, roughly between latitudes 50° and 60°N, then their journeys, varying in length from about 1,100 to 1,400 miles (1,770–2,253 km), could have been accomplished in about 18–22 hours, allowing a flight-speed of 30–35 mph (48–56 km/ph) plus a roughly equal wind speed. A late afternoon departure from Greenland would be not inconsistent with a mid-afternoon appearance at Fair Isle next day.

The big invasions of *coburni* in October 1956 pose a problem. The synoptic charts clearly point to nocturnal migration over a vast seascape beset by stormy weather, the westerly winds frequently at or above gale force. Yet unless these birds possessed the power to orientate their flight in a south-east 'standard direction', making continuous correction for the drifting effect of strong beam winds, Iceland could have been their immediate place of origin. If we insist that it was, then we are bound to acknowledge that the two races begin and perform their migrations in very different kinds of weather. If, on the other hand, we concede that the factors influencing departure and performance are much the same in the two races, then we must satisfy ourselves that it is not improbable that large numbers of

FIGURES 68a & 68b
*Weather charts
showing cyclonic
drift movements,
probably from
southern
Greenland, of
500 Icelandic
Redwings at Fair
Island on 24th
October 1956
(*68a *mid-day
map, 23rd,* 68b
*midnight map,
23rd–24th
October 1956).
(After Williamson
1958b)*

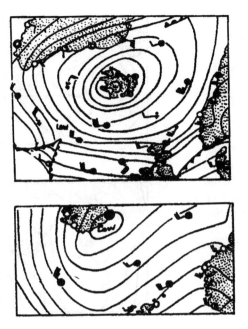

Icelandic Redwings reach the Greenland coast from time to time on the first stage of their autumn migration, and then fly a cyclonic route from the south of that country to Britain.

Meteorological developments which would favour this postulated first stage are not rare. Iceland is much troubled by depressions, and the weather charts for 1956 show that in a stormy autumn the best weather comes in the brief intervals between the passing of one low and the arrival of the next. The Greenland ice-cap supports an anticyclone, and intermittently a ridge has the chance to penetrate to Iceland between gales, so that the complementary airflow between it and the next advancing depression is easterly. This happened on 29–30 September, and again on 4–5 October, when a small high reached Iceland from east Greenland, the east wind persisting as a new low came up from the Atlantic on the 6–7 October. Calms and light easterlies were also prevalent from 15–17 October, when a rapidly filling low lay at the entrance to Denmark Strait; and it may not be a coincidence that this quiet period so closely preceded the biggest invasion of Icelandic Redwings at Fair Isle (Figure 69). This ice-cap anticyclone also ensures that there are more fair days for departure from the south of Greenland than from Iceland: in October 1956 there were twenty-three days with winds of forces 0–4, little cloud and no precipitation in southern Greenland (as shown on the mid-day charts), against seventeen similar days in southern Iceland—on many of which, as already indicated, the wind was from between south and east. Hørring and Salomonsen (1941) say that *coburni* is 'casual' in Greenland in October and the winter months, but one wonders if

FIGURES 69a & 69b
*Downwind
cyclonic drift
movements of
Icelandic
Redwings to Fair
Isle in autumn.
Weather condi-
tions in Iceland
and south-eastern
Greenland from
mid-day maps for
15th and 16th
October 1956.
(After Williamson
1958b)*

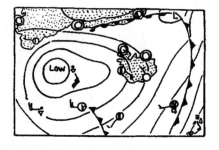
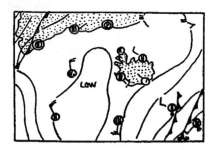

this opinion might not be subject to radical review if there were more intensive observation along this desolate coast.

A prophetic example of what might happen to the Icelandic Redwings if
the mean track of the depressions should shift to a more southerly latitude
(as has happened in the past) took place in 1959. It is an example, too,
which drives home the point that fitness and strength are of vital importance
to the continuance of the race. During the climatic amelioration many
Icelandic Redwings got into the habit of delaying their migration, preferring
to stay in the fields and gardens around Reykjavik and other settlements
until forced out by frost and snow. In November 1959, a vast low-pressure
system, centred over the English Midlands, carried flocks of late departing
birds to the Biscay coast (Figure 70). On three successive days, 15–17
November, three Iceland-ringed birds were found dead or dying from
exhaustion in southern France and northern Spain. Others survived, since a
fourth ringed bird was found in France on 9 December, and a fifth in
Portugal on 3 January. These were the first records from the Biscay region
of a subspecies which normally winters in Ireland and western Scotland, and
they had had to travel twice as far as usual—a feat in excess of the capabil-
ity of many of the birds.

Lapland and Snow Buntings
The Lapland Bunting is predominantly a low-Arctic breeder in Greenland.
The Snow Bunting is a common summer visitor to high-Arctic as well as
low-Arctic regions and also nests in Iceland, where it is designated *insulae*

on account of its darker plumage (Salomonsen 1931). In Greenland both seek the warmer, drier anticyclonic climate of the interior fjord country, and are relatively scarce in the less benign coastal areas.

Greenland's Snow Buntings cannot be separated from those of Canada and northern Europe. Greenland's Lapland Buntings, however, are just about

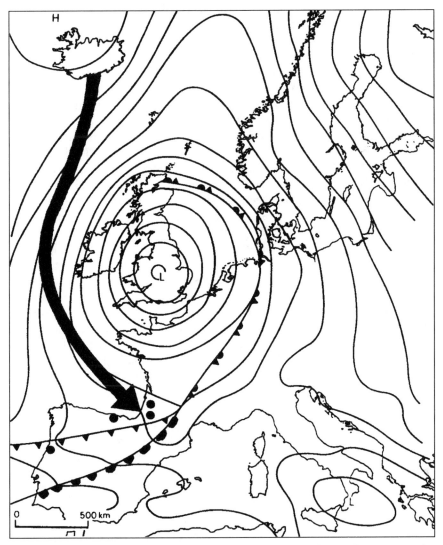

FIGURE 70 *Weather chart showing arrival of Icelandic Redwings on the Biscay coasts of France and Spain, as indicated by ringing recoveries (black spots), following 'cyclonic approach' around western half of a deep depression centred over Britain on 14th/15th November 1959 (redrawn from Williamson 1975).*

353

distinguishable from those of northern Europe, and are noticeably darker than more eastern birds; in other words, there is a 'cline' in plumage coloration with the darkest race, *subcalcaratus*, at the western end (Williamson and Davis 1956). The bill is more robust, and the wing is slightly longer on average, Hørring (1937) giving for fifty-two Greenland males 91–101 mm, and for twenty Lapland males 90–96 mm. At best *subcalcaratus* is a poorly defined form, especially when compared with the Wheatear *leucorrhoa* and the Redwing *coburni*.

A very successful bird-marking programme carried out in Greenland by Salomonsen's Eskimo helpers has shown that whereas Snow Buntings from the western part of the country cross the Davis Strait to winter in the Great Lakes region of the USA and Canada, those from the north-east make a direct flight across the Polar Sea to north Russia. These groups are thus widely segregated in winter, but this is not necessarily true of Snow Buntings inhabiting the southern parts of the country. There is a remarkable case of a spring migrant ringed at Fair Isle on 7 April 1959 being found on Fogo Island, Newfoundland, on 1 May 1960. This is on the migration route of south-west Greenland birds, and indeed one marked in Godthåb District at 64°N on 18 July 1958 had been recovered on Fogo Island only a week before the Fair Isle bird. It looks as though the Fair Isle migrant wintered in the Old World one year and the New World the next—and this is unlikely to be unique. It may well be the case that Snow and Lapland Buntings from southern Greenland normally travel to the American continent, the shorter journey under suitable weather conditions, but that in some years passing depressions whisk them off to Europe. The low-Arctic Lapland Buntings, having the longest sea crossing, would be particularly exposed to eastward displacement in cyclonic weather, and this subspecies might be expected to develop a size differential compared with continental birds. But if the bulk of both Snow and Lapland Bunting populations normally move to America, the need for greater energy storage is not so pressing as in the Greenland Wheatear and Icelandic Redwing.

Redpolls

The Redpolls have a Holarctic distribution, forming an 'emergent inter-species' which can be conveniently divided into several ecogeographical forms (Salomonsen 1928; Williamson 1961*b*). They comprise an Arctic group of very pale, white-rumped birds, *hornemanni*, in Greenland and the east Canadian Arctic, and the smaller but otherwise similar Coues's Redpoll, *exilipes*, in the most northerly parts of the continental regions; a subarctic group of much darker birds usually lacking the white rump, which comprises *rostrata* in Greenland and *islandicus* in Iceland; and a Boreal and Temperate zone group, the Mealy and Lesser Redpolls. The Lesser Redpolls are isolated in the wooded country of the south European alpine system (*cabaret*) and the British Isles (*disruptis*), although in recent years *disruptis* has established breeding colonies in the Netherlands and Denmark. The

Mealy Redpoll, *flammea*, extends across Eurasia and North America as a forest bird generally to the south of the near-tundra zone of birch and willow scrub occupied by the white-rumped *exilipes*; but in late springs many of the latter breed to the south of their normal range, and in Scandinavia hybrids—which have been named *pallescens*—between this and the Mealy Redpoll are not uncommon (Payn 1947; Harris *et al.* 1965). The American Mealy, pushing eastwards, colonised parts of Baffin Island and southern Greenland, giving rise to the large, dark and thick-billed *rostrata*, usually called the Greater or Greenland Redpoll. The Icelandic population appears to be an intermediate one in which the extremes are barely separable from the very pale Hornemann's Redpoll on the one hand, and the very dark Greenland Redpoll on the other. Possibly the latter invaded Iceland from low-Arctic Greenland at the close of some colder phase of climate when *hornemanni* was the established breeding bird, and a 'hybrid swarm' has developed. This may perhaps have been in the relatively warm Viking era when *hornemanni* was in retreat (Williamson 1961*b*). Something similar began to happen in Greenland during the recent climatic amelioration, when *rostrata* pushed its breeding range further north into the areas already occupied by Hornemann's Redpoll, so that interbreeding occurred where they met (Salomonsen 1951; Waterston and Waterston 1970).

Hornemann's Redpoll moves no further south than the low Arctic for the winter, and returns to its breeding places some ten to fourteen days ahead of the return from abroad of the slightly smaller *rostrata* (Salomonsen 1951; Wynne-Edwards 1952). In this case Hemmingsen's and Salomonsen's Rules both apply, though *rostrata*'s migratory behaviour ensures that the size gap is minimal. The extent to which Icelandic birds migrate is not known; certainly many, perhaps most, stay in Iceland throughout the year, although occasional specimens have been taken in the Hebrides (Williamson 1956).

The Mealy Redpoll and its derivative the Greenland Redpoll are the only forms which are at all strongly migratory; they nest at much the same latitudes, enjoying a similar warm continental-type climate. The Greenland Redpoll is the bigger bird, Salomonsen giving the wing length as 77–83 mm for males, 75–82 mm for females in *rostrata*, compared with 71–78 mm for males, 69–75 mm for females in *flammea*. The Mealy has only a short overland migration, wintering for the most part in temperate Eurasia and in some years in fair strength in the British Isles. Greenland Redpolls from the west coast make a back-track migration to the eastern provinces of Canada, but east coast birds travel to Iceland, Faeroes, Scotland and Ireland. Thus, in general, the large form enjoys the milder maritime climate, and the smaller one a colder continental winter climate, so that on this count there is a reversal of the expected size trend. An average weight of 14.4 gm (range 12.0–17.5 gm) was given for migrant Mealy Redpolls at Heligoland by Weigold (1926). Table IV, taken from Williamson (1956), gives some idea of the variations in Greenland birds, whose normal departure weight in autumn may well be around 20 gm.

TABLE IV *Gains in weight of migrant Greenland Redpolls*

| Date | | Interval | Weight in gm | | % |
1st capture	Recapture	(days)	1st capture	Recapture	Gain
21.10.54	25.10.54	4	17.09	17.26	1
27.08.55	05.09.55	9	14.40	20.23	40
06.09.55	16.09.55	10	13.80	16.55	20
06.09.55	23.09.55	17	16.35	20.46	25
16.09.55	17.09.55	1	16.55	14.61	-
29.09.55	01.10.55	2	10.77	11.77	1

These birds invade Fair Isle in large numbers at rather rare intervals; such invasions are known to have taken place in 1905, 1925 and 1955. In 1955 there were unusual numbers also at Foula and elsewhere in Shetland, and in western Ireland (Williamson 1956). Since they leave early, in late August and early September, they are often able to travel south in anticyclonic weather *via* Iceland, Faeroes and Shetland; but many are caught in cyclonic situations and, like the Wheatears and Lapland Buntings, travel round the depressions. It is possible that, as with the buntings, some may winter in different continents in different years; nevertheless, their migratory performance far outstrips that of the Mealy Redpoll and is a sufficient explanation of their greater size.

Meadow Pipit and White Wagtail

Two other common breeding birds of Iceland and northern Europe which are strongly migratory are the Meadow Pipit and White Wagtail. The former species winters from the British Isles south to the Mediterranean basin, while the White Wagtail (*alba*) leap-frogs the range of its close relative the Pied Wagtail (*yarrellii*) of Britain to join continental *alba* in tropical Africa. Icelandic Meadow Pipits are more richly coloured than birds of the typical race (*pratensis*), being a deeper brownish-olive above and strongly suffused with buff below; most British breeders appear to be of the typical race, but the exact distribution of the colourful Atlantic race (*theresae*) has not been worked out (Williamson 1959). All that can be said is that winter specimens collected in southern France, Iberia and Morocco are of *pratensis* type; that such birds as winter in Britain and Ireland are predominantly *theresae*; and that there are no winter recoveries in the British Isles of home-bred young. It may be that there is a more or less complete change-over of stocks, the breeders going to southern Europe and north Africa, where (as with continental birds) there are numerous recoveries of ringed birds, and the Icelandic population taking their place.

The Icelandic breeding population must therefore make an obligatory overseas flight, and big influxes coinciding with both cyclonic and anticyclonic weather have been recorded at Fair Isle and elsewhere in the north and west. Over 7,000 Meadow Pipits appeared at St. Kilda on 14–15 September 1958 after journeying round the periphery of a depression, and several exhausted birds came aboard an army LCT which was sailing to St. Kilda from the Outer Hebrides. So far as is known there is no size difference between Icelandic and continental populations, and therefore neither fits Salomonsen's Rule. The warmer winter influence—tropical in the wagtail, Mediterranean in the pipit—should have ensured a smaller size for continental as opposed to Atlantic Meadow Pipits (a differential which the latter's overseas flight should have exaggerated), and for White Wagtails as opposed to the British resident Pied Wagtails. Perhaps the recent spread of the Icelandic populations of both species to south-east Greenland is indicative of a comparatively new and continuing expansion of range, and the time-lapse since their arrival in the north-east Atlantic sector is too brief for subspecific differences in size to have evolved.

Waders

Among wading birds, there are slight differences in length of wing between Icelandic and continental Whimbrels, Oystercatchers and Purple Sandpipers, but none provides a good case for the hypothesis that the birds with a mandatory overseas migration have evolved greater robustness. The race *islandica* of the Black-tailed Godwit is separable from the typical race *limosa* of temperate Europe on account of its redder plumage in the nesting season and its shorter bill (Salomonsen 1935); and as it obeys ecogeographical rules in both these respects, one may conclude that it has been established in Iceland for a long time. The Black-tailed Godwit does not nest on the continent north of the Kattegat and Skagerrak, so there is a considerable latitude difference in the respective breeding ranges. Continental birds should show a further tendency towards smaller size since they spend the winter in Africa, much further south than Icelandic birds which visit Ireland and the English south coast estuaries. Yet the wing-length range is much the same in the two races, and the lack of greater dimensions in *islandica*, despite its more northerly breeding and wintering areas and its obligatory overseas migration, makes it a puzzling exception to the hypothesis we have developed.

Greenland White-fronted Geese and Icelandic Redshanks and Merlins

We may conclude with three cases which do seem to fit the bill, supporting the view that a trans-oceanic migration demands a bigger body to contain the necessary energy stores. The Greenland White-fronted Goose (*flavirostris*) resorts to Ireland and south-west Scotland (and an outpost in central Wales) after its post-nuptial moult, as the numerous recoveries of ringed birds show. Their exodus from the low-Arctic breeding grounds is delayed until middle and late October, when most flocks are compelled to travel round the Atlantic depressions (Ruttledge and Williamson 1952). According to Salomonsen (1950–51) the wing length of *flavirostris* has a range of

420–455 mm, compared with 368–440 mm for the typical race *albifrons*, which has a largely overland migration to much the same latitudes in Britain and western Europe.

The Icelandic Redshank (*robusta*) is not only longer in the wing than the typical race (*totanus*) whose range extends to northern Norway, but also has a bigger sternum (Nørrevang 1954) and stouter legs and feet (Harrison 1944), indicating a greater body weight. Although Fair Isle weighings support this (average 133 gm compared with 119 gm, in favour of *robusta*) the records are very few. The whole of Britain and Ireland lie within the winter range of the Icelandic race, and there are some records from further south. Only a small part of the British stock (*britannica*) ventures across the English Channel, but some birds from northern localities go to Ireland. The winter distribution of the European populations has been worked out by Salomonsen (1954), who provided maps based on ringing recoveries. Danish birds travel overland to the central Mediterranean; birds from western Germany and the Netherlands go to the Iberian peninsula and north Africa; and Scandinavian birds leap-frog them on the way to and from tropical Africa. They have the shortest average wing length despite their more northerly home, a response to the warm winter environment, as in the *tundrae* Ringed Plover. British and west European birds, though enjoying Atlantic maritime and Mediterranean influences respectively, do not differ significantly in size. The synhiemic British and Icelandic Redshanks, however, show a mean difference of 10 mm in wing length in favour of the latter, best explained as a response to a mandatory overseas migration.

The Icelandic Merlin (*subaesalon*) breeds between the same latitudes as a good many continental birds (*aesalon*). In an examination of museum specimens I found a mean wing length for Icelandic birds of 209 mm in males and 227 mm in females (this being the more powerful sex, as in many birds of prey), compared with 197 mm and 215 mm respectively for European birds. A difference in wing length between the two races is found in 96% of the males and 94% of the females (Butterfield 1954). There are some limitations in drawing comparisons between the weights of the two races, since any individual bird of prey can be expected to vary within a fairly wide range. Obviously, a newly arrived migrant at Fair Isle would register much less than an off-passage bird which had made a recent kill, especially as the prey is sometimes as big as a Redwing. When all the available weights are averaged, sex for sex, including records of the same individuals if recaptured, it is probable that these inequalities are reduced. If the figures are comparable, then it would appear that *subaesalon* is considerably heavier than *aesalon*—i.e. males 182 gm, females 255 gm in the former, and males 175 gm, females 210 gm in the latter. Kluz in 1943 gave for *aesalon* males 150–180 gm, females 188–210 gm, the upper limit in each case being below the average for Icelandic immigrants at Fair Isle.

The sedentary British population shows a male wing length close to the mean for continental birds, but a female wing length nearer that of Icelandic females. The continental race winters over most of Europe south to the Mediterranean, while the Icelandic population spends the off-season mostly in western Britain, with some slight overspill on to the continent between Norway and France. It is therefore synhiemic with the British population and a considerable part of the continental one in a mild maritime climate, and the remarkable size discrepancy would appear to present a clear case in which selection pressure has been strongest during the actual migration periods, when the Icelandic birds have a compulsory sea crossing of several hundred miles (Williamson 1954; Williamson and Butterfield 1955).

REFERENCES

Butterfield, A. 1954. *Falco columbarius subaesalon* Brehm: a valid race. *British Birds* 47:342–347.

Hall, B.P. & Moreau, R.E. 1970. *An Atlas of Speciation in African Passerine Birds.* London (British Museum).

Harris, M.P., Norman, F.I. & McColl, R.H.S. 1965. A mixed population of redpolls in northern Norway. *British Birds* 58:288–294.

Harrison, J.M. 1944. Some remarks upon the western Palaearctic races of *Tringa totanus* (Linnaeus). *Ibis* 86:493–503.

Hemmingsen, A.M. 1951. The relation of bird migration in north-eastern China to body weight and its bearing on Bergmann's Rule. *Proceedings of the Xth International Ornithological Congress,* Uppsala 1950, pp.289–294.

Hørring, R. 1937. *Report of the Fifth Thule Expedition 1921–24: Zoology.* Vol. 2 (Birds). Copenhagen.

Hørring, R. & Salomonsen, F. 1941. Further records of rare or new Greenland birds. *Meddelelser om Grønland* 131(5):1–86.

Huxley, J. 1942. *Evolution: the Modern Synthesis.* London (Allen & Unwin).

Mayr, E. 1956. Geographical character gradients and climatic adaptation. *Evolution* 10:105–108.

Nørrevang, A. 1954. Sternal measurements as a mean in identifying the Iceland Redshank. *Dansk Ornithologisk Forenings Tidsskrift* 48:235–236.

Payn, W.A. 1947. Redpolls from Norway. *Bulletin of the British Ornithologists' Club* 67:41–42.

Rensch, B. 1939. Klimatische Auslese von Grössenvarianten. *Archive Naturgeschichte Leipzig* 8:89–129.

Ruttledge, R.F. & Williamson, K. 1952. Early arrival of White-fronted Geese into Ireland. *Irish Naturalists' Journal* 10:263–264.

Salomonsen, F. 1928. Bemerkungen über die Verbreitung der *Carduelis linaria* Gruppe und ihre Variationen. *Vidensk. Medd. Dansk Naturh. Foren* 86:123–202.

Salomonsen, F. 1931. On the geographical variation of the Snow-Bunting (*Plectrophenax nivalis*). *Ibis* (13)1:57–70.

Salomonsen, F. 1935. *Zoology of the Faroes. Aves.* Copenhagen.

Salomonsen, F. 1950–51. *Grønlands Fugle: The Birds of Greenland.* Copenhagen (Munksgaard).

Salomonsen, F. 1951. The immigration and breeding of the Fieldfare (*Turdus pilaris* L.) in Greenland. *Proceedings of the Xth International Ornithological Congress,* Uppsala 1950, pp.515–526.

Salomonsen, F. 1954. The migration of the European Redshanks (*Tringa totanus* (L.)). *Dansk Ornithologisk Forenings Tidsskrift* 48:94–122.

Salomonsen, F. 1955. The evolutionary significance of bird migration. *Dan. Biol. Medd.* 22(6):1–62.

Scholander, P.F. 1955. Evolution of climatic adaptation in homeotherms. *Evolution* 9:15–26.

Ticehurst, N.F. 1909. *A History of the Birds of Kent.* London (Witherby).

Waterston, G. & Waterston, I. 1970. Greenland Redpoll breeding in High Arctic Region. *Dansk Ornithologisk Forenings Tidsskrift* 64:93–94.

Weigold, H. 1926. *Masse, Gewichte und Zug nach Alter und Geschlecht bei Helgoländer Zugvögeln.* Oldenburg.

Williamson, K. 1954. The migration of the Iceland Merlin. *British Birds* 47:434–441.

Williamson, K. 1956. The autumn immigration of the Greenland Redpoll (*Carduelis flammea rostrata* (Coues)) into Scotland. *Dansk Ornithologisk Forenings Tidsskrift* 50:125–133.

Williamson, K. 1958a. Bergmann's rule and obligatory overseas migration. *British Birds* 51:209–232.

Williamson, K. 1958b. Autumn migration of Redwings *Turdus musicus* into Fair Isle. *Ibis* 100:582–604.

Williamson, K. 1959. Meadow Pipit migration. *Bird Migration* 1:88–91.

Williamson, K. 1961a. The concept of 'cyclonic approach'. *Bird Migration* 1:235–240.

Williamson, K. 1961b. The taxonomy of the redpolls. *British Birds* 54:238–241.

Williamson, K. & Butterfield, A. 1955. Merlin migration at Fair Isle in 1954. *Fair Isle Bird Observatory Bulletin* 2:265–268.

Williamson, K. & Davis, P. 1956. The autumn 1953 invasion of Lapland Buntings and its source. *British Birds* 49:6–25.

Wynne-Edwards, V.C. 1952. Zoology of the Baird expedition (1950). I. The birds observed in central and south-east Baffin Island. *Auk* 69:353–391.

BIBLIOGRAPHY

Axell, H.E. 1966. Eruptions of Bearded Tits during 1959–65. *British Birds* 59:513–543.

Balfour-Browne, F. 1958. The origin of our British Swallowtail and our Large Copper butterflies. *Entomologist's Record and Journal of Variation* 70:33–36.

Barthel, P.H. 1990. Hinweise zur Bestimmung der Zitronenstelze *Motacilla citreola*. *Limicola* 4:149–182.

Batten, L.A., Bibby, C.J., Clement, P., Elliott, G.D. & Porter, R.F. (eds.) 1990. *Red Data Birds in Britain*. London (Poyser).

Batten, L.A., Dennis, R.H., Prestt, I. and the Rare Breeding Birds Panel 1979. Rare breeding birds in the United Kingdom in 1977. *British Birds* 72:363–381.

Beirne, B.P. 1952. *The Origin and History of the British Fauna*. London (Methuen).

Beresford, J. (ed.) 1935. *The Diary of a Country Parson 1758–1802 by James Woodforde*. London (Oxford University Press).

Bergthórsson, P. 1969. An estimate of drift ice and temperature in Iceland in 1,000 years. *Jökull* 19:94–101.

Berthold, P. 1974. Die gegenwärtige Bestandsentwicklung der Dorngrasmücke (*Sylvia communis*) und anderer Singvögelarten im westlichen Europa bis 1973. *Vogelwelt* 95:170–183.

Berthold, P., Fliege, G., Querner, U. & Winkler, H. 1986. Die Bestandsentwicklung von Kleinvögeln in Mitteleuropa: Analyse von Fangzahlen. *Journal für Ornithologie* 127:397–437.

Bewick, T. 1797–1804. *History of British Birds*, Vols. 1–2. Newcastle-upon-Tyne.

Bijleveld, M. 1974. *Birds of Prey in Europe*. London (Macmillan).

Bolin, B., Döös, B., Jäger, J. & Warrick, R. (eds.) 1986. *The Greenhouse Effect, Climatic Change and Ecosystems*. Chichester (Wiley).

Bonham, P.F. & Robertson, J.C.M. 1975. The spread of Cetti's Warbler in north-west Europe. *British Birds* 68:393–408.

Boswall, J. 1974. Midsummer field notes on the birds of coastal south-east Iceland. *Bristol Ornithology* 7:51–66.

Braaksma, S. 1960. [The spread of the Curlew *Numenius arquata* as a breeding bird.] *Ardea* 48:65–90. (In Dutch, with English summary.)

Brenchley, A. 1986. The breeding distribution and abundance of the Rook (*Corvus frugilegus*) in Great Britain since the 1920s. *Journal of Zoology, London.* (A) 210:261–278.

Breuer, G. 1980. *Air in Danger*. Cambridge (Cambridge University Press).

Broecker, W.S. 1975. *Science* 189:460–463.

Brooks, C.E.P. 1949. *Climate through the Ages*. (2nd ed.) London.

Brown, P. and Waterston, G. 1962 *The Return of the Osprey*. London (Collins).

Bryson, R.A. 1975. The lessons of climatic history. *Environmental Conservation* 2:163–170.

Buckley, T.E. & Harvie-Brown, J.A. 1891. *A Vertebrate Fauna of the Orkney Islands*. Edinburgh (David Douglas).

Burton, J.F. 1974a. Subtropical species, Kerry air: Ireland's Mediterranean Wildlife. *Country Life* (23 May, 1974), pp.1270–1271.

Burton, J.F. 1974b. Friend or foe? Collared Doves and agriculture. *Country Life* (27 June, 1974), pp.1674–1675.

Burton, J.F. 1974c. The paradoxical wanderer: Pallas's Sandgrouse. *Country Life* (25 July, 1974), pp.216–217.

Burton, J.F. 1975. The effects of recent climatic changes on British insects. *Bird Study* 22:203–204.

Burton, J.F. 1976. *Field and Moor*. Bath (Kingsmead Press).

Burton, J.F. 1981. Half north, half south: Britain's rich wildlife. *New Scientist* 90:16–18.

Burton, J.F. 1984. Wildlife and our changing climate. *The Living Countryside* 14:3346–51.

Burton, J.F. 1991. British grasshoppers and bush-crickets may be responding to the "Greenhouse" warming. *Country-Side* 27:29–31.

Burton, J.F. & French, R.A. 1969. Monarch butterflies coinciding with American passerines in Britain and Ireland in 1968. *British Birds* 62:493–494.

Calder, N. 1974a. Arithmetic of ice ages. *Nature* 252:216–218.

Calder, N. 1974b. *The Weather Machine*. London (BBC Publications).

Campbell, B. 1954–55. The breeding distribution and habitats of the Pied Flycatcher (*Muscicapa hypoleuca*) in Britain. *Bird Study* 1:81–101; 2:24–32, 179–191.

Campbell, B. 1965. The British breeding distribution of the Pied Flycatcher, 1953–62. *Bird Study* 12:305–318.

Chancellor, R.D. (ed.) 1977. *Report of Proceedings of the World Conference on Birds of Prey, Vienna, 1–3 October, 1975*. International Council for Bird Preservation.

Christie, D.A. 1975. Studies of less familiar birds. 176: Barred Warbler. *British Birds* 68:108–114.

Clark, J.G.D.C. 1954. *Excavations at Star Carr*. Cambridge.

Clark, J.M. 1984. *Birds of the Hants/Surrey Border*. Aldershot (Hobby Books).

Cook, A. 1975. Changes in the Carrion/Hooded Crow hybrid zone and the possible importance of climate. *Bird Study* 22:165–168.

Coope, G.R. 1965. The response of the British insect fauna to late Quaternary climatic oscillations. In Freeman, P. (ed.), pp.454–455.

Coope, G.R. 1975. Climatic fluctuations in northwest

Europe since the Last Interglacial, indicated by fossil assemblages of Coleoptera. In Wright, A.E. and Moseley, F. (eds.), pp.153–168.

Cowdy, S. 1973. Ants as a major food source of the Chough. *Bird Study* 20:117–120.

Cramp, S. *et al.* (eds.) 1977–1994. *Handbook of the Birds of Europe, the Middle East and North Africa (The Birds of the Western Palearctic)*: Vols. 1–9. Oxford (Oxford University Press).

Crawford, O.G.S. & Keiller, A. 1928. *Wessex from the Air.* Oxford.

Crick, H., Dudley, C. & Glue, D. 1993. Breeding birds in 1992. *BTO News* 189:14–16.

Cumming, I.G. 1979. Lapland Buntings breeding in Scotland. *British Birds* 72:53–59.

Curry-Lindahl, K. 1959–63. *Våra Fåglar i Norden,* Vols. 1–4. Stockholm (Natur och Kultur).

Dadds, M. 1993. Icterine warbler at Brentry, Bristol. *Avon Bird Report 1992*: 100–101.

Dansgaard, W., Johnsen, S.J., Reeh, N., Gundestrup, N., Clausen, H.B. & Hammer, C.U. 1975. Climatic changes, Norsemen and modern man. *Nature* 255:24–28.

Dennis, R.H. 1983. Purple Sandpipers breeding in Scotland. *British Birds* 76:563–566.

Dennis, R.L.H. 1977. *The British Butterflies: Their Origin and Establishment.* Faringdon (Classey).

Donald, P. 1993a. The fat bird thins out. *Birds (RSPB)* 14(5):23–24.

Donald, P. 1993b. Corn Buntings in winter: are there better times ahead? *BTO News* 189:12–13.

Dorling, E. 1992. Record breeding season. *Birds (RSPB)* 14(4):20–22.

Durango, S. 1946. (The Roller (*Coracias g. garrulus* L.) in Sweden.) *Vår Fågelvärld* 5:145–90. (In Swedish, with English summary.)

Durango, S. 1948. (Observations on the distribution and breeding of the Icterine Warbler in Sweden). *Fauna och Flora* 43:186–220. (In Swedish, with English summary.)

Dybbro, T. 1976. *De Danske Ynglefugles Udbredelse.* Copenhagen (Dansk Ornithologisk Forening).

Evans, I.M. & Pienkowski, M.W. 1991. World status of the Red Kite. *British Birds* 84:171–187.

Evans, P.R. & Flower, W.U. 1967. The birds of the Small Isles. *Scottish Birds* 4:404–445.

Everett, M.J. 1971. Breeding status of Red-necked Phalaropes in Britain and Ireland. *British Birds* 64:293–302.

Ferguson-Lees, I.J. 1970. Red-breasted Flycatcher. In Gooders, J. (1969–71), pp.2248–2249.

Ferguson-Lees, I.J. 1971. Serin. In Gooders, J. (1969–71), pp.2615–2617.

Ferguson-Lees, I.J., Hockcliffe, Q. & Zweeres, K. 1975. *A Guide to Bird-Watching in Europe.* London (The Bodley Head).

Ferguson-Lees and the Rare Breeding Birds Panel. 1977. Rare breeding birds in the United Kingdom in 1975. *British Birds* 70:2–23.

Ferguson-Lees, I.J. & Sharrock, J.T.R. 1977. When will the Fan-tailed Warbler colonise Britain? *British Birds* 70:152–159.

Fisher, J. 1952. *The Fulmar.* London (Collins).

Fisher, J. 1953. The Collared Turtle Dove in Europe. *British Birds* 46:153–181.

Fisher, J. 1966a. The Fulmar population of Britain and Ireland, 1959. *Bird Study* 13:5–76.

Fisher, J. 1966b. *The Shell Bird Book.* London (Ebury Press and Michael Joseph).

Fitter, R.S.R. 1959. *The Ark in our Midst.* London (Collins).

Fitter, R.S.R. 1976. Black Redstarts breeding in Britain in 1969–73. *British Birds* 69:9–15.

Fjeldså, J. 1973. Distribution and geographical variation of the Horned Grebe *Podiceps auritus* (Linnaeus, 1758). *Ornis Scandinavica* 4:55–86.

Flegg, J.M. 1970. Savi's Warbler. In Gooders, J.

(1969–71), pp.2085–2086.

Flohn, H. 1980. *Possible Climatic Consequences of a Man-Made Global Warming.* Laxenburg (IIASA).

Ford, M.J. 1978. The variety of biological effects of climatic change. In Frydendahl, K. (ed.), pp.98–102.

Ford, M.J. 1982. *The Changing Climate: Responses of the Natural Fauna and Flora.* London (Allen & Unwin).

Francis, P. 1975. Fire and ice. *New Scientist* 67:19–22.

Freeman, P. (ed.) 1965. *Proceedings of XIIth International Congress of Entomology*, London, 1964. London.

Frydendahl, K. (ed.) 1978. *Proceedings of the Nordic Symposium on Climatic Changes and Related Problems.* Copenhagen (Danske Meteorologiske Institut).

Gibbons, D.W., Reid, J.B. & Chapman, R.A. 1993. *The New Atlas of Breeding Birds in Britain and Ireland: 1988–1991.* London (T. & A.D. Poyser).

Gimpel, J. 1973. Population and environment in the Middle Ages. *Environment and Change* (December 1973): 233–242.

Godfrey, W.E. 1966, 1986 (revised edition). *The Birds of Canada.* Ottawa (National Museum of Canada).

Gooders, J. 1969–71. *Birds of the World.* London (IPC).

Greenoak, F. (ed.) 1986–89. *The Journals of Gilbert White, 1751–1793.* London (Century Hutchinson).

Greenwood, J.J.D. 1968. Bluethroat nesting in Scotland. *British Birds* 61:524–525.

Gribbin, J. 1975a. *Our Changing Climate.* London (Faber & Faber).

Gribbin, J. 1975b. Climatic change down under. *New Scientist* 68:523–525.

Gribbin, J. 1976a. *Our Changing Universe.* London (Macmillan).

Gribbin, J. 1976b. *Forecasts, Famines and Freezes.* London (Wildwood House).

Gribbin, J. 1979. Disappearing threat to ozone. *New Scientist* 81:474–476.

Gribbin, J. 1990. *Hothouse Earth: the Greenhouse Effect and Gaia.* London (Bantam Press).

Gribbin, J. & Kelly, M. 1989. *Winds of Change.* London (Headway).

Grzimek, B. 1971. *Grzimeks Tierleben, Enzyklopädie des Tierreiches,* Vol. 6. Zurich (Kindler Verlag).

Gudmundsson, F. 1951. The effects of the recent climatic changes on the birdlife of Iceland. *Proceedings of the Xth International Ornithological Congress,* Uppsala, 1950, pp.502–514.

Haartman, L. von. 1973. In Farner, D.S. (ed.) *Breeding biology of birds.* (Proc. of a symposium on breeding behaviour and reproductive physiology in birds, Denver, 1972), pp.448–481. Washington (National Academy of Sciences).

Haftorn, S. 1958. (Population changes, especially geographical changes, in the Norwegian avifauna during the last 100 years.) *Sterna* 3:105–137. (In Norwegian, with English summary.)

Harris, G. 1964. Climatic changes since 1860 affecting European birds. *Weather* 19:70–79.

Harris, M.P. 1976. The present status of the Puffin in Britain and Ireland. *British Birds* 69:239–264.

Harris, M.P. & Murray, S. 1977. Puffins on St. Kilda. *British Birds* 70:50–65.

Harrison, C. 1982. *An Atlas of the Birds of the Western Palaearctic.* London (Collins).

Harrison, G.R., Dean, A.R., Richards, A.J. & Smallshire, D. 1982. *The Birds of the West Midlands.* Studley (West Midlands Bird Club).

Harrison, J.M. & Harrison J.G. 1965. The juvenile plumage of the Icelandic Black-tailed Godwit and further occurrences of this race in England. *British Birds* 58:10–14.

Harvey, W.G. 1977. Cetti's Warblers in east Kent in 1975. *British Birds* 70:89–96.

Hawksworth, D.L. (ed.) 1974. *The Changing Flora and Fauna of Britain.* London (Academic Press).

Heinzel, H., Fitter, R. & Parslow, J. 1972. *The Birds of Britain and Europe with North Africa and the Middle East*. London (Collins).

Heppleston, P.B. 1972. The comparative breeding ecology of Oystercatchers (*Haematopus ostralegus* L.) in inland and coastal habitats. *Journal of Animal Ecology* 41:23–51.

Hildén, O. 1989. The effects of severe winters on the bird fauna of Finland. *Memoranda Soc. Fauna Flora Fennica* 65:59–66.

Hildén, O. & Sharrock, J.T.R. 1985. A summary of recent avian range changes in Europe. *Proceedings of the XVIIIth International Ornithological Congress*, Moscow 1982, pp.716–736.

Hoyle, F. & Lyttleton, R.A. 1939. The effect of interstellar matter on climatic variations. *Proceedings of the Cambridge Philosophical Society, Biological Sciences* 35:405–415.

Hudson, R. 1973. *Early and Late Dates for Summer Migrants*. BTO Field Guide No. 15. Tring.

Hustich, L. (ed.) 1952. *The Recent Climatic Fluctuation in Finland and its Consequences; a Symposium. Fennia* 75. Helsinki.

Hustings, F. 1988. *European Monitoring Studies of Breeding Birds*. Beek, Netherlands (SOVON).

Hutchins, L.W. 1947. The bases for temperature zonation in geographical distribution. *Ecological Monographs* 11:325–335.

Hutchinson, C.D. 1989. *Birds in Ireland*. Calton (Poyser).

Ijzendoorn, A.L.J. van 1950. *The Breeding Birds of the Netherlands*. Leiden (Brill).

Jardine, Sir W. 1850. *Contributions to Ornithology for 1850*.

Jespersen, P. 1946. *The Breeding Birds of Denmark*. Copenhagen (Munksgaard).

John, B.S. 1977. *The Ice Age*. London (Collins).

Johnson, C.G. & Smith, L.P. (eds.) 1965. *The Biological Significance of Climatic Changes in Britain*. London (Academic Press).

Kalela, O. 1949. Changes in geographic ranges in the avifauna of northern and central Europe in relation to recent changes in climate. *Bird Banding* 20:77–103.

Kalela, O. 1952. Changes in the geographic distribution of Finnish birds and mammals in relation to recent changes in climate. *Fennia* 75:38–51.

Keve, A. 1963. Peculiarities of range expansion of three European bird species. *Proceedings of the XIIIth International Ornithological Congress*, Ithaca 1962, pp.1124–1127.

Lack, D. 1954. *The Natural Regulation of Animal Numbers*. Oxford (Clarendon Press).

Lack, P. (ed.) 1986. *The Atlas of Wintering Birds in Britain and Ireland*. Calton (Poyser).

Lamb, H.H. 1963. What can we find out about the trend of our climate? *Weather* 18:194–216.

Lamb, H.H. 1966. *The Changing Climate*. London (Methuen).

Lamb, H.H. 1972–77. *Climate: Present, Past and Future*, Vols. 1–2. London (Methuen).

Lamb, H.H. 1975. Our understanding of the global wind circulation and climatic variations. *Bird Study* 22:121–141.

Lea, D. 1974. Black Redstarts breeding in Orkney. *Scottish Birds* 8:80–81.

Lockley, R.M. 1953. *Puffins*. London (Dent).

Lyall, I.T. 1970. Recent trends in spring weather. *Weather* 25:163–165.

Macdonald, D. 1965. Notes on the Corn Bunting in Sutherland. *Scottish Birds* 3:235–246.

Makatsch, W. 1966. *Wir Bestimmen die Vögel Europas*. Melsungen (Neumann-Neudamm).

Manley, G. 1972. 5th impress. *Climate and the British Scene*. London (Collins).

Marchant, J.H. & Balmer, D. 1993. Common Birds Census: 1991–92 index report. *BTO News* 187:9–12.

Marchant, J.H., Hudson, R., Carter, S.P. &

Whittington, P. 1990. *Population Trends in British Breeding Birds*. Tring (BTO).

Marshall, N. 1968. The ice-fields around Iceland in Spring 1968. *Weather* 23:368.

Matthews, S.W. 1976. What's happening to our climate? *National Geographic Magazine*. 150:576–615.

Matvejev, S.D. 1985. Expansion of areas by 15 bird species in Balkan peninsula. *Proceedings of the XVIIIth International Ornithological Congress*, Moscow 1982, pp.763–768.

May, R.M. 1979. Arctic animals and climatic changes. *Nature* 281:177–178.

Mayr, E. 1926. Die Ausbreitung des Girlitz (*Serinus canaria serinus* L.). *Journal für Ornithologie* 74:571–671.

McCrea, W.H. 1975. Ice ages and the galaxy. *Nature* 255:607–609.

Mead, C.J. 1973. Movements of British raptors. *Bird Study* 20:259–286.

Mead, C.J. & Hudson, R. 1985. Report on bird-ringing for 1984. *Ringing and Migration* 6:125–172.

Mead, C.J. & Pearson, D.J. 1974. Bearded Reedling populations in England and Holland. *Bird Study* 21:211–214.

Meek, E.R. & Little, B. 1977. The spread of the Goosander in Britain and Ireland. *British Birds* 70:229–237.

Merikallio, E. 1951. Der Einfluss der letzten Wärmeperiode (1930–49) auf die Vogelfauna Nordfinnlands. *Proceedings of the Xth International Ornithological Congress*, Uppsala 1950, pp.484–493.

Merikallio, E. 1958. *Finnish Birds. Their Distribution and numbers*. Fauna Fennica 5. Helsinki.

Merrett, C. 1666. *Pinax rerum Naturalium Britannicarum*.

Mikkola, H. 1973. The Red-flanked Bluetail and its spread to the west. *British Birds* 66:3–12.

Møller, A.P. 1989. Population dynamics of a declining Swallow *Hirundo rustica* population. *Journal of Animal Ecology* 58:1051–1063.

Monk, J.F. 1963. The past and present status of the Wryneck in the British Isles. *Bird Study* 10:112–132.

Moreau, R.E. 1955. The bird-geography of Europe in the Last Glaciation. *Proceedings of the XIth International Ornithological Congress*, Basel 1954, pp.401–405.

Morris, A. 1993. The 1992 Nightjar survey—a light at the end of the tunnel for this threatened species? *BTO News* 185:8–9.

Morris, P.A. 1990. The preserved Hawfinch attributed to Gilbert White. *Archives of Natural History* 17:361–366.

Murton, R.K. 1965. *The Woodpigeon*. London (Collins, New Naturalist).

Murton, R.K., Westwood, N.J. & Isaacson, A.J. 1964. The feeding habits of the Woodpigeon *Columba palumbus*, Stock Dove *C. oenas* and Turtle Dove *Streptopelia turtur*. *Ibis* 106:174–188.

Nagell, B. & Frycklund, I. 1965. (Irruptions of the Snowy Owl (*Nyctea scandiaca*) in Scandinavia in the winters of 1960–63.) *Vår Fågelvärld* 24:26–55. (In Swedish, with English summary.)

Nethersole-Thompson, D. 1966. *The Snow Bunting*. Edinburgh and London (Oliver & Boyd).

Nethersole-Thompson, D. 1973. *The Dotterel*. London (Collins).

Newton, I. 1972. *Finches*. London (Collins, New Naturalist).

Nicholson, E.M. 1948. Poets as bird observers. *Bird Notes* (RSPB) 23:192–196.

Nørrevang, A. 1955. (Changes in the birdlife of the Faeroes in relation to the climatic changes in the North Atlantic area.) *Dansk Ornithologisk Forenings Tidsskrift* 49:206–229. (In Danish, with English summary.)

Norris, C.A. 1947. Report on the distribution and status of the Corn-Crake, Part Two. *British Birds*

40:226–244.

Norris, C.A. 1960. The breeding distribution of thirty bird species in 1952. *Bird Study* 7:129–184.

O'Connor, R.J. & Shrubb, M. 1986. *Farming and Birds*. Cambridge (Cambridge University Press).

Olney, P.J.S. 1963. The food and feeding habits of Tufted Duck *Aythya fuligula*. *Ibis* 105:55–62.

Olsson, V. 1969. Die Expansion des Girlitzes (*Serinus serinus*) in Nordeuropa in den letzten Jahrzehnten. *Vogelwarte* 25:147–156.

Olsson, V. 1971. Studies of less familiar birds. 165: Serin. *British Birds* 64:213–223.

Olsson, V. 1975. Bearded Reedling populations in Scandinavia. *Bird Study* 22:116–118.

Orford, N. 1973. Breeding distribution of the Twite in central Britain. *Bird Study* 20:51–62, 121–126.

Osborne, P.J. 1965. The effect of forest clearance on the distribution of the British insect fauna. In Freeman, P. (ed.), pp.456–457.

Osborne, P.J. 1969. An insect fauna of late Bronze Age date from Wilsford, Wiltshire. *Journal of Animal Ecology* 38:555–66.

O'Sullivan, J.M. 1976. Bearded Tits in Britain and Ireland, 1966–74. *British Birds* 69:473–489.

O'Sullivan, J. and the Rarities Committee. 1977. Report on rare birds in Great Britain in 1976. *British Birds* 70:405–453.

Owen, M. & Salmon, D.G. 1988. Feral Greylag Geese *Anser anser* in Britain and Ireland, 1960–86. *Bird Study* 35:37–45.

Palmer, R.S. 1962–76. *Handbook of North American Birds*, Vols. 1–3. New Haven & London (Yale University Press).

Parkin, D.W. 1976. Solar constant during a glaciation. *Nature* 260:28–31.

Parslow, J.L.F. 1967. Changes in status among breeding birds in Britain and Ireland. *British Birds* 60:261–285.

Parslow, J.L.F. 1973. *Breeding Birds of Britain and Ireland: a Historical Survey*. Berkhamsted (Poyser).

Peakall, D.B. 1962. The past and present status of the Red-backed Shrike in Great Britain. *Bird Study* 9:198–216.

Peal, R.E.F. 1968. The distribution of the Wryneck in the British Isles, 1964–66. *Bird Study* 15:111–126.

Pearson, R. 1978. *Climate and Evolution*. London (Academic Press).

Pennie, I.D. 1962. A century of bird-watching in Sutherland. *Scottish Birds* 2:167–192.

Ratcliffe, D.A. 1976. Observations on the breeding of the Golden Plover in Great Britain. *Bird Study* 23:63–116.

Rheinwald, G. 1977. *Atlas der Brutverbreitung westdeutscher Vogelarten. Kartierung 1975*. Bonn (Dachverband Deutscher Avifaunisten).

Rheinwald, G. 1993. *Atlas der Verbreitung und Häufigkeit der Brutvögel Deutschlands—Kartierung um 1985*. Bonn (Dachverband Deutscher Avifaunisten).

Richmond, W.K. 1959. *British Birds of Prey*. London (Lutterworth).

Robbins, C.S., Bruun, B. & Zim, H.S. 1983. *A Guide to Field Identification: Birds of North America*. New York (Golden Press).

Robbins, C.S. 1985. Recent changes in the ranges of North American birds. *Proceedings of the XVIIIth International Ornithological Congress*, Moscow 1982, pp.737–742.

Roberts, P.J. 1982. Foods of the Chough on Bardsey Island, Wales. *Bird Study* 29:155–161.

Rolfe, R. 1966. The status of the Chough in the British Isles. *Bird Study* 13:221–236.

Ryder, J.P. & Cooke, F. 1973. Ross's Geese nesting in Manitoba. *Auk* 90:691–692.

Sage, B.L. & Vernon, J.D.R. 1978. The 1975 national survey of rookeries. *Bird Study* 25:64–86.

Sage, B. & Whittington, P.A. 1985. The 1980 sample

survey of rookeries. *Bird Study* 32:77–81.

Salinger, M.J. 1976. New Zealand temperatures since AD 1300. *Nature* 260:310–311.

Salinger, M.J. & Gunn, J.M. 1975. Recent climatic warming around New Zealand. *Nature* 256:396–398.

Salomonsen, F. 1948. The distribution of birds and the recent climatic change in the North Atlantic area. *Dansk Ornithologisk Forenings Tidsskrift* 42:85–99.

Salomonsen, F. 1951. The immigration and breeding of the Fieldfare (*Turdus pilaris* L.) in Greenland. *Proceedings of the Xth International Ornithological Congress*, Uppsala 1950, pp.515–526.

Salomonsen, F. 1965. The geographical variation of the Fulmar (*Fulmarus glacialis*) and the zones of marine environment in the North Atlantic. *Auk* 82:327–355.

Schifferli, A., Géroudet, P. & Winkler R. 1980. *Atlas des Oiseaux Nicheurs de Suisse*. Schweizerische Vogelwarte Sempach. (Published in German and French.)

Sharrock, J.T.R. 1976. *The Atlas of Breeding Birds in Britain and Ireland*. Tring (BTO).

Sharrock, J.T.R. and the Rare Breeding Birds Panel. 1978. Rare breeding birds in the United Kingdom in 1976. *British Birds* 71:11–33.

Sharrock, J.T.R. and the Rare Breeding Birds Panel. 1980. Rare breeding birds in the United Kingdon in 1978. *British Birds* 73:5–26.

Shotton, F.W. 1965. The geological background to European Pleistocene entomology. In Freeman, P. (ed.), pp.452–454.

Simms, E. 1971. *Woodland Birds*. London (Collins, New Naturalist).

Simms, E. 1985. *British Warblers*. London (Collins, New Naturalist).

Simms, E. 1992. *British Larks, Pipits and Wagtails*. London (Collins, New Naturalist).

Simpson, G.C. 1957. Further studies in world climate. *Quarterly Journal, Royal Meteorological Society* 83:459–485.

Sitters, H.P. 1982. The decline of the Cirl Bunting in Britain, 1968–80. *British Birds* 75:105–108.

Sitters, H.P. 1986. Woodlarks in Britain, 1968–83. *British Birds* 79:105–116.

Sovinen, M. 1952. The Red-flanked Bluetail, *Tarsiger cyanurus* (Pall.), spreading into Finland. *Ornis fennica* 29:27–35.

Spencer, R. 1975. Changes in the distribution of recoveries of ringed Blackbirds. *Bird Study* 22:177–190.

Spencer, R. and the Rare Breeding Birds Panel. 1989. Rare breeding birds in the United Kingdom in 1987. *British Birds* 82:477–504.

Spencer, R. and the Rare Breeding Birds Panel. 1990. Rare breeding birds in the United Kingdom in 1988. *British Birds* 83:353–390.

Spencer, R. and the Rare Breeding Birds Panel. 1991. Rare breeding birds in the United Kingdom in 1989. *British Birds* 84:349–370, 379–392.

Spencer, R. and the Rare Breeding Birds Panel. 1993. Rare breeding birds in the United Kingdom in 1990. *British Birds* 86:62–90.

Spitzer, G. 1972. Jahreszeitliche Aspekte der Biologie der Bartmeise (*Panurus biarmicus*). *Journal für Ornithologie* 113:241–275.

Stafford, J. 1956. The wintering of Blackcaps in the British Isles. *Bird Study* 3:251–257.

Stamp, L.D. 1946. *Britain's Structure and Scenery*. London (Collins, New Naturalist).

Stegmann, B. 1938. Grundzüge der Ornithogeographischen Gliederung des paläarktischen Gebietes. *Faune de l'URSS, Oiseaux* 1:77–156.

Stenton, D.M. 1951. *English Society in the Early Middle Ages*. London (Penguin).

Stjernberg, T. 1985. Recent expansion of the Scarlet Rosefinch (*Carpodacus erythrinus*) in Europe. *Proceedings of the XVIIIth International Ornithological Congress*, Moscow 1982, pp.743–753.

Stuart, A.J. 1974. Pleistocene history of the British

vertebrate fauna. *Biological Reviews* 49:225–266.

Summers-Smith, J.D. 1988. *The Sparrows*. Calton (Poyser).

Summers-Smith, J.D. 1989. A history of the status of the Tree Sparrow *Passer montanus* in the British Isles. *Bird Study* 36:23–31.

Tansley, Sir A. 1968. *Britain's Green Mantle* (2nd edn. revised by Proctor, M.C.F.). London (Allen & Unwin).

Terasse, J.H. 1964. *Report on Working Conference on Birds of Prey*. Caen.

Terry, J.H. 1986. Corn Bunting in Hertfordshire. *Transactions of the Hertfordshire Natural History Society* 29:303–312.

Thiede, W. 1975a. Bemerkenswerte faunistische Feststellungen 1970/71 in Europa. Non-Passeriformes. *Vogelwelt* 96:29–36.

Thiede, W. 1975b. Bemerkenswerte faunistische Feststellungen 1970/71 in Europa. Passeriformes. *Vogelwelt* 96:71–77.

Thom, V.M. 1986. *Birds in Scotland*. Calton (Poyser).

Thompson, D., Evans, A. & Galbraith, C. 1992. The fat bird of the barley. *BTO News* 178:8–9.

Thompson, D.B.A. & Gribbin, S. 1986. Ecology of Corn Buntings (*Miliaria calandra*) in NW England. *Bulletin of the British Ecological Society* 17:69–75.

Tomkovich, P.S. 1992. Breeding-range and population changes of waders in the former Soviet Union. *British Birds* 85:344–365.

Turner, W. 1544. *Avium Praecipuarum*. Cologne.

Väisänen, R.A., Hildén, O. & Pulliainen, E. 1989. (Monitoring of Finnish land bird populations in 1979–88.) *Lintumies* 24:60–67. (In Finnish, with English summary.)

Venables, L.S.V. & Venables, U.M. 1955. *Birds and Mammals of Shetland*. Edinburgh (Oliver & Boyd).

Voous, K.H. 1960. *Atlas of European Birds*. London (Nelson).

Waag, A. 1975. Iceland and the Faeroes. In Ferguson-Lees, I.J., Hockcliffe, Q & Zweeres, K. (eds.), pp.37–50.

Wahlstedt, J. 1970. Collared Flycatcher. In Gooders, J. (1969–71), pp.2245–2247.

Wallace, D.I.M., Cobb, F.K. & Tubbs, C.R. 1977. Trumpeter Finches: new to Britain and Ireland. *British Birds* 70:45–49.

Wallace, James. 1791–1797. In *The Statistical Account of Scotland*.

Wanless, S. 1987. A survey of the numbers and breeding distribution of the North Atlantic Gannet *Sula bassana* and an assessment of the changes which have occurred since Operation Seafarer 1969–70. Nature Conservancy Council. Research and survey in Nature conservation Series No. 4. Peterborough.

Watson, D. 1977. *The Hen Harrier*. Berkhamsted (Poyser).

Weertman, J. 1976. Milankovitch solar radiation variations and ice age ice sheet sizes. *Nature* 261:17–20.

White, G. 1789. *The Natural History of Selborne*. London.

Williamson, K. 1951. The moorland birds of Unst, Shetland. *Scottish Naturalist* 63:37–44.

Williamson, K. 1970. *The Atlantic Islands—A study of the Faeroe Life and Scene*. London (Routledge & Kegan Paul).

Williamson, K. 1974. New bird species admitted to the British and Irish lists since 1800. In Hawksworth, D.L. (ed.), pp.221–227.

Williamson, K. 1975. Birds and climatic change. *Bird Study* 22:143–164.

Williamson, K. 1976. Recent climatic influences on the status and distribution of some British birds. *Weather* 31:362–384.

Williamson, K. 1977a. Birds and farmland. *Country Life* 161:192–93.

Williamson, K. 1977b. Unpublished draft chapters for proposed book on the effects of climatic change on birds.

Willughby, F. & Ray, J. 1676. *Ornithologia* (English edition 1678). London (John Martyn).

Wilson, M. 1979. Further range expansion by Citrine Wagtail. *British Birds* 72:42–43.

Winstanley, D., Spencer, R. & Williamson, K. 1974. Where have all the Whitethroats gone? *Bird Study* 21:1–14.

Wolley, J. 1850. Some observations on the birds of the Faeroe Islands. In Jardine, Sir W., (1850). *Contributions to Ornithology for 1850*.

Wright, A.E. & Moseley, F. (eds.) 1975. *Ice Ages: Ancient and Modern*. Geological Journal, Special Issue, No. 6. Liverpool (Seel House Press).

Yapp, B. 1981. *Birds in Medieval Manuscripts*. London (The British Library).

Yeatman, L.J. 1971. *Histoire des Oiseaux d'Europe*. Paris (Bordas).

Yeatman, L.J. 1976. *Atlas des Oiseaux Nicheurs de France de 1970 à 1975*. Paris (Ministère de la Qualité de la Vie—Environnement).

INDEX

LFS

INDEX OF SCIENTIFIC NAMES